MICROHARDNESS OF METALS AND SEMICONDUCTORS

MIKROTVERDOST' METALLOV I POLUPROVODNIKOV

МИКРОТВЕРДОСТЬ МЕТАЛЛОВ И ПОЛУПРОВОДНИКОВ

MICROHARDNESS OF METALS AND SEMICONDUCTORS

V. M. Glazov and V. N. Vigdorovich

A. A. Baikov Institute of Metallurgy
Academy of Sciences of the USSR
Moscow, USSR

Translated from Russian by G. D. Archard

$\left(\frac{c}{b}\right)$ CONSULTANTS BUREAU · NEW YORK–LONDON · 1971

Vilenin Naumovich Vigdorovich was born in 1931 and graduated from the Moscow Institute of Nonferrous Metals and Gold in 1954, specializing in metallography and heat treatment. From 1954 to 1960 he worked as a research fellow at the Moscow Institute of Nonferrous Metals, and from 1960 to 1969 at the State Scientific-Research and Design Institute of the Rare-Metal Industry. In 1958 he defended a dissertation in pursuit of the scientific degree of Candidate of Technical Sciences, and in 1966 that of Doctor of Technical Sciences. Since 1969 he has been the Director of the Faculty of Material Science and Professor of the Moscow Institute of Electronic Technology. His principal research work is devoted to questions of chemical thermodynamics, the physicochemical analysis of metallic alloys and semiconducting systems, and crystallization meth   for the intensive purification of metals and semiconductors.

Vasilii Mikhailovich Glazov was born in 1931. In 1954 he was graduated from the Moscow Institute of Nonferrous Metals and Gold, specializing in metallography. From 1954 to 1963 he worked at the A. A. Baikov Institute of Metallurgy of the Academy of Sciences of the USSR, eventually as a Senior Research Fellow. In 1959 he defended a dissertation in pursuit of the scientific degree of Candidate of Technical Sciences, and in 1966 that of Doctor of Chemical Sciences. From 1963 to 1968 he taught at the Moscow Institute of Steel and Alloys. At the present time he heads the Faculty of Physical Chemistry at the Moscow Institute of Electronic Technology. His principal research is devoted to problems in the physicochemical analysis of semiconductors in the liquid phase, the laws of doping semiconductors, the phase diagrams of metallic systems, and also the structure and crystallization of metallic alloys. A large number of his papers, systematized in the monograph "Microhardness of Metals and Semiconductors," are devoted to the application of the microhardness method to the solution of a number of problems of physicochemical analysis and also the physics of metals and semiconductors.

The original Russian text, published in Moscow by Metallurgiya Press in 1969, has been corrected and enlarged by the authors for the present edition. The English translation is published under an agreement with Mezhdunarodnaya Kniga, the Soviet book export agency.

ГЛАЗОВ Василий Михайлович
ВИГДОРОВИЧ Виленин Наумович

МИКРОТВЕРДОСТЬ МЕТАЛЛОВ И ПОЛУПРОВОДНИКОВ

Library of Congress Catalog Card Number 70-128504

ISBN-13: 978-1-4684-8248-5   e-ISBN-13: 978-1-4684-8246-1
DOI: 10.1007/978-1-4684-8246-1

© 1971 Consultants Bureau, New York
A Division of Plenum Publishing Corporation
227 West 17th Street, New York, N.Y. 10011

United Kingdom edition published by Consultants Bureau, London
A Division of Plenum Publishing Company, Ltd.
Donington House, 30 Norfolk Street, London, W.C. 2, England

All rights reserved

No part of this publication may be reproduced in any
form without written permission from the publisher

FOREWORD TO THE ENGLISH EDITION

The use of refined methods of physical experimentation in studying the properties of matter in very small volumes has enabled us to discover some new and interesting laws and phenomena revealing the physical nature of the materials studied. The method of microhardness measurement is one of these. The application of this method to the study of metals, semiconductors, and their alloys has led to the development of a whole series of problems associated with the fine structure of matter. In this we primarily mean such phenomena as dendritic liquation and microheterogenization accompanying nonequilibrium crystallization. A study of the laws associated with these phenomena is of general importance both for metals and for semiconductors, since the formation of crystallites in the course of nonequilibrium crystallization with the suppression of diffusion in the solid phase is based on the same laws as those governing the production of single crystals by the methods of Czochralski, Bridgman, and others. The microhardness method has been widely accepted in studying phase equilibria and plotting the phase diagrams of binary, ternary, and quaternary metallic and semiconductor systems. The method may be used both for plotting the boundaries between phase regions and establishing the positions of conodes, and also for identifying individual phases and structural constituents in complex alloys. A justification for the use of this method in such investigations based on the principles of physicochemical analysis enunciated by N. S. Kurnakov is presented for the first time in this monograph.

Naturally, success in the solution of problems of solid-state physics, material science, and physicochemical analysis will depend substantially on the corrections of the method chosen for preparing the samples and carrying out the research. For this reason a number of sections in the book are devoted to methodical problems. The methodical techniques are developed in connection with the Soviet-built PMT-3 hardness tester; however, the laws discussed are of more general significance and may also be employed in connection with other types of apparatus and pyramids of different configurations (for example, Knoop pyramids).

It has proved extremely fruitful to apply the microhardness method to a study of the fine structure of semiconducting crystals. The revelation of microinhomogeneities in the distribution of impurities along the length and across the diameter of single crystals, the effect of dislocation density on the mechanical properties of a crystal, the anisotropy of the properties, the polarity of the crystallographic planes of semiconducting compounds of the $A^{III}B^V$ group, all these are new fields in which advances have been made with the help of the microhardness method. All recent investigations carried out in these fields are reflected in this monograph. In the Soviet Union, "Microhardness of Metals and Semiconductors" constitutes the second edition of our earlier (1962) monograph; it has been considerably extended by sections devoted to semiconductors. The book furthermore extends beyond the realm of the purely methodical, since the majority of its sections are of independent scientific interest from the point of view of discovering the physical essence of the phenomena under consideration. It should be noted that certain problems and principles, for example, those concerning diffusionless crystallization, the plotting of the solidus by analyzing the microhardness of alloys crystallized under non-equilibrium conditions, the effect of superheating the melt on liquations, and so on, bear a controversial character; however, despite this, the experimental material presented retains its value. The American edition of "Microhardness of Metals and Semiconductors" differs little from the Russian. Corrections have been made for errors and inexactitudes noticed after publication of the original. An additional section has been added to Chapter 5; this relates to the establishment of the ranges of homogeneity of solid solutions based on semiconducting compounds (using as examples silver selenide and germanium and tin tellurides).

The authors wish to thank Plenum Publishing Corporation for facilitating participation in the preparation of the book for the American edition; they welcome the publication of the monograph in the United States of America and hope that their labors will be favorably received by the American reader.

The authors would be extremely grateful for any comments on shortcomings in this book or suggestions for its improvement, or indeed for the addition of further material.

Professor V. Glazov
Professor V. Vigdorovich

FOREWORD

In our earlier monograph "Microhardness of Metals" [1], published in 1962, we attempted to correlate a variety of investigations scattered throughout a number of journals with the general theme of solving problems of physicochemical analysis and metallography by microhardness measurements.

The publication of "Microhardness of Metals" promoted the widespread use of this technique in studying physicochemical phenomena in various materials. In recent years the microhardness method has been used most extensively in studying semiconducting materials, and this has necessitated the revision of the monograph with the aim of incorporating new experimental data relating to both metals and semiconductors.

The greater proportion of the material presented in this book reflects the authors' own investigations. Other investigations associated with the use of the microhardness method in the physicochemical analysis of metallic and semiconducting systems are also taken into account.

The authors are extremely indebted to Academician G. G. Urazov for his great interest in this work at its inception and for a number of valuable comments regarding the possibility of using the microhardness technique in physicochemical analysis.

Sincere thanks are extended to Academician A. A. Bochvar and Professors A. N. Krestovnikov, M. M. Khrushchov, M. V. Mal'tsev, M. V. Zakharov, and I. I. Novikov for their interest in these investigations.

V. Glazov
V. Vigdorovich

INTRODUCTION

Measuring the hardness of materials is generally accepted as being one of the easiest and most rapidly executed forms of mechanical test. By using this method the quality of parts and materials may be rapidly and accurately tested and many physicochemical investigations associated with the recognition of materials and the study of their properties, functions, and structural transformations may be carried out. A number of important uses of this method occur in connection with the possibility of making an indirect estimate of other mechanical characteristics of materials having a specific correlation with their hardness.

The hardness-measuring method first became widely applied for purposes of physicochemical analysis after the investigations of N. S. Kurnakov and his school.

The hardness depends not only on the properties of the material under test but also largely on the conditions of measurement. In the most general case the hardness constitutes an integral property, being governed by a wide selection of the mechanical characteristics of the test sample (ductility, elastic limit, strength, and so forth). It is thus particularly important to choose measuring conditions in which the results will depend in a uniform manner on the same basic mechanical characteristics of the material under consideration.

At the present time there are a number of methods of determining hardness.

The qualitatively different method of microhardness arose as a result of an investigation into the measuring conditions of these earlier methods when using very small loads.

The microhardness method may well be employed in conjunction with a study of the microstructure of the test material. This important qualitative distinction of the method offers excellent prospects for further investigations, the limits of which are as yet hard to discern.

In 1936 Lips (of the Netherlands) constructed the first apparatus capable of measuring microhardness with loads of 35 to 100 g for metallographic purposes. The diamond pyramid used for the impression had a base in the form of a square, while the angle between the faces at the vertex of the pyramid was 136°. Low-load hardness measurements with a Knoop diamond pyramid having a base in the form of an extended rhombus have developed widely in the United States since 1939. The angles between the faces at the vertex of the Knoop pyramid are 130 and 172°30'. Devices later created for measuring microhardness constituted either attachments to metallographic microscopes or individual specialized instruments. Other types of indentors in addition to the square pyramid with a vertical angle of 136° and the Knoop pyramid already mentioned have not received such general acclaim.

In the USSR the development of instruments for measuring microhardness started in 1940. The basic idea was the most reasonable one according to which the apparatus was founded on a vertical microscope with a revolving head having one socket for the objective lens and another for the indentor. The first Soviet instruments were created in the Institute of Engineering of the Academy of Sciences of the USSR.

The most widely used instruments for measuring microhardness are the PMT-2 and PMT-3 derived by M. M. Khrushchov and E. S. Berkovich [2, 3].

The PMT-2 and PMT-3 are very original as regards their mode of operation, and in construction and simplicity of application they have proved much better than many devices of the same type produced in other countries.

The development of the essential apparatus has advanced the study of microhardness in a wide range of test objects. Microhardness tests have been applied to the fine components of clock and instrument mechanisms, thin metal strip, foils, wires, metallic fibers, thin galvanic coatings, artificial oxide films, etc., as well as the thin surface layers of metals which change their properties as a result of mechanical treatment (machining), rolling, friction, and other effects. The microhardness method is widely used for studying the individual structural constituent elements of metallic alloys, minerals, glasses, enamels, and artificial abrasives.

Soviet achievements in this field were reported in the Transactions of the Conferences on Microhardness held in Moscow in November 1950 [4] and October 1963 [5].

The possibility of using the microhardness method for physicochemical analysis is based on the classical investigations into the relation between hardness and the composition of solid solutions carried out by N. S. Kurnakov and his colleagues [6, 7].

However, as originally noted by N. S. Kurnakov and investigated further by A. A. Bochvar [8-10], in the study of hardness, microhardness, and other properties as functions of composition, a great part is played by the actual structure of the alloys under examination.

The first work carried out by Bochvar and his colleagues on the basis of microhardness measurements showed that the structure of the crystals formed by solid solutions of two-phase alloys, crystallized under nonequilibrium conditions, was distinguished by a complex internal substructure and microinhomogeneities of a liquation type, as well as heterogeneity of the second order. This type of structure will presumably lead to deviations from the theoretical relationships arising from Kurnakov's principles and hence complicate the prospects of using the microhardness method in studying the phase equilibrium in binary, ternary, and quaternary systems.

It clearly became essential to make a detailed study of the laws governing the relationship between microhardness and the structure of the alloys in order to evaluate the prospects of using this simple and convenient method more widely for physicochemical analysis.

The investigations of Kurnakov and Bochvar serve as a foundation for subsequent work in this direction.

In 1953, we started a systematic investigation [1] into the relation between the microhardness of individual phases and the structural constituents of binary and ternary alloys and their structures and compositions at various temperatures.

We succeeding in establishing a number of laws relating microhardness to composition and were able to discover the characteristic structural features of real alloys. This facilitated the application of the microhardness method to purposes of physicochemical analysis.

The initiation of these investigations was preceded by the development of a number of aspects relating to methods of measuring microhardness when using this technique for the purposes under consideration.

The aim of this book is to correlate the results of all these investigations (earlier published as individual articles in various journals) as well as those of other research workers in the same field.

A consideration of these matters should give a reasonable idea of the prospects of using the microhardness method and ways of applying it to physicochemical analysis; this should facilitate the choice of optimum procedures for presenting and solving a large number of material-science and physicochemical problems.

CONTENTS

xiii

CHAPTER 1

APPARATUS AND METHOD OF MEASURING MICROHARDNESS

The PMT-3 Apparatus

Main Components, Operating Principles, and Construction. The principal instrument for testing the microhardness of materials used in modern laboratories of various types is the Soviet-made PMT-3, the latest model of which was designed by M. M. Khrushchov and E. S. Berkovich [2, 3].

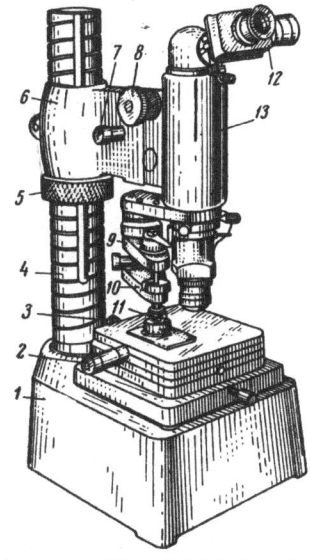

Fig. 1. The PMT-3: 1) Stand; 2, 3) object-table screws; 4) support; 5) ring nut; 6) bracket; 7) microdrive mechanism; 8) macrodrive mechanism; 9, 10) loading-mechanism brackets; 11) indentor (diamond pyramid); 12) micrometer eyepiece; 13) tube.

The PMT-3 and other Soviet models are usually employed for testing the hardness of materials by impression under a load of between 2 (5) and 200 g. As impressing tool (indentor) a diamond pyramid with a square base and an angle of 136° between opposite faces at the vertex is employed. In the tests the diagonal of the impression is measured and the hardness number is calculated by dividing the area of the impression into the value of the applied load. In other words, microhardness tests are carried out in the same way as hardness tests in a Vickers instrument; to some extent, however, the two kinds of test only take on the same physical meaning when studying fairly homogeneous, single-crystal materials.

In addition to this the use of small loads and the necessity of measuring small impressions conditions all the remaining characteristics of instruments for testing microhardness and makes them completely different from macrohardness testers.

Figure 1 shows a PMT-3 instrument with its main components: a microscope tube 13 carrying a micrometer eyepiece 12, an objective, a condenser, and a loading mechanism attached to it by a special bracket. The tube moves in the guides of a bracket fixed to the support 4 of the stand 1 by means of a rack and pinion 8 ("gear") for effecting coarse movement (macroscopic drive) and a special multistage gear mechanism 7 for effecting small displacements (microscopic drive). Under the tube is the object table turning through approximately 180° around its central axis from stop to stop and having two forward motions of the upper part.

1

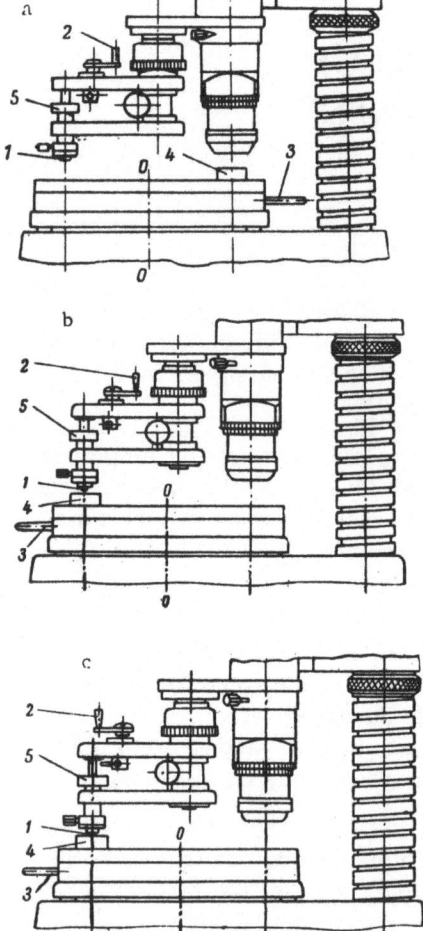

Fig. 2. Arrangement for making hardness tests with the PMT-3: a) Object under objective, choice of point for test, measurement of diagonal after making impression; b) object under diamond pyramid before dropping indentor and after making impression and lifting indentor; c) diamond pyramid on object, producing impression.

Figure 2 illustrates the operation of the device. Thus the microscope tube and loading mechanism are rigidly fixed together and move together when focusing. This ensures that the diamond pyramid is set exactly in the same working position relative to the surface of the test sample, irrespective of the motion of the tube. In order to make an impression, first of all the test object 4 is moved about on the object table and a suitable point for an impression is chosen under the microscope (Fig. 2a). Then the object is moved, bringing the point chosen on its surface under the point of the diamond pyramid 1. This is achieved by rotating the whole table 180° around the axis O−O by means of the handle (Fig. 2b). In the PMT-3 an adjustment may be made to ensure that the axis of rotation of the table O−O is at an equal distance from the optical axis of the microscope and the axis of the indentor with the diamond pyramid. In order to ensure that after the rotation of the object table around the axis O−O the point chosen for test on the object always lies strictly opposite to the point of the diamond, stops limiting the rotation are provided, and the table only turns from one stop to the other.

After the test object has been arranged under the indentor, the impression is made. The rod is released by a half turn of the stop 2 (Fig. 2c). Under the influence of the load 5 the rod falls until the diamond pyramid comes into contact with the test object, and is driven into the latter by the same load. In order to lift the rod back into its original position the handle of the stop 2 is reversed (Fig. 2b).

Then, on turning the table by the handle 3 to the stop, the table with the object 4 is returned to the initial position and the diagonal is measured.

Loading Mechanism. The operating principles and adjustment of the loading mechanism of the PMT-3 are shown in Fig. 3. The illustration indicates how the rod with the diamond pyramid 1 is suspended from the free ends of two elastic steel plates 3 and 4, which are fixed rigidly in the body 5 of the mechanism and balance the weights of the parts of the indentor. The total rigidity of the plates usually corresponds to a motion of 6 or 7 μ in the rod under a load of 0.5 g.

The indentor is lifted and let down by a half rotation of the handle 7, which corresponds to a 1-mm movement of the rod.

Microscope. As indicated in Fig. 4, the bracket carrying the microscope is able to move vertically along the column 2 of the massive base 1 of the microscope stand by means of the nut 3. The extent of this motion depends on the height of the test object and is established by means of the screw 4. The necessity of this motion arises from the comparatively short travel of the microscope tube in the guides of the bracket as facilitated by the macrodrive 5.

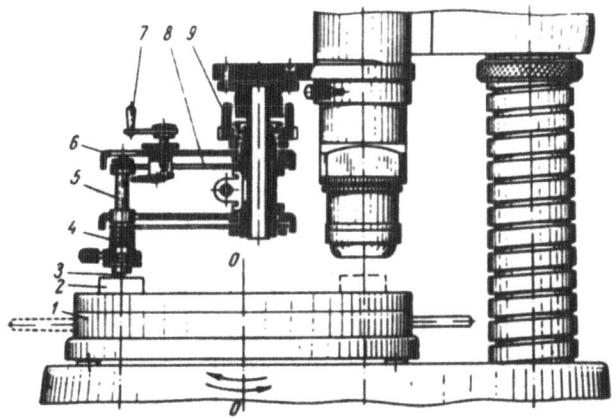

Fig. 3. Operating conditions and adjustment of the loading mechanism of the PMT-3: 1) Rod; 2) nut for raising the loading mechanism; 3, 4) lower and upper elastic plates; 5) body; 6) one of the two centering screws (right); 7) handle; 8) point of diamond; 9) test sample.

Fig. 4. External view of the PMT-3.

In order to ensure that the microscope tube does not move downward during observation and measurement under the weight of the attached parts, the handle 6 is capable of fixing the tube in the bracket. Subsequent fine adjustment of the tube for focusing can only be effected by means of the microdrive 7.

The course of the rays in the microscope of the PMT-3 is illustrated in Fig. 5. The PMT-3 has a complete illuminating system, so that the object surface may be viewed with an episcopic objective in "bright field" (Fig. 5, left) and in "dark field" (Fig. 5, right). For this purpose the holder of the mirror 1 is rotated around the axis AB from one stop to the other. Then the plate 5 is placed in the path of the rays for bright field or plate 7 for dark.

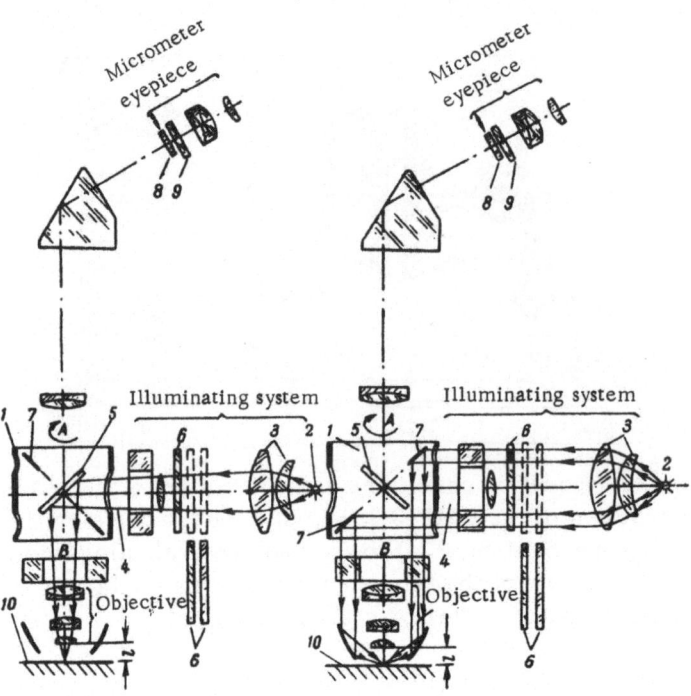

Fig. 5. Optical system of the PMT-3: 1) Mirror holder;
2) light source; 3) condenser; 4) diaphragm; 5) reflecting
plate; 6) light filters; 7) mirror; 8) movable grid; 9) stationary grid; 10) object; l) free distance between the object and objective.

The total magnification of the microscope in the PMT-3 for visual observations and measurements by means of a 40-times episcopic objective (OE-6) with an aperture A = 0.65 (focal
length F = 6.16) and the screw eyepiece of a 15-times AM9-3 microscope equals 485 to 487.

The intensity of the light source of the PMT-3 is regulated by means of a transformer
and rheostat which reduce the voltage from 127 to 8 V. The observation conditions improve
with the use of light filters.

The sharpness of the image may be increased by varying the aperture of the diaphragm 4;
however, this is only possible when viewing the object in bright field (Fig. 5, left).

In order to center the optical axis of the microscope, which involves making the axis of
the optical system coincide with that of the indentor, an additional 8-times episcopic objective
OE-23 (focal length F = 23.17, aperture A = 0.17) is added to the instrument; by means of this
the linear field of view of the microscope is magnified to 1.5 instead of 0.2 mm (the latter relating to the ordinary OE-6 objective), and this makes it easier to find the pyramid impressions
on the surface of the test sample.

In the upper part of the tube of the PMT-3 provision is made for replacing the inclined
eyepiece tube with a straight one. The firing ring of the fitting is placed on this and 7-, 10-, or
15-times standard eyepieces are placed in the eyepiece tube. If necessary, a camera is fixed
to the end of the fitting and the surface of the object is observed either on the mat glass of the
camera or through the eyepiece tube of the photographic attachment.

Industry produces photographic attachments differing only in the size of the photographic
plates: a universal microphotographic attachment (MFN-1) with a mat glass 6.5 × 9 cm in size

and a photographic attachment (MFN-2) with a mat glass 9 × 12 cm in size. The ordinary photographic material for microphotographs of medium sensitivity are used for taking the pictures.

Object Table. In the PMT-3, the object table is placed directly under the microscope tube and loading-mechanism indentor. The table has two motions: coordinate (linear) by virtue of two micrometer screws, and half circular, by rotating the whole table from stop to stop around an axis equidistant from the optic axis of the microscope and the axis of the indentor.

The coordinate motion of the table enables a place suitable for testing to be chosen (within certain limits) and also facilitates exact and accurate placing of impressions on the sample surface at specified distances from one another. The maximum travel of the object table is 10 to 12 mm. The scale division on the drums of the micrometer screws is 0.01 mm.

The half-circular motion serves to bring the test object alternately under the microscope and under the point of the diamond pyramid.

The sample of material to be tested is set and fixed on the object table by means of spring clamps or a thin layer of modeling clay.

During operation it is important to observe the correct sequence of movements and not to allow the object table to rotate with the indentor lowered, since this would damage the diamond pyramid.

Determining the Scale of Microscope Magnification

Determining the scale of magnification of the microscope is extremely important.

First of all, in order to eliminate parallax leading to error in determining the diagonal of the impression, it is essential to be able to see the crosswires and sample surface equally sharply. The eyepiece mounting has a screw thread for focusing the crosswires sharply relative to the eye of the observer. After adjusting the wires to sharpness the microscope is focused on the object.

The crosswire and two lines are moved by a screw nut mechanism with a pitch of 1 mm. The screw has a drum divided into a hundred divisions at the end to record fractional rotations. Thus the smallest division on the drum corresponds to a 0.01-mm movement of the crosswires, which is reckoned relative to the stationary eight-division scale visible in the eyepiece. Thus the whole travel of the screw (with the crosswire) is 8 mm, or eight full turns.

In order to make the magnification of the microscope known, it is essential to establish what real distance corresponds to a specified motion of the crosswires. For this purpose an objective micrometer (an accurate small scale 1 mm long divided into 100 parts) is substituted for the object on the table.

The linear magnification of the objective is the ratio of the difference of the readings on the scale of the micrometer eyepiece to the product of the number of divisions of the objective micrometer and the actual value of the scale division of the latter (0.01 mm).

From the known magnification of the objective it is easy to determine the scale division of the drum on the screw eyepiece. For this purpose the movement of the eyepiece crosswires on rotating the screw through one scale division of the drum (0.01 mm) is to be divided by the linear magnification of the objective.

In PMT-3 instruments furnished with an AM9-2 [or (AM9-3)] eyepiece and an episcopic objective OE-6, the ocular drum division equals about 0.3 μ.

We may then measure the length of any object in any part of the microscope field of view. However, it is best to measure the diagonal of the impression in the center of the microscope field of view, since in this case certain possible shortcomings in the optical system of the microscope which manifest themselves most at the edges of the field will be eliminated.

The diagonal of the impression is determined as the product of the scale division of the eyepiece drum times the difference in the drum readings on moving the crosswires from one corner of the impression to the opposite corner.

Micrometer eyepiece measurements may be made to an accuracy of ±0.5 divisions in the PMT-3, or on allowing for the magnification to an accuracy of 0.15 μ.

Adjusting and Calibrating the Apparatus

In view of the very severe demands made on the accuracy of matching the optical axis of the microscope to the loading axis on rotating the object table, provision is made in the PMT-3 for a special centering device, which enables the objective (with the illuminating system) to be shifted slightly in a horizontal plane, i.e., in directions perpendicular to the two coinciding axes. The centering device is brought into action by means of screws 8 and 9 (Fig. 4).

We use this device when first preparing the instrument for operations and also when testing the microhardness of objects of different heights or an object with a considerable surface "relief."

In measuring samples differing in height the loading axis may not coincide with the optical axis owing to the fact that the axis of rotation of the object table may not be quite parallel to the direction of travel of the microscope tube. Then a slight correction will be needed, and this is called the centering of adjusting of the instrument.

The sequence of operations in centering is indicated in Fig. 6. Let us suppose that the instrument is not yet centered. Then we set the crosswires of the screw micrometer eyepiece so as to be exactly in the center of the field of view of the microscope. In so doing the moving double line in the eyepiece should appear against the figure 4 of the stationary scale, and the zero on the drum scale exactly opposite the graduation line. Then, by moving the object table with the help of the micrometer screws, we bring the point chosen for testing under the crosswires (Fig. 6, position I). After this we carry out the test. However, in view of the fact that the device is not centered, the impression a will have been obtained to one side of the crosswires and the intended test point (position II). Then we move the objective by operating with the centering screws until the crosswires coincide with the center of the impression a (position III). After this we move the table so that the crosswires are again brought to the place at which it is desired to make the impression (position IV). The renewed impression b is exactly in the right place (position V).

Sometimes in order to achieve exact centering this operation has to be repeated several times. The centering will be completely adequate if the impressions are situated at a distance of no more than 0.5 to 2.0 μ from the crosswires.

It may chance that the impression is not only not on the crosswires but not in the field of view of the microscope at all. Hence in this case there is a considerable difference between the distances from the axis of rotation of the table to the axis of the objective and from the axis of rotation of the table to the axis of the diamond pyramid. For this reason an additional operation of preliminary centering will be required. Instead of the main objective we in fact use an auxiliary one with a lower magnification and carry out all the centering operations. Then we restore the episcopic objective and carry out a more accurate centering.

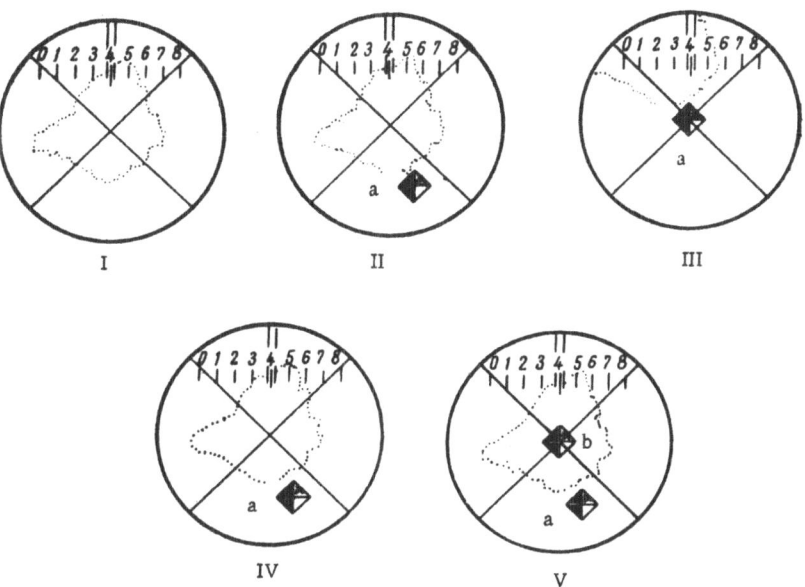

Fig. 6. Diagrams to illustrate the successive processes involved in centering the objective of the PMT-3.

The centering mechanism should be used as rarely as possible, in fact only in those cases in which the impression has to be at an exact spot on the object. Rather than centering, it is better to check carefully whether an observed displacement between the crosswires and the impressions might not arise from other causes. These might be incorrect rotation of the object table (not from stop to stop), imprecise setting of the measuring mechanism of the screw micrometer eyepiece to zero, unstable fixing of the sample on the object table, and various others.

In preparing the instrument for study the whole loading mechanism is set in height in such a way that the point of the diamond 8 (see Fig. 3) should touch the surface of the test sample 9, while the microscope should be focused on this surface. This setting is effected by means of the nut 2, moving the whole loading mechanism upward or downward until the point of the diamond in the free, unloaded mechanism touches the surface and leaves hardly any perceptible mark on a soft metal with a well-visible surface such as aluminum or tin on viewing in the microscope. We may use the following rule: For a load of 0.5 g hardly any visible mark should be seen in the microscope, and without any load no mark should be seen at all.

In addition to this we may use the results of A. D. Kuritsyn [4], who showed that the microhardness of common salt crystals remained constant on keeping for a long time and was independent of the size and mode of preparation of the crystals, although polishing the cleavage planes might raise the microhardness by a factor of 1.5. For calibrating the loading mechanism we may consider that, on measuring the microhardness in the natural cleavage plane or the plane of growth obtained from solutions, common salt gives a constant value within the narrow limits of 20 to 22 kg/mm^2. It is also important that the microhardness of common salt remains almost constant on varying the applied load.

After completing the calibration of the instrument the position of the loading mechanism is fixed by the handle stopper and periodically checked.

In the PMT-3 loading is effected by means of special weights in the form of a split disc 5, 10, 20, 50, 100, and 200 g in weight. The weights are placed on a flat space provided for the purpose in the gap between two flat springs, which retain the rod. The permissible deviations in the value of the applied load are not more than ±0.1% for loads up to 10 g and not more than ±1% for larger loads. The error in the load due to the elasticity of the spring plates themselves during the impression is slight (for example, in an impression with a diagonal of 140 μ it is about 0.3 g).

The adjusted instrument should ensure smooth application of the load on the pyramid, without shocks or excessive friction of the parts of the loading mechanism, since all these would introduce error into the value of the applied load. The pyramid should be impressed under the influence of an inertia-free load.

Special requirements are imposed on one of the most important parts of the loading mechanism, the mounting with the diamond pyramid.

The working faces of the diamond should be carefully polished and should be free from cracks and scratches. The shape of the working surface of the diamond should correspond to a regular tetragonal pyramid with a vertical angle between opposite faces of 136°, the greatest permissible deviation being ±20'. The edges and vertex of the pyramid should not have any rounding or chipping.

The quality of the sharpened pyramid may affect the determination of microhardness. In practice the vertex of the pyramid will be a "comb" (ridge) rather than a point. The comb at the vertex of the pyramid should not be longer than 0.5 μ.

All the parameters and quality characteristics of the diamond pyramid mentioned may easily be verified by inspecting either the diamond itself or the impression which it makes under the microscope with a fairly high magnification (not less than 950).

Measuring the Diagonal of the Impression and

Determining the Value of the Microhardness

First of all it is important to set the micrometer eyepiece and mounting with the diamond pyramid in such a position that on rotating the drum of the eyepiece the crosswires move strictly along one of the diagonals of the impression. For this purpose, on first setting the pyramid mounting, its position is fixed exactly with the screw, as far as possible making the mark on the pyramid mounting agree with that on the indentor rod. The diamond pyramid mounting fixed on the rod then occupies a specific position, so that all the impression diagonals will be oriented in the same way. Then the fixing screw 10 is released (Fig. 4) and the micrometer eyepiece is rotated through an appropriate angle, only being fixed when the direction of motion of the crosswires agrees with the direction of one of the impression diagonals.

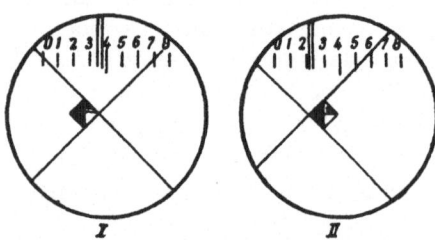

Fig. 7. Diagrams explaining the successive execution of procedures employed in measuring the diagonals of the impressions in the PMT-3 with an AM9-2 or AM9-1 micrometer eyepiece.

The diagonal of the impression may be measured in several ways. We ourselves consider that the best method is the following (this is suitable when using the AM9-2 or AM9-1 micrometer eyepieces).

Using the micrometer screws for the coordinate motion of the object tables 11 and 12 (Fig. 4) and observing in the eyepiece, we bring the impression to the left-hand side of the crosswires. Then with the drum we bring the crosswires to the corner of the impression. The operation must always be carried out in the same way, for example, from right to left, so that the wires will always stop on the right of the contour of the impression, thus eliminating the effects of gaps in the screw

TABLE 1. Hardness Numbers for Testing a Square (Tetragonal)
Diamond Pyramid Having a Vertical Dihedral Angle
of 136° and a Load of 20 g

Diagonal of impression, μ	Hardness number, kg/mm²									
	0	1	2	3	4	5	6	7	8	9
0	—	—	—	—	—	3710	2575	1890	1450	1145
10	925	765	645	550	473	417	362	321	286	257
20	232	210	192	175	161	148	137	127	118	110.5
30	103	96.5	90.5	85.0	80.0	75.5	71.5	67.5	64.0	61.0
40	58.0	55.0	52.5	50.0	47.9	45.8	43.8	42.0	40.25	38.6
50	37.1	35.65	34.3	33.0	31.8	30.65	29.55	28.55	27.55	26 65
60	25.75	24.9	23.9	23.35	22.65	21.95	21.3	20.65	20.05	19.5
70	18.90	18.40	17.90	17.40	16.95	16.50	16.05	15.65	15.25	14.85
80	14.50	14.15	13.80	13.45	13.15	12.85	12.55	12.25	12.00	11.7
90	11.45	11.20	10.95	10.70	10.50	10.25	10.05	9.85	9.65	9.45
100	9.25	9.10	8.90	8.75	8.55	8.4	8.25	8.10	7.95	7.8
110	7.65	7.55	7.40	7.25	7.45	7.00	6.90	6.75	6.65	6.55
120	6.45	6.35	6.25	6.15	6.05	5.95	5.85	5.75	5.65	5.55
130	5.50	5.40	5.30	5.25	5.15	5.10	5.00	4.94	4.87	4.80
140	4.73	4.66	4.60	4.53	4.47	4.41	4.35	4.29	4.24	2.18
150	4.12	4.07	4.02	3.95	3.91	3.86	3.81	3.76	3.72	3.67
160	3.62	3.58	3.54	3.49	3.45	3.41	3.37	3.32	3.28	3.25
170	3.21	3.17	3.14	3.10	3.07	3.03	3.00	2.96	2.93	2.90
180	2.86	2.83	2.80	2.77	2.74	2.71	2.68	2.65	2.63	2.60
190	2.57	2.54	2.52	2.49	2.47	2.44	2.41	2.39	2.36	2.34
200	2.32	2.29	2.27	2.25	2.23	2.21	2.19	2.16	2.14	2.12
210	2.10	2.08	2.06	2.04	2.03	2.01	1.99	1.97	1.95	1.93
220	1.92	1.90	1.88	1.87	1.85	1.83	1.82	1.80	1.78	1.77
230	1.75	1.74	1.72	1.71	1.69	1.68	1.67	1.65	1.64	1.62
240	1.61	1.60	1.59	1.57	1.55	1.54	1 53	1.52	1.51	1.50
250	1.48	1.47	1.46	1.45	1.44	1.43	1.42	1.41	1.40	1.38
260	1.37	1.36	1.35	1.34	1.33	1.32	1.31	1.30	1.29	1.28
270	1.27	1.26	1.25	1.24	1.23	1.22	1.215	1.21	1.20	1.19
280	1.180	1.174	1.166	1.16	1.15	1.140	1.13	1.125	1.120	1.11
290	1.105	1.10	1.090	1.08	1.07	1.065	1.06	1.050	1.04	1.035
300	1.030	—	—	—	—	—	—	—	—	—

of the reading mechanism of the micrometer eyepiece. This facilitates reading on the scales, since the edge of the wire is first brought up (from right to left) to the right-hand corner of the impression and the scales are read, and then the same edge of the wire is brought up to the left-hand corner of the impression and again the scale readings are taken from right to left (Fig. 7, I and II). Then the difference between the two readings (N) is obtained, and the result is multiplied by the value of the scale division in microns (C), giving the length of the diagonal in microns:

$$d = CN. \tag{1}$$

When using an episcopic objective of aperture A = 0.65 the scale division of the microm-eter-eyepiece drum is C ~ 0.3 μ.

In microhardness testing the numerical value of the result is expressed as the ratio of the load P (kg) to the lateral area of the impression F (mm²)* on the assumption that the angles

---

* According to V. G. Grigorovich [11], in calculating the microhardness it is more accurate to relate the load not to the area of the surface of the impression but to the area of its projec-tion. However, the difference thus obtained is insignificant. In the majority of cases, up to the present, microhardness data have been calculated on the basis of Eq. (2).

TABLE 2. Hardness Numbers for Testing with a Square (Tetragonal) Diamond Pyramid Having a Vertical Dihedral Angle of 136° and a Load of 50 g

Diagonal of impression, μ	Hardness number, kg/mm^2									
	0	1	2	3	4	5	6	7	8	9
0	—	—	—	—.	—	1484	1030	756	580	458
10	370	306	258	220	189.2	164.8	144.8	128.4	114.4	102.8
20	92.8	84.1	76.6	70.1	64.4	59.4	54.8	50.8	47.3	44.2
30	41.2	38.6	36.2	34.0	32.0	30.2	28.6	27.0	25.6	24.2
40	23.2	22.0	21.0	20.0	19.16	18.32	17.52	16.80	16.10	15.44
50	14.84	14.26	13.72	13.20	12.72	12.26	11.82	11.42	11.02	10.66
60	10.30	9.96	9.56	9.34	9.06	8.78	8.52	8.26	8.02	7.80
70	7.56	7.36	7.16	6.96	6.78	6.60	6.42	6.26	6.10	5.94
80	5.80	5.66	5.52	5.38	5.26	5.14	5.26	4.90	4.80	4.68
90	4.58	4.48	4.38	4.28	4.20	4.10	4.02	3.94	3.86	3.78
100	3.70	3.64	3.56	3.50	3.42	3.36	3.30	3.24	3.18	3.12
110	3.06	3.02	2.96	2.90	2.86	2.80	2.76	2.70	2.66	2.62
120	2.58	2.54	2.50	2.46	2.42	2.38	2.34	2.30	2.26	2.22
130	2.20	2.16	2.12	2.10	2.06	2.04	2.00	1.976	1.948	1.920
140	1.892	1.866	1.840	1.814	1.788	1.764	1.740	1.746	1.694	1.670
150	1.648	1.626	1.606	1.584	1.564	1.544	1.524	1.504	1.486	1.468
160	1.448	1.430	1.414	1.396	1.380	1.362	1.346	1.330	1.314	1.298
170	1.284	1.268	1.254	1.240	1.226	1.212	1.198	1.184	1.170	1.158
180	1.144	1.132	1.120	1.108	1.096	1.084	1.072	1.060	1.050	1.038
190	1.028	1.016	1.006	0.996	0.986	0.976	0.966	0.956	0.946	0.936
200	0.928	0.916	0.908	0.900	0.892	0.884	0.876	0.864	0.856	0.848
210	0.841	0.832	0.824	0.816	0.812	0.804	0.796	0.788	0.780	0.772
220	0.766	0.760	0.752	0.748	0.740	0.732	0.728	0.720	0.712	0.708
230	0.701	0.696	0.688	0.684	0.676	0.672	0.668	0.660	0.656	0.648
240	0.644	0.638	0.634	0.628	0.622	0.618	0.612	0.608	0.604	0.598
250	0.594	0.588	0.584	0.580	0.574	0.570	0.566	0.562	0.558	0.552
260	0.548	0.544	0.540	0.536	0.532	0.528	0.524	0.520	0.516	0.512
270	0.508	0.506	0.502	0.498	0.494	0.490	0.486	0.484	0.480	0.476
280	0.473	0.470	0.466	0.464	0.460	0.456	0.454	0.450	0.448	0.444
290	0.442	0.438	0.436	0.432	0.430	0.426	0.424	0.420	0.418	0.414
300	0.412	—	—	—	—	—	—	—	—	—

in the impression are the same as in the pyramid itself:

$$H_\mu = \frac{P}{F} = \frac{2P \sin \alpha/2}{d^2} = \frac{1.854\,P}{d^2},$$ (2)

where α is the vertical angle of the diamond pyramid (136°, or 2.47 rad).

If P is expressed in grams and d in microns, the formula for calculating the microhardness has the form

$$H_\mu = \frac{1854P}{d^2} \text{ kg/mm}^2.$$ (3)

Under the same conditions, for calculating the depth of the impression on perfectly ductile materials in which there is no elastic restitution, we may use the formula

$$t = \frac{d}{2\sqrt{2}\tan\frac{\alpha}{2}},$$ (4)

which for $\alpha = 136°$ takes the form

$$t \approx \frac{d}{7}.$$ (5)

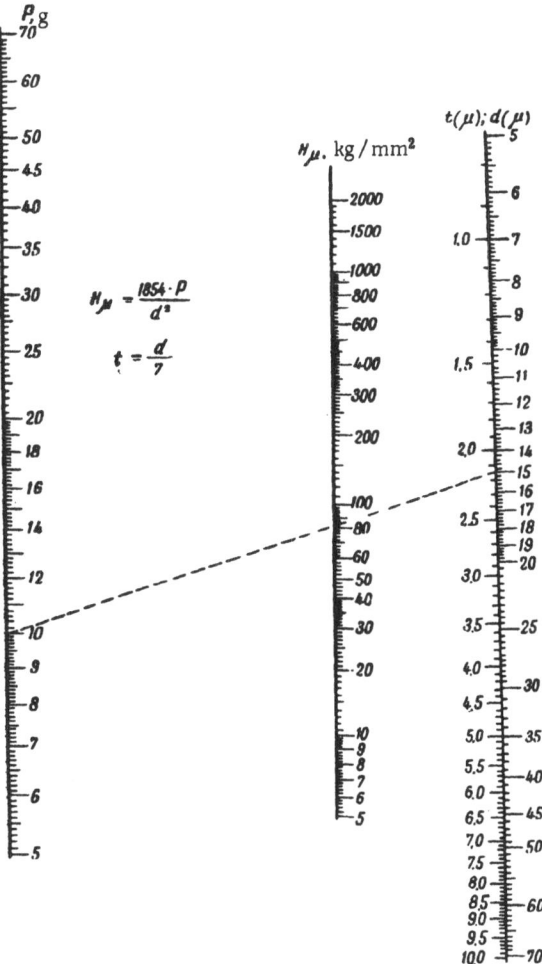

Fig. 8. Nomogram for determining hardness numbers from the load P, the diagonal d, and the depth of penetration of the diamond pyramid t when measuring microhardness on the PMT-3.

For the sake of convenience and also in order to accelerate the calculations Tables 1, 2, and 3 have been calculated for loads of 20, 50, and 100 g. The same tables may be used for loads ten times smaller (5 and 10 g) or ten times larger (200, 500, and 1000 g). In so doing the numbers found from the tables are reduced or increased by a factor of ten as necessary.*

In addition to this, a nomogram may be employed in order to calculate microhardness values.

Figure 8 shows a nomogram borrowed from the handbook by M. M. Khrushchov and E. S. Berkovich [3], enabling us to calculate the hardness numbers from the known value of the load and the diagonal of the impression or the depth of penetration of the diamond pyramid for the PMT-2 and PMT-3 instruments.

---

*Composite tables for loads of 5, 10, 20, 50, 100, 200, and 500 g are provided in All-Union State Standard 9450-60.

TABLE 3. Hardness Numbers (kg/mm^2) for Testing with a Square
(Tetragonal) Diamond Pyramid Having a Vertical Dihedral
Angle of 136° and a Load of 100 g

Diagonals of impression, μ	Hardness number, kg/mm^2									
	0	1	2	3	4	5	6	7	8	9
0						7420	5150	3780	2900	2290
10	1850	1530	1290	1100	946	824	724	642	572	514
20	464	420	383	350	322	297	274	254	236	221
30	206	193	181	170	160	151	143	135	128	122
40	116	110	105	100	95.8	91.6	87.6	84.0	80.5	77.2
50	74.2	71.3	68.6	66.0	63.6	61.3	59.1	57.1	55.1	53.3
60	51.5	49.8	47.8	46.7	45.3	43.9	42.6	41.3	10.1	39.0
70	37.8	36.8	35.8	34.8	33.9	33.0	32.1	31.3	30.5	29.7
80	29.0	28.3	27.6	26.9	26.3	25.7	25.1	24.5	24.0	23.4
90	22.9	22.4	21.9	21.4	21.0	20.5	20.1	19.7	19.3	18.9
100	18.5	18.2	17.8	17.5	17.1	16.8	16.3	16.2	15.9	15.6
110	15.3	15.1	14.8	14.5	14.3	14.0	13.8	13.5	13.3	13.1
120	12.9	12.7	12.5	12.3	12.1	11.9	11.7	11.5	11.3	11.1
130	11.0	10.8	10.6	10.5	10.3	10.2	10.0	9.88	9.74	9.60
140	9.46	9.33	9.20	9.07	8.94	8.82	8.70	8 58	8.47	8.35
150	8.24	8.13	8.03	7.92	7.82	7.72	7.62	7.52	7.43	7.34
160	7.24	7.15	7.07	6.98	6.90	6.81	6.73	6.65	6.57	6.49
170	6.42	6.34	6.27	6.20	6.13	6.06	5.99	5.92	5.85	5.79
180	5.72	5.66	5.60	5.54	5.48	5.42	5.36	5.30	5.25	5.19
190	5.14	5.08	5.03	4.98	4.93	4.88	4.83	4.78	4.73	4.68
200	4.64	4.58	4.54	4.50	4.46	4.42	4.38	4.32	4.28	4.24
210	4.20	4.16	4.12	4.08	4.06	4.02	3.98	3.94	3.90	3.86
220	3.83	3.80	3.76	3.74	3.70	3.66	3.64	3.60	3.56	3.54
230	3.50	3.48	3.44	3.42	3.38	3.36	3.34	3.30	3.28	3.24
240	3.22	3.19	3.17	3.14	3.11	3.09	3.06	3.04	3.02	2.99
250	2.97	2.94	2.92	2.90	2.87	2.85	2.83	2.81	2.79	2.76
260	2.74	2.72	2.70	2.68	2.66	2.64	2.62	2.60	2.58	2.58
270	2.54	2.53	2.52	2.40	2.47	2.45	2.43	2.42	2.40	2.38
280	2.36	2.35	2.33	2.32	2.30	2.28	2.27	2.25	2.24	2.22
290	2.21	2.19	2.18	2.16	2.15	2.13	2.12	2.10	2.09	2.07
300	2.06	—	—	—	—	—	—	—	—	—

G. A. Il'inskii [12] proposed a graph for determining the microhardness number by reference to the diagonal of the impression expressed in divisions of the AM9-2 micrometer eyepiece of the PMT-3 for a load of 20 g only. In view of the fact that the graph had a hyperbolic
nature, the accuracy of the determinations was low for small dimensions of the diagonal.

A nomogram proposed by L. G. Kharitonov is free from this failing, since it is constructed on the basis of a logarithmic transformation of the equations, and calculations may
be carried out for loads of 5, 10,..., and 1000 g by using the diagonals of the impressions
measured in the micrometer eyepiece (Fig. 9).

In order to construct the nomogram a line corresponding to Eq. (1) is drawn on a logarithmic mesh at an angle of 45° to the axes through a point with coordinates d = 30 and N = 100.
Lines corresponding to Eq. (2) are drawn through the points with coordinates d = 10 and H_μ =
18.54 P; the slope of the lines is obtained by geometrical construction. Setting off an arbitrary
section downward along the line d = 10 from the points in question, a section half as long is
taken parallel to the d axis; a second point is found, also belonging to the line representing Eq.
(2) for the specified load. Then a straight line is drawn between the two points.

The use of the nomogram is explained by the example shown in Fig. 9 (dotted and dashed
line) (N = 300, A = 0.65, P = 200 g, and H_μ = 46 kg/mm^2). The error in calculations carried
out by means of the nomogram is no greater than 1%.

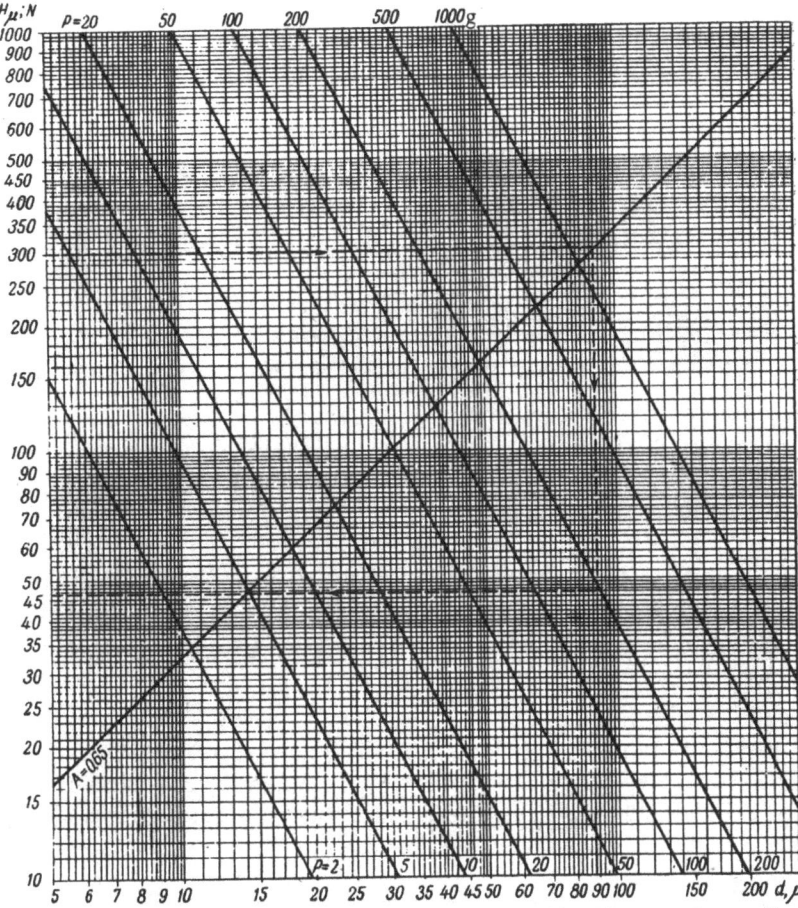

Fig. 9. Nomogram for determining microhardness H_μ on the
PMT-3.

At least three impressions should be obtained, when determining microhardness. If the results of the measurements differ substantially and constitute a consequence of random testing errors rather than errors in the method and apparatus, then the number of tests should be increased.

The impression obtained may deviate from the square shape, perhaps, if the axis of the diamond pyramid is not perpendicular to the surface on which the test is being made, or possibly if the elastic properties of the test material are anisotropic.

In the first case the test must be regarded as invalid and another carried out.

In the second case the lengths of the two perpendicular diagonals of the impression are measured and the average taken.

Some Special Cases of Using the Devices for Measuring

Microhardness

Measuring the Microhardness of Large Samples. Since it proved difficult to measure the microhardnesses of large parts and objects in the PMT-2 and PMT-3, N. A. Sologub [14] proposed combining the tube of the PMT-3 with the base of a UIM-21 universal measuring microscope. For this purpose the tube of the PMT-3 is put in place of the eyepiece head on the column of the UIM-21 microscope by means of a special bracket fixed in the appropriate position with a stop screw. The sample is set in the center of the microscope or if necessary fixed to the object table with screw clamps forming part of the microscope kit.

This arrangement enables the microhardness of cylindrical objects 12 mm in diameter and up to 300 mm long to be measured.

Measuring the Microhardness of Metals in the Stressed and Strained State. V. D. Lisitsyn [15] used a PMT-3 as the basis of a special apparatus for simultaneously deforming metal samples and measuring the deforming forces. The microhardness may be measured in this apparatus without removing the deforming load. Results obtained with aluminum, brass, and steel samples subjected to various degrees of strain verified various aspects of the physics of metals and the theory of plastic deformation.

Apparatus for Taking Samples from Microscopic Regions of a Sample by Drilling. Following a suggestion by Academician G. G. Urazov in 1949 E. S. Berkovich and A. D. Kuritsyna [16] constructed a modified version of the PMT-2 and PMT-3 facilitating the drill-sampling of metals in order to carry out a microchemical analysis of microscopic regions in the surface of metallographic samples. The modification of the PMT-2 and PMT-3 for this purpose took the following form. The diamond-pyramid mounting was removed and through the hollow rod of the loading mechanism a rod with a special diamond drill at the lower end was passed. This drill facilitated the removal of a sample from a precisely specified part of the object.

Use of the PMT-3 for Microspectral Analysis. Since existing methods of local chemical (micro)analysis are very difficult or poorly located, the idea of overcoming these difficulties by using the potentials of the PMT-3 arose. A method of microspectral analysis using the microhardness-measuring apparatus was developed by I. L. Mirkin and E. P. Rikman [17].

A "point" source of rectified high-frequency current was used. The high-frequency current obtained by means of a PS-39 generator was rectified with the help of a kenotron rectifier. A voltage of 6 V on the electrodes and a current of 0.007 A in the spark circuit was obtained.

The sample was connected as cathode; the anode was a needle arranged perpendicular or parallel to the spectrograph slit, depending on the problem in hand.

The microhardness-measuring device formed the universal stand for fixing the steel needle. The mounting of the diamond pyramid was replaced by an insertion piece in which the steel needle serving as anode was fixed with a screw. The layer of plasticene fixing the sample to the table of the apparatus formed a reliable insulator. The sample was furnished with a special clamp. The insertion piece and clamp were connected to the rectifier terminals with flexible leads.

The microspectral analyzer thus constructed may be used to measure the analytical gap to an accuracy of 0.002 mm and the diameter of the crater and the size of inclusions to an accuracy of 0.005 mm (magnification 100). In addition to this, the rupture of microscopic regions taking place during the discharges and the corresponding changes in the appearance and microstructure of the surface of the crater may be followed.

I. L. Mirkin and E. P. Rikman tested the method thus proposed by studying the magnesium distribution of high-strength nodular cast iron [18]. The apparatus in question also facilitated the determination of the amounts of a number of elements present in the center and at the periphery of the grains in samples of complex-alloyed austenitic steel.

Study of Microhardness at High Temperatures. In order to study micro-hardness at high temperatures, a special miniature electric furnace may be placed on the tables of the PMT-2 and PMT-3 for heating the test samples. If the diamond pyramid is calked into the steel mounting and not lined with soft solder, heating to 750°C is permissible.

The first attempt at using the PMT-2 for measuring microhardness at temperatures up to 350°C was mentioned in a book by M. M. Khrushchov and E. S. Berkovich [3]. The modification required for this purpose was very slight. In testing, first a point was chosen for impression on the cold sample, then the sample was heated and the diamond pyramid was pressed in, and finally the dimensions of the impressions were measured after cooling. The work was accelerated by imprinting a number of impressions arranged in a straight line in a single heating.

G. V. Boguchava [19] described an original device which may be numbered among other instruments for measuring microhardness at high temperatures. The device was intended for determining the microhardness of abrasive materials on vacuum heating to 1300°C under loads of 50 to 300 g. However, the same apparatus may be used without modification for testing metal samples.

Several devices were designed under the direction of M. G. Lozinskii [4, 5, 20-22] for studying hardness in vacuum over a wide temperature range, from room temperature to 1100°C. The same instruments may be used for determining the microhardness of samples subjected to high-temperature vacuum heating.

Study of Microhardness at Low Temperatures. The PMT-3 was used for microhardness tests at temperatures down to −55°C in connection with a study of the properties of ice by M. M. Khrushchov and E. S. Berkovich [23]. The experiments were carried out in a refrigerator room. In order to even out the temperature the PMT-3 and test sample are held for two hours at the test temperature before starting the experiments. In order to prevent the sample from being heated, the illuminating system is deliberately taken from the microscope and removed from the apparatus. Experiments at not too low a temperature are conducted in the open air in freezing conditions. Loads of 10 to 200 g are used and the time under load is 5 to 300 sec.

V. I. Trefilov and Yu. V. Mil'man [24] proposed determining the microhardness of metals at low temperatures under a layer of cooling liquid. The proposed method ensures equality of temperature in the sample and indentor at the moment of applying the load and in the intervals between applications of the load. As cooling liquids the following are suitable: ethyl alcohol, petroleum ether, pure aviation gasoline, and mixtures of organic liquids (14.5% chloroform + 25.3% methylene hydrochloride + 33.4% ethyl bromide + 10.4% transdichloroethylene + 16.4% trichloroethylene). The required temperature is reached by adding liquid nitrogen to the cooling liquid. Microhardness at high temperatures may be measured in a similar way by using glycerin as heating liquid.

By way of example, the results of some measurements of microhardness for pure and commercial bismuth are presented in [24] for the temperature range between −200 and +200°C.

Use of the Microhardness Method with Small Impressions (under 1 μ). One of the limitations of the field suitable for microhardness tests is the minimum size of the impression. When using the PMT-2 and PMT-3 in particular, considerable errors accrue from measuring impression diagonals of the order of 5 μ. For smaller impressions the measuring error increases rapidly and this makes the results insufficiently reliable. However, measurements with short diagonals are extremely necessary when determining the micro-hardness of individual particles or the structural constituents of alloys having sizes of less than

10 μ in cross section, the peripheral layers of crystals and the material at their boundaries, surface layers of the order of 3.5 μ thick, and so on.

In order to solve this problem an electron microscope may be used rather than an optical microscope [4, 25].

The microhardness method was used in the field of small impressions by M. M. Khrush-chov and E. S. Berkovich [3, 25] when studying the microhardness of synthetic ruby and quenched ShKh15 steel. The impressions were made in the PMT-3 with a load of 1 g, using diamond pyramids of two types:

1) A tetrahedral pyramid with a square base and a vertical angle of 136° between opposite faces;

2) a trihedral regular pyramid with a vertical angle of 65° between the side and the height.

In the case of a trihedral pyramid a sharp vertex is always obtained, although this is generally not realized for the tetrahedral type. This important difference between the two types gives the trihedral pyramid a considerable advantage in the case of very small impressions.

A lacquer replica was taken from the surface of the ruby on which the impressions had been made; this replica was shadowed with chromium and photographed in the electron microscope. The dimensions of the impressions were determined from the photographs.

For impressions made by the trihedral pyramid the microhardness was determined from the formula

$$H_\triangle = \frac{1.570P}{l^2} \ \text{kg/mm}^2, \tag{6}$$

where P is the load in kg, l is the height of the triangular impression on the test surface in mm, determined as the average of three measurements.

Measurements for pyramids of both types were made with a load of 1 and 100 g. The results indicate that the microhardness is independent of the size of the impression (or load) in the range studied. It was further noted that the small impressions (1 to 10 μ) appeared more sharply in the replicas that those made by the tetrahedral pyramid and this facilitated their measurement.

Study of the Thermoelectric Properties of Materials in Small Volumes. In studying the structures of metals and alloys under the microscope it is often also important to study the physicochemical properties of individual phases or structural constituents.

In 1946 G. V. Akimov [26] suggested the possibility of making precisely directed ("aimed") measurements of the thermo-emf of individual structural constituents. Provided that they were not too fine, individual structural constituents could be pinpointed in Akimov's apparatus with a bent, heated needle at 100- to 150-times magnification. The author recommended his method for identifying individual phases and structural constituents in alloys. Then E. S. Berkovich also tested the possible use of the PMT-3 for this purpose.

V. N. Novogrudskii and I. G. Fakidov [27] and E. M. Strug and E. V. Panchenko [28] advocated Akimov's method again.

The possibility of using microthermo-emf as a means of physicochemical analysis for solving various metallographic problems was demonstrated in [29].

An apparatus was constructed for measuring the thermoelectric properties of materials in very small volumes. In this case "very small" (micro) means of the same order of volumes as in the measurement of microhardness.

An attachment was made for operation together with the optical system of the PMT-3; this enabled "aimed" (precisely directed) measurements of thermo-emf to be made at any point on the microsection of test material. Needles were fixed in the mounting by means of two ebonite insertion pieces and stop screws. To one side of the mounting a steel bushing was screwed; this had a socket for fixing a miniature furnace with the aid of stop screws. The length of the needle protruding from the furnace could be regulated. The material of the needles was chosen so as to obtain a thermo-emf of the desired magnitude. The miniature furnace was made by winding a Nichrome wire 100 μ in diameter on a quartz tube, the internal diameter of which was taken in accordance with the diameter of the needle in such a way that there should be only a small gap between them. A layer of thermally insulating lubricant was deposited from the outside. The external diameter of the furnace was no greater than 5 or 6 mm.

The furnace was heated with alternating current controlled by a rheostat. The heating was monitored by means of an ammeter and also by measuring the temperature of the needle surface with a thermocouple at some distance from the end (3 to 4 mm). Of course the thermo-couple readings were not the same as the temperature at the end of the needle, but they gave a fair indication of this.

The apparatus in question was fixed in the PMT-3 instead of the bushing usually holding the diamond pyramid for measuring microhardness. After this the system was adjusted.

In order to let down the mounting carrying the needle and furnace, the same mechanism was used as in the case of microhardness measurements. A thermo-emf developed as soon as the heated needle touched the sample.

The throw of the galvanometer scale needle could be regulated for a particular choice of material and dimensions of the electrode by varying the temperature at the end of the needle and also by using the rheostat in the galvanometer circuit. Other conditions being equal, the value of the thermo-emf is determined by the properties of the very small volume with which the heated end of the needle comes into contact.

The distinguishing feature of the method is the useful combination of microscope analysis and thermo-emf measurement, which offers the possibility of making measurements at any point of a microsection selected under the microscope.

The order of operation when measuring thermo-emf is the same as in measuring micro-hardness. The method is best used for comparative measurements by expressing the thermo-emf in millimeters of scale. Absolute measurements may nevertheless also be made by choosing an appropriate needle/sample pair giving a known thermo-emf at a specified temperature; the apparatus may accordingly be calibrated. It is then essential to make a thermal calculation allowing for the parameters of the sample and needle materials and to estimate the temperature at the point of contact more exactly.

It should be noted that in thermo-emf measurements the needle penetrates to a specific depth in the sample, depending on the relative hardnesses of the sample and needle materials for the load employed. At the same time, the character of the temperature field, the resistance at the point of contact, and hence the value of the current flowing (as measured by the galvanometer) depend on the area of contact; this must be allowed for in comparative and absolute measurements by varying the load so as to secure identical sizes of impression from the needle.

When studying solids with a comparatively soft needle, it is recommended that smallish loads (not over 5 to 10 g) should be used and the tip of the needle slightly blunted.

The microthermo-emf method was used in [29] for studying the liquation inhomogeneity in the germanium−silicon and bismuth−antimony systems and the chemical interaction between the components in copper−chromium−zirconium alloys.

Use of Microhardness Meters when Studying Plastic-Deformation Processes and the Motion of Dislocations. V. Z. Benguz-Olevskii and V. I. Startsev [30] used the PMT-3 for the local deformation of crystals and analyzed the results with a view to elucidating the nature of the deformation processes. The experiments were carried out with calcite crystals. By studying the character of the breaks developing in the crystal as seen in the polarization microscope on reversing the contrast, one may judge as to whether the deformation is effected by twinning (on changing the contrast in the polarization microscope twin layers become light instead of dark) or by the formation of slip lines (these are not visible either in polarized light, or in indirect illumination, in the so-called "dark-field" condition). By comparing the lengths of the slip lines (smaller) and twins (greater) originating in the impressions, it was concluded that there was a considerably greater resistance to the motion of complete dislocations in calcite than to the motion of twinning dislocations.

V. Z. Bengus-Olevskii* gradually reduced the stress on the indentor (a steel sphere was driven into the surface of the calcite) and achieved purely elastic contact between the sphere and the crystal, for which no dislocations appeared. This corresponded to the maximum tangential stress

$$\tau_{max} = 1.86 \frac{P}{\pi d^2},$$

where P is the load; d is the diameter of the sphere.

It was found that

$$\tau_{max} = 0.01G,$$

where G is the shear modulus; i.e., the homogeneous generation of dislocations does not take place in calcite even at stresses exceeding the yield stress by one or two orders.

In the same paper the PMT-3 was used in order to study the motion of twinning dislocations under the influence of a mechanical load. This time the distance between the impression and the twin layer in the calcite was varied. The velocity of the dislocations along the interlayer varied in accordance with this distance. Calculations carried out without allowing for the stress fields of neighboring dislocations (coarse approximation) showed that the starting stress was 10 to 60 g/mm^2. In order of magnitude this agrees with the macroscopic yield stress of calcite.

Similar experiments may be set up for metal and semiconductor crystals.

A slightly modified method of measuring microhardness was used in [31] for studying the plastic deformation of germanium single crystals. Pobedit points (diameter of curvature 280 μ) were driven in at 550 and 780°C in a nitrogen atmosphere. For a load of 200 g the number of slip bands was very low; however, on increasing to 350 this rose and thereafter changed little, in complete agreement with the germanium hardening curve. First elastic strain took place; then a large number of sources of dislocations with approximately equal activation energies came into play; as a result of the multiplication of these, hardening occurred and a great increase in stress was required for further multiplication of the dislocations.

---

*V. Z. Benguz-Olevskii, Dissertation, Institute of Crystallography of the Academy of Sciences of the USSR, Moscow (1963).

It was noted that in the case of large deformations (strains) there were three regions of dislocation density: the central region (or "cloud") with a density of the order of 10^8 cm^{-2}, a region with clearly-visible, separated slip lines (density 10^7 to 10^5 cm^{-2}), and a region of lower density (10^4 to 10^3 cm^{-2}). The length of the slip lines was almost independent of the distance to the point at which the indentor was applied.

It is also interesting to use microhardness meters when studying the results of irradiation, with the subsequent development of etch figures near the impression [32]. The hardness or stressed state of the material may be estimated by reference to its relative tendency toward the generation and motion of dislocations as a result of the application of the impression.

A lithium fluoride single crystal was irradiated with x rays, part of the sample being screened. After this a series of impressions were made, the line intersecting the boundary between the irradiated and screened regions. Both edge and screw dislocations were revealed near the impressions by etching.

On passing from one region to the other, the microhardness undergoes little change, whereas the size of the etch-figure pattern around the impression changes substantially. Whereas in the screened section the impression from a tetrahedral diamond pyramid with a vertical angle of 136° under a load of 100 g is 1.2 times greater than in the irradiated section, as regards the size of the pattern these sections differ by a factor of 4.3. The asymmetry of the patterns also indicates the direction in which the hardness or stress vary in the test sample.

Studying the Properties of Crystals by Scratching Their Faces and Examining the Scratches. Recently the method of measuring microhardness by scratching has become widely employed [33]. A PMT-3 with a tetrahedral diamond pyramid and a load of 5 to 15 g is used for this. Special means are engaged to create a scratch at a velocity of about 0.01 cm/sec [34-42].* In this case the microhardness is estimated from the width of the resultant scratch. It was shown in [34] that different microhardness rosettes were obtained on the same faces of single crystals for different scratching methods. It was suggested and later confirmed in [37, 38] that the anisotropy of microhardness observed on using a standard PMT-3 pyramid instrument for scratching was associated with the disposition of the slip elements in the crystals rather than with cleavage.

The use of PMT-3 instruments for testing by the scratch method has proved most fruitful when studying the anisotropy of the properties of the faces in both semiconducting and metallic crystals.

Anisotropy has been studied in detail for beryllium [43, pp. 23-28], antimony [43, pp. 29-34; 44], bismuth [39, p. 131], tellurium [39, p. 64], InSb [45], In$_2$Te and InBi [5], Sb$_2$Se$_3$ [46], and also solid solutions of the Te−Se and Te−S systems [47].

---

* Yu. S. Boyarskaya, Dissertation, Kishinev State University (1954).

CHAPTER 2

TECHNIQUES OF DETERMINING MICROHARDNESS

Evaluating the Object and Choosing the Point

for Studying Microhardness

The selection of samples of the test material is an extremely vital operation, the neglect of which may lead the research worker to false conclusions.

The sample to be studied must reflect the properties of the test metal or alloy correctly and not possess random deviations from these properties. The cutting of the sample must be carried out with due regard to the problem in hand, being determined by the particular requirements in each individual case.

After the cutting of the sample and its appropriate treatment (deformation, quenching, annealing, tempering, etc.) we have a no less important operation, the preparation of the surface for the microhardness measurements. Questions of preparing the surface are considered in detail in Chapter 3. Here we simply note the great importance of the orientation of the plane of the microsection to be tested in the material. The choice of the position of this plane necessarily involves consideration of structural features such as the crystallization or machining texture, various forms of edge damage, the arrangement of crystals or dendrites, and so on. The plane in question must be situated so as to enable tests to be carried out under conditions eliminating the possible influence of such structural characteristics on the results.

Direct measurement of the microhardness is usually preceded by a microscope study of the structure of the material, during which the test object is evaluated visually and places are chosen for making the impressions. This operation demands special attention to the possible effects of various factors associated with the structural features of the material as a whole and also the shape, size, and relative disposition of the phases and structural constituents under investigation.

The simplicity of the actual microhardness test in the technological respect may easily lead one into error, and neglect of all the preliminary work in the selection of the sample, the preparation of the test object itself and its surface, the evaluation of the material, and the choice of a spot for impression may invalidate the research. The method of measuring microhardness is extremely sensitive to various factors, and this sensitivity is decisive when the method is used for certain particular purposes. Hence as in any other sensitive experimental method a certain amount of experience in the work and an exact knowledge of its characteristic features are required.

A knowledge of the structure of metals and alloys is particularly important. In addition to this, one needs experience in handling the optical and simple but accurate mechanical control and measuring systems. Hence the servicing of microhardness meters and the carrying out of experimental work should only be entrusted to personnel with special qualifications.

Choice of Load

According to the earlier-established and accepted point of view, the hardness determined by the impression of a cone or pyramid is independent of the load. This follows from the law of similarity; it was established for a cone by Ludwick [48] and confirmed for a pyramid by Smith and Sandland [49], using loads between 10 and 100 kg. However, in measuring microhardness, i.e., using very small loads (order of grams or tens of grams), from the very outset deviations from this law were observed; the discrepancies lay in both directions and obeyed no obvious laws. Hence the first question presenting itself for solution was whether the law of similarity were valid in the case of impressions having diagonals of the order of 1 to 10 μ and a volume of the order of 10^{-11} to 10^{-8} cm^3. From this one might be able to draw conclusions regarding the comparability of micro- and macrohardness results.

Investigations showed that the law of similarity lost its significance for such small impressions. This was due to a number of causes, some of which still remain uncertain.

Thibault and Niquist [50], using the Knoop method of measuring microhardness, found a sharp rise in the hardness of nonmetallic materials under small loads.

Campbell, Henderson, and Donlevy [51], using a tetrahedral pyramid and loads of 5 to 100 g in an apparatus of their own construction, obtained a sharp fall in hardness with falling load, starting at 50 g.

A. A. Bochvar and O. S. Zhadaeva [52] made measurements on carefully prepared surfaces. The measurements showed that with increasing load the microhardness first increased and then after passing through a maximum starting falling slowly. Qualitatively similar results were obtained by D. B. Gogoberidze and N. A. Kopatskii on a PMT-2 and V. A. Egorov on a PMT-3 [4].

D. B. Gogoberidze and N. A. Kopatskii proposed [53] that this dependence of microhardness on load had a universal character. On the other hand, B. I. Kostetskii and P. K. Topakha [54] asserted that the microhardness was independent of a certain quantity which they used as a measure of the load.

Hanemann and Bernhardt [55], and Schultz and Hanemann [56] made measurements on various materials and came to the conclusion that in certain materials, for example, sodium fluoride, the microhardness increased with increasing load, while in others, such as metals, it diminished.

Thus the experiments which have been carried out give no unique answer to the question as to the effect of load on microhardness. Such an answer can hardly be obtained by direct methods, since up to the present time there has been no generally accepted explanation for the dependence of microhardness on load. It is entirely possible that the establishment of a law reflects not so much the dependence of the microhardness on the load as its dependence on the dimensions (chiefly the depth) of the impression and the properties of the layer of metal bearing the impression. Owing to the cracking and low strength of the surface layer (physical factor [57]), its subjection to intensive erosion or decomposition (chemical factor [58]), and also owing to the possibility of surface work-hardening, a change in the load used may lead to the penetration of the pyramid into layers differing in physical or chemical properties. The resultant value of the microhardness will reflect the characteristic properties of these layers.

All this imposes special demands on the quality of the surfaces of objects studied by the microhardness technique.

In choosing the value of the load for microhardness tests it is important to allow for two further circumstances, one of which calls for as large loads as possible and the other for small loads.

Analyzing Eqs. (2) and (3) in Chapter 1, it is easy to see that, if we keep all other parameters considered when calculating the microhardness constant, the relative computing error is proportional to the relative error in the applied load. If we consider that the permissible deviation is about ±1%, then the greater the load used the less will this error affect the results of the microhardness measurements (for a load of 10 g the relative measuring error is ±0.1%, for 100 g it is ±0.01%). However, when carrying out microhardness tests for purposes of physicochemical analysis, using the same load, the absolute error introduced by any deviation of the weight of the load from its norm may be neglected.

The size of the impression increases at the same time as the load. If other parameters affecting the accuracy of calculations based on Eqs. (2) and (3) are kept constant, the relative error in determining the impression diagonal is not doubled but is transferred to the final result. Hence if we consider that the accuracy in determining the diagonal is about ±0.15 μ, then the greater the size of the impression the less will this error affect the results of the microhardness measurements (for a diagonal of 5 μ the accuracy of the measurement is ±3%, for 20 μ it is ±0.75%).

For example Brown and Ineson [59] established that the values of microhardness obtained by research workers in 13 laboratories using 10 instruments with loads of 10 to 20 g deviated by up to 50% from the norm, while for loads between 100 and 200 g the deviation was 15%.

On the other hand, the size of the crystals and structural constituents tested always limit the permissible loads and the size of the impressions. In addition to this, for certain relationships between the shape and ductility of the structural constituents, on greatly increasing the load these constituents may move relatively to each other under the impress of the indentor, and this will introduce an additional error into the results of the microhardness measurement. In individual cases, raising the load leads to the rupture of the structural constituents under test, and this will invalidate the measurements.

In addition to this, as we shall show later, on increasing the load the influence of factors such as the rate of loading, the period spent under load, and so on, will appear to a greater extent.

Taking all this into consideration, as regards the general effect on load on microhardness, in any particular case we must choose the load empirically, seeking an optimum compromise.

In order to make this easier, we may draw up some practical recommendations based on earlier measurements (Table 4).

Choice of Loading Time and Period under Load

The correctness of the result obtained from a microhardness test largely depends on the speed at which the test is carried out. For a detailed analysis of the factors at work, the process of making the impression may conveniently be divided into three stages: loading, holding under load, and unloading.

It is evident that the rate of unloading will have the least effect on the result of the measurement. However, remembering that to some extent the existence of forces of cohesion between the test material and the diamond may enter or friction may arise, it is best to remove

TABLE 4. Examples of Loads Used for Measuring
the Microhardness of Certain Metals
and Semiconductors

Object	Load, g	Order of magnitude of the diagonal, μ
Bismuth	10	20-30
Cadmium	10	20-30
Aluminum.......	10	30-32
	20	40-45
Magnesium	20	25-28
	50	40-43
Copper	20	18-22
	50	30-35
Silicon	50	10-20
Germanium	50	10-20
$A^{III}B^{V}$ compounds..	50	10-20
Borides	100	8-10
Carbides	100	8-15
Nitrides	150	10-20

the load reasonably slowly. The time of unloading should be roughly equal to the time of loading.

A change in the duration of the loading process and the period under load has a considerable effect on the result of microhardness measurements.

Increasing the speed of loading leads to a fall in the value of microhardness obtained. This is associated with the fact that deformation takes place not only under the static effect of the weight chosen for the test but also under the associated dynamic action, equal to the component of kinetic energy involved. As a result of this there is a greater deformation of the material, the impression becomes larger, and the calculated microhardness diminishes.

As the values of the applied load falls, the relative effect of the mass of the moving part of the apparatus increases rapidly [60]. In addition to this, loading at high speed inevitably involves a large amount of vibration and shaking. The loading conditions should accordingly be made as static as possible.

However, too slow loading will extend the time corresponding to the intermediate position of the indentor, when the pyramid is only partly loaded and the material is not under the action of the nominal test load. If the time under load is short, an undue increase in loading time introduces error of another kind, and the value of microhardness obtained may be either too low or too high.

Thus in order to ensure that the loading conditions should have the least effect it is important to choose the optimum loading time and to carry out the loading uniformly.

Experience shows that on working with the PMT-2 and PMT-3 it is entirely sufficient to carry out the loading in a period of 5 to 8 sec.

Still more rigorous demands are made on choosing the time spent under load and keeping this constant for all tests. Of the factors most liable to distort the results of the microhardness measurement as a result of changes in the time spent under load, we may primarily mention creep and vibration.

In a number of cases in which a material with a low creep limit is being studied, the pyramid may penetrate progressively into the material during the period under load until the

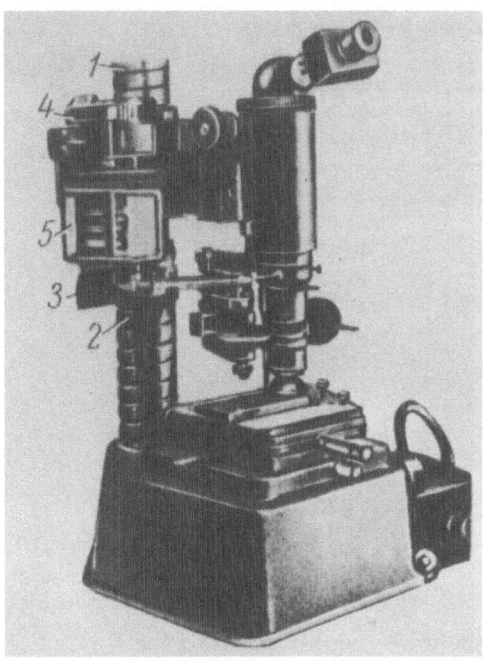

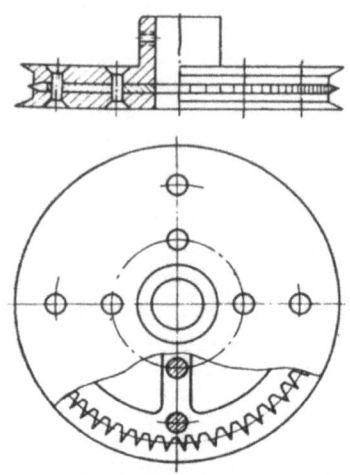

Fig. 10. General view of the PMT-3
with the attachment for automatic loading [63]: 1) Stand; 2) tie; 3) bracket; 4)
motor; 5) reducer.

Fig. 11. Construction of the
gear for the belt drive for the
attachment of [63].

average pressure evens out or falls below the creep limit. However, for the majority of metals
under small loads at room temperature the effect of creep is not too serious.

Bergsmann (see [61]) showed that for a load of 300 g the microhardness of steel after
holding under a load for 15 and 25 sec fell by 0.5 and 1.0% respectively as compared with the
value obtained after holding for 5 sec; he recommended that a holding time of about 5 sec should
be used for steels.

Hendey (see [61]) studied the dependence of the microhardness on time under load and
found that creep occurred in annealed copper even after 30 sec, whereas in worked copper no
such effect was observed.

Knoop, Peters, and Emerson (see [61]) recommend holding under load for 20 sec on the
basis of the fact that after this time the microhardness of most solids changes very little.

Any undue increase in time under load, however desirable in the interests of securing a
stable creep tendency of the material in question, should be avoided owing to the possible effects of shaking and vibration.

Shaking and vibration, although not seriously affecting the shape of the impression and remaining small in magnitude, may nevertheless introduce appreciable errors into the results.
The error introduced is particularly serious for small loads [60]. Experience shows that careful shock protection of the apparatus still fails to eliminate the effects of shaking completely.
Hence in order to eliminate this influence the time spent under load should be reduced to a minimum.

In the majority of cases quite satisfactory results may be obtained with the PMT-2 and
PMT-3 for holding periods of 5 to 10 sec.

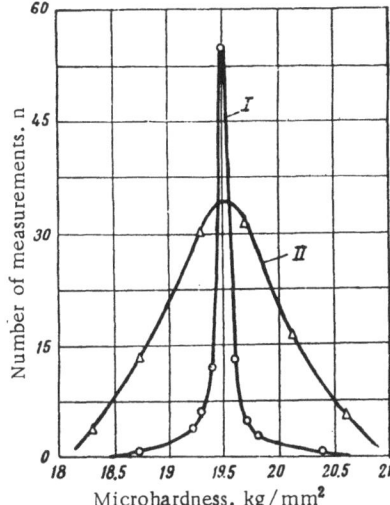

Fig. 12. Effect of the method of loading on the scatter in microhardness data for a pure aluminum single crystal: I) Automatic loading [63]; II) manual loading.

Special attention should be directed not only to the necessity of specifying the optimum loading time and time under load but also to the preservation of identical conditions throughout all the test measurements. This may best be achieved by automating the loading, holding, and unloading processes.

Measuring Microhardness with Automatic Sample Loading

In the PMT-3 microhardness meters now being produced by industry [2-4] and also in foreign microhardness meters [61] the indentor is let down manually when loading the sample at the moment of testing and then removed manually.

B. Ya. Petrenko's attachment [62] to the PMT-3 for automating the loading process has not been widely accepted, as it demands a rather radical modification to the loading system.

V. M. Glazov and V. A. Borisov [63] designed and constructed a simpler attachment to the PMT-3, facilitating uniform raising and lowering of the diamond pyramid in a strictly regulated time during the loading and unloading of the sample. Figure 10 shows a photograph of the PMT-3 with this attachment.

A bracket holding a reversible motor with a reducer is fixed to the central support with a tie coupling. Rotational motion is imparted to the indentor from the reducer by means of a belt drive, for which purpose two special gears are sited on the axles of the indentor and reducer (Fig. 11), giving uniform transmission of the motion with no slipping. This arrangement enables the sample to be loaded for microhardness measurements in 1.5, 5, 15, and 60 sec by putting the sliding gear in the reducer into the correct position.

A study of the working of the system revealed that no vibrations occurred either in loading or unloading the device.

Automation of the loading process when measuring microhardness greatly reduces the scatter in the results. For illustration, some comparative measurements were made on the aluminum of the AV000 type (99.998% Al) with manual and automatic loading of the sample.

In order to avoid the effect of any side factors associated with the presence of grain boundaries in the matrix, the tests were carried out on a single-crystal sample obtained by zone crystallization and vacuum-annealed at 400°C for 3 h in order to eliminate initial stresses which might arise on cooling after passage of the molten zone. The surface of the sample was prepared in two ways: electropolishing and mechanical polishing. In order to eliminate the influence of surface cold hardening due to the grinding and polishing, the sample was again annealed in vacuum at 400°C for 1 h. The effect of the method of loading samples so prepared on the scatter of the resultant microhardness data was studied.

One hundred tests were made on the same sample by both manual and automatic loading. In automatic loading the loading, resting, and unloading time amounted to 5 sec. In manual loading an attempt was made at preserving the corresponding timing. The load was 10 g.

TABLE 5. Comparison of Microhardness Measurements with
Manual and Automatic Loading on the PMT-3 Equipped with
the Attachment of [63]

P, g	Loading	Serial No.										Mean of 10
		1	2	3	4	5	6	7	8	9	10	
10	Manual	19.2	18.3	18.7	17.5	17.6	17.8	21.5	20.0	19.2	21.5	19.0
	Automatic	18.3	19.2	20.0	18.7	19.2	20.0	20.5	20.0	19.2	18.7	19.4
5	Manual	18.0	21.3	17.2	16.5	18.7	20.3	17.2	18.0	18.7	19.5	18.5
	Automatic	20.3	19.5	20.3	19.5	18.7	18.7	18.7	21.3	19.5	19.5	19.5
1	Manual	18.2	17.1	21.1	18.2	17.6	18.7	17.6	17.1	20.0	21.1	18.8
	Automatic	18.7	19.4	19.4	18.2	20.0	18.7	20.0	19.4	18.2	19.4	19.1
0.5	Manual	17.8	16.6	17.8	21.3	15.3	19.5	14.3	23.6	17.8	17.8	18.2
	Automatic	17.8	18.6	18.6	20.4	17.8	19.5	20.4	19.5	18.6	17.8	18.9

The results of the measurements were analyzed statistically and converted into the frequency curves of Fig. 12. This shows that in automatic loading the scatter was much lower than in manual. More than 90% of all the measurements in automatic loading lay within ± 0.2 kg/mm^2 of the true value and 80% within ± 0.1 kg/mm^2.

From this we may conclude that, in order to obtain reliable values of microhardness, which are usually taken as the average of a small number of measurements (often 15 to 20), much less time is needed with automatic loading.

It should also be noted that with automatic loading there is a considerably smaller scatter in the microhardness data when using small weights. Table 5 shows the results of comparative tests using the same aluminum single crystals with manual and automatic loading based on different loads, starting from 0.5 g.

We see from Table 5 that with automatic loading the scatter in the results with the small loads is much smaller and the average results for different loads agree better with each other than in the case of manual loading.

Another version of a device for automating the loading when determining microhardness on the PMT-3 was proposed in [64].

The loading and unloading of the weights was here carried out smoothly, the part of the apparatus bracketed to the central support (microscope, illuminating system, and loading device) not being overloaded and the apparatus as a whole not experiencing additional vibrations.

The device was made in the form of an attachment to the apparatus. This constitutes an individual unit, comprising a master mechanism, the regulating shoe of an SD-2 motor, a VME-609005 microswitch, an RPT-100 relay, and a starting button, mounted on the body of the lighting transformer independently of the PMT-3 (Fig. 13).

The master loading mechanism consists of four profiled working discs and one contact disc 3 mm thick, possibly made of vinyl plastic or organic glass. The discs are mounted on a single axle at a distance of 3 mm and are fixed together in such a way that when the motor is stopped by the contact disc all the working discs are in the position "load off."

The working discs have a shape ensuring smooth and uniform loading and unloading of the weight. The profile of the working discs is formed by circles: 1) of diameter d = 70 mm on an arc of 60 $\omega\tau$ where ω is the angular velocity of the disc in rpm and τ is the required holding time under load (sec), and 2) of diameter d_1 = 65 mm, on a section of free path, for which the indentor remains in the arrested state. The transition from d to d_1 is smooth (along an Archimedes spiral). The working discs are calculated for holding periods under load of 3, 5, 10, and 15 sec. The loading time is 6 sec.

Fig. 13. General view of the PMT-3 with the attachment for automatically operating the loading mechanism [5, 64].

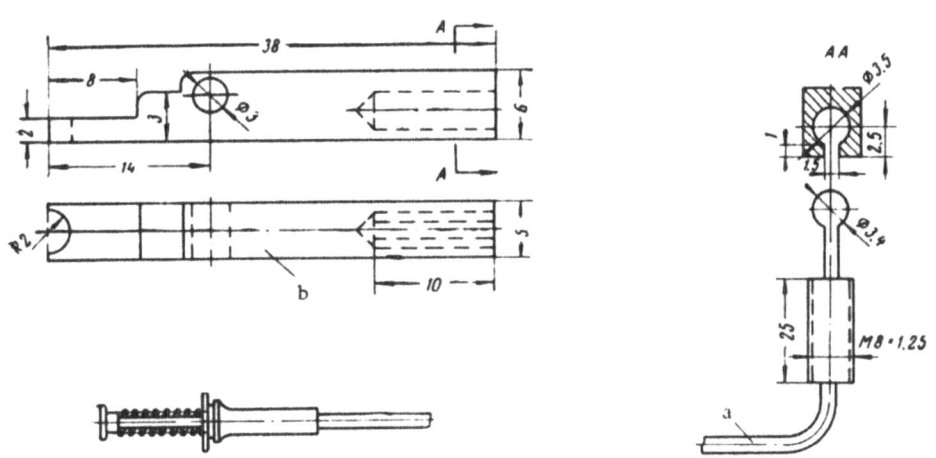

Fig. 14. Parts of the attachment to the PMT-3 for automatically operating the loading mechanism [5, 64]: a) Cable; b) lever of stopping device.

The indentor with the diamond pyramid is let down and lifted up by an executive mechanism (Fig. 14) consisting of a stop lever and a cable from a photographic shutter, similar to those in the MIM-7 and MIM-8 metallographical microscopes. The cable has an end piece with a conical thread, replacing the previous end piece with a cylindrical thread (M8×1.25). The striker of the cable is attached to the head of a spherical universal joint (diameter 3.5 mm) sited in a special socket in the stop lever. An extra spring is placed on the rod of the cable control. For operating the attachment, the end of the cable furnished with the universal head

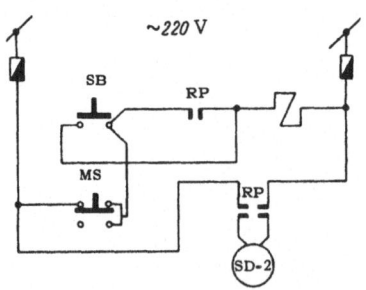

Fig. 15. Electrical circuit of
the attachment [5, 64].

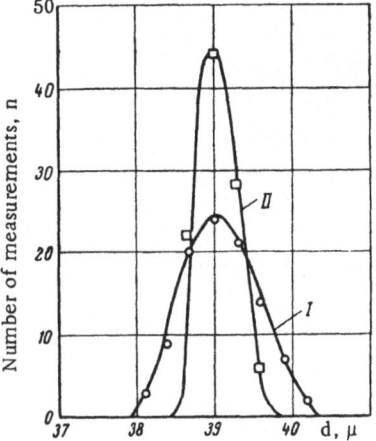

Fig. 16. Comparison of microhardness measurements with
manual (I) and automatic (II)
control of loading on the PMT-3,
obtained when testing the attachment described in [5, 64].

is introduced into a drilled aperture (diameter 8.5 mm)
in the lower bracket of the loading mechanism and fixed
with two nuts. The universal head is introduced into the
socket of the stop lever. The other end of the cable is
fixed in the shoe clamp, which may be moved by means
of a screw nut and stopped against any of the working
discs on the motor axle.

Figure 15 shows the electrical circuit of the attachment. The start button SB operates the relay RP (RPT-
100), which feeds the motor SD-2. When the motor (with
the master mechanism on its axle) rotates, the profiled
working disc bears on the cable control, and a forward
motion is imparted to the striker with the spherical universal joint and the stop lever. The lever, rotating about
its axis, frees the indentor with the diamond pyramid.
After the holding period under load, the striker returns
to its original position under the influence of the spring,
stopping the indentor. After the removal of the load, the
contact disc breaks the circuit of the relay coil by means
of the microswitch MS and the motor stops.

The proposed method of moving the lever of the
stopping device has a certain advantage, as it allows the
arm of the lever opposite to the indentor to be lengthened.
This in turn reduces the harmful effect of nonuniform motion of the drive mechanism on the motion of the indentor
while being raised or lowered. With the existing arrangement of the stopping mechanism, an increase in this arm
of the lever was clearly limited owing to the necessity
that the thread on the axle of the stop mechanism should
be self-locking under the action of the load, i.e., the necessity of increasing the pitch or number of turns of the thread.

Table 6 presents the results of microhardness
measurements on an aluminum single crystal under a load
of 20 g obtained with manual and automatic loading and un-
loading on the PMT-3. We see that with both methods of loading the average microhardness is
24.4 kg/mm^2. However, the scatter in the values obtained falls sharply on automating the loading process. This is indicated by the frequency curves of Fig. 16 based on 100 impressions,
characterizing the accuracy of the microhardness measurements.

E. S. Berkovich [5, 65] proposed an unmotorized drive for automatic loading and unloading in microhardness tests. The proposed construction is based on the electrical heating of a
thermal bimetal plate, which stops the suspension of the indentor in the PMT-3. There is a
possible version of this system suitable for automation; in this the thermal bimetal strip moves
a lever, the sensitivity of the latter being achieved by the use of hollow-cast pivots in corundum bearings. In addition to the stopping of the indentor, provision is made for the automatic
movement of the coordinate table.

Thus the automatic system of the Berkovich device [5] controls the motion of the table
and the progress of the experiment itself (loading, holding under load, and raising the indentor).
The motion of the object table between individual positions covered a distance of 50 μ. The
duration of a single test cycle was 45 to 50 sec, the holding time under the load being 20 sec.

TABLE 6. Microhardness Measurements of an Aluminum Single Crystal with Manual and Automatic Loading Using the Device of [5, 64] (Load 20 g, Loading Time 6 sec, Time under Load 6 sec) *

Manual loading			Automatic loading		
d, μ	H_μ, kg/mm^2	n	d, μ	H_μ, kg/mm^2	n
38.1	25.5	3	—	—	—
38.4	25.1	9	—	—	—
38.7	24.7	20	38.7	24.7	22
39.0	24.4	24	39.0	24.4	44
39.3	24.0	21	39.3	24.0	28
39.6	23.7	14	39.6	23.7	6
39.9	23.3	7	—	—	—
40.1	22.9	2	—	—	—

* d) diagonal of impression, H_μ) microhardness, n) number of measurements.

However, tests with the proposed apparatus showed that the value of the load had a considerable effect on the time taken for loading and unloading the sample.

M. S. Ablova and A. A. Averkin [66, 67] proposed the most highly perfected construction of attachments for the PMT-3. The system recommended by these authors provided for: 1) automatically loading and unloading the sample; 2) changing the rate of loading between 28 and 5 sec; 3) varying the time under load between 1 sec and several hours (so as to be able to study local creep in the materials); 4) making a series of impressions at arbitrary loads (25 to 50 impressions in each series). The attachment required no rebuilding of the apparatus; most of it was assembled in the form of a separate unit.

The general form of the apparatus with the automatic attachment appears in Fig. 17. Two motors are attached to the system; one of them loads and unloads the sample and the second moves the sample table. The control part of the apparatus (the electrical circuit) is made in the form of individual units connected to the apparatus by plugs and sockets.

A brass plate 2 with an SRD-2 motor is fixed to the central support of the system 1; two brass discs 3 and 4 (with projections) and a pulley 5 are fitted to the motor axle. The plate 2 carries contacts 6 (K_2 and K_3 in Fig. 18) transmitting the signal from the apparatus to the electrical circuit. A pulley 7 is fitted to the indentor axle under the handle provided for manual loading. Motion is imparted by the motor to the indentor through a belt drive between pulleys 5 and 7. In order to avoid pulling of the belt and also in order to ensure uniformity of transmission, a notch is made inside the pulleys and a satellite pulley is arranged on plate 2 to regulate the tension in the belt.

Ablova and Averkin [66, 67] made two pulleys for the axle of the indentor, 14, 28, and 42 mm in diameter. By changing the pulleys on the motor and indentor axles, six combinations could be obtained and hence six possible loading times, 28, 15, 14, 9, 7, and 5.5 sec (supposing the motor to rotate at 2 rpm).

The electrical circuit of the apparatus is shown in Fig. 18. The left-hand side of the circuit (see also 8 in Fig. 17), providing loading and unloading of the sample, has three relays P_1, P_2, and P_3 of the KRD-2 type, one relay P_4 of the RP-4 type, and a neon lamp L_1 of the MN-3 type. After pressing the start button BK$_1$ the relay P_1 operates (the armature of this relay remains attracted even after the button has been released), and this starts the SRD-2 motor. The sample is then loaded.

When the motor starts turning, the discs with the projections 3 and 4 are rotated. At the instant at which the projection on disc 3 closes the plates of the contacts K_2 the relay P_2 operates;

Fig. 17. General view of the PMT-3 together with the device for automatic
measurement [66, 67].

this disconnects P_1 and thus stops the motor, and also connects the circuit for evaluating the
time spent by the sample under load. The holding time is determined by the rate of charging
the capacity C_1 through the resistance R_1 and the potentiometer R_2. When the voltage on the
plates of the condenser C_1 reached the ignition potential of the lamp L_5, the condenser discharges through the neon lamp and the winding of relay P_4. When the plates of P_4 close, relay
P_3 operates (its armature remains attracted even after relay P_4 has been disconnected); this
connects the motor in the opposite direction. Thereupon the contacts K_2 are opened and relay
P_2 is disconnected. Relay P_3 is disconnected when contact K_3 is opened by the projection on
disc 4. The motor stops and the load−hold−unload cycle ends. The holding time may be varied
widely by changing the charging time of the capacity C_1 (by varying the resistance R_2).

Because of the way in which the plates of relays P_2 and P_3 are connected, the start button
BK_1 cannot reconnect relay P_1 while the load −hold −unload cycle is proceeding. The two discs
with projections 3 and 4 on the motor axle enable the required angle α between the projections
on these discs to be set in accordance with the pair of pulleys selected. The angle must be
changed on changing the loading time. Since for different pulley diameters (i.e., different
transmission factors) a motor rotation angle differing from 180° (and differing for different
pulleys) will be required for a 180° rotation of the indentor, an angle α suited to the specified
transmission factor must be set up. For example, with 28-mm diameter pulleys on the motor
and 14-mm diameter on the indentor, i.e., a loading time of 7.5 sec, the angle between the
projections on the discs is 90°; with 42- and 14-mm diameter pulleys respectively (loading time
5.5 sec) the angle is 60°, and so on.

We see from Fig. 17 that in order to make a series of impressions a motor 9 of the RD-09
type is fixed to the base of the apparatus; the rotation of this is imparted through a micrometer
screw to the object table of the apparatus by means of a reducing gear. The reducer comprises four gears creating a transmission factor of 1:5. One of the gears is mounted on the

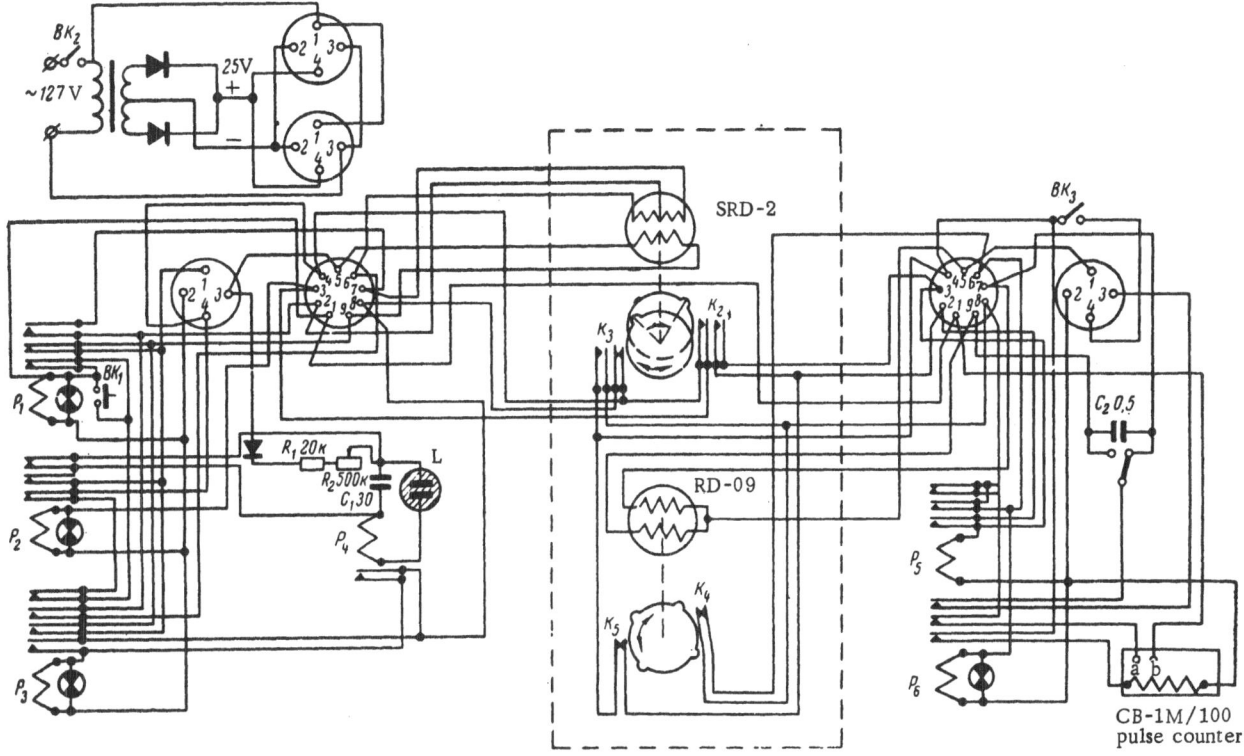

Fig. 18. Electrical circuit of the Ablova−Averkin system [66, 67].

micrometer screw of the apparatus. There are two contacts K_4 and K_5 (Fig. 18) on the plate holding the reducer, and a disc with projections mounted on the axle of the RD-09 motor. The number of projections on the disc determines the distance between the centers of the impressions. We made three discs with one, two, and four projections, giving distances of 120, 60, and 30 μ between the impressions respectively.

The right-hand side of the circuit consists of two relays P_5 and P_6 of the KDR-2 type. After pressing the start button BK_1 the rotation of the motor (as indicated earlier) causes the protection on disc 3 to close the second contact of K_2. As a result of this the relay P_5 operates, and the armature of this remains attracted even after the contact K_2 has been released. The relay P_5 is introduced into the circuit in order to prepare for the operation of relay P_6. At the instant at which the cycle ends, the projection on the disc 4 closes the second contact of K_3. Then relay P_6 operates, and this connects the RD-09 motor. The sample on the object tables moves. The projection on the disc on the axle of the RD-09 motor, which in the original position opened the contact K_4, moves away from this contact and allows it to close; then as it moves it opens the contact of K_5 which connects relay P_5.

In subsequent rotation one of the projections on the disc opens the contact of K_4 and thus disconnects relay P_6. The motor stops. With relays P_5 and P_6 disconnected the supply passes to relay P_1 without the participation of the button BK_1, and the sample-loading process recommences. The supply only passes to P_1 when P_5 and P_6 are simultaneously disconnected. In series with P_5 and P_6 a pulse counter of the SB-1M 100 type is connected into the same circuit through terminals a and b. In addition to counting the number of impressions made (the operator sees the number of these on a scale), the counter breaks the circuit after 25 or 50 impressions. In order to achieve this, some modifications were made to the counter. A brass disc 2 with insulated strips was fitted to the axle of gear 1 (which counted the pulses); the disc cor-

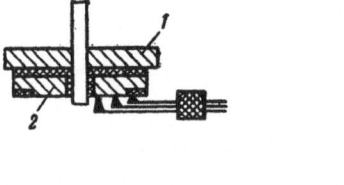

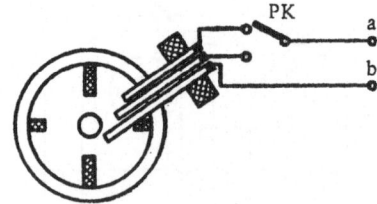

Fig. 19. Reading part of the SB-1M/100 counter with supplementary attachment [66, 67].

responded to three contacts (Fig. 19) connected to terminals a and b. When a contact falls on an insulated strip, the circuit feeding relay P_1 (see Fig. 18) breaks, so that the process of making the impressions stops automatically after 25 or 50 have been introduced. The switch BK_3 serves to connect the part of the circuit producing the series of impressions. The switch PK is provided for reversing the RD-09 motor. A signal lamp is connected as each relay operates.

Using the automatic apparatus just described, in conjunction with the PMT-3, the microhardness of rock salt, silicon, and germanium single crystals was measured. The measurements showed that the apparatus was convenient in operation and gave a smaller scatter than in manual operation. On working with such brittle materials as germanium and silicon, no additional cracks or cleavages were noticed around the impressions. This bears witness to the importance of uniformity of sample loading and experimental conditions when carrying out microhardness tests. Automation of the loading process yields more reliable data in a shorter time.

Automation of the loading process in the PMT-3 greatly eases the use of the microhardness method in physicochemical analysis, since all the measuring errors associated with the operation of the loading mechanism become systematic.

Effect of Vibrations on Microhardness

The harmful effects of vibrations on the relation between the microhardness and the load have been noticed by a number of research workers. This question is directly related to the problem of automated loading with motorized drive and to the study of microhardness in the field of small loads, in which the influence of vibrations increases very sharply.

Kalei [68]* studied the effect of the vibrations created by a motor of the SD-2 type on the microhardness over a load range of 1 to 200 g when testing with the PMT-3. In order to reduce the effect of the dynamic characteristics of the load for small load values, the indentor was let down manually at a slow speed (100 μ/sec).

For recording the oscillograms of the indentor vibrations, a profilograph (model 201) with a magnification of 20,000 to 200,000 was used; the fixing arrangements for the sensing device of this system are shown in Fig. 20. The main body of the sensing device is placed in a massive plate 1, which is fixed rigidly to the table of the PMT-3. The diamond needle 4 of the sensing device is set on the polished surface of the holder of the indentor 6, the vibrations of which are transmitted directly to the needle of the profilograph sensor 5. In order to eliminate the relative motion of the sample and the plate, the sample 3 is fixed to the plate by means of the low-melting-point alloy 2. Analysis of published constructions of automatic loading systems with motorized drive leads to the conclusion that the most typical points for siting the SD-2 motor are: 1) on the table supporting the apparatus, 2) on the support of the PMT-3, 3) on the actual loading mechanism. These forms of fixing were studied by carrying out experiments on 1Kh18N9 austenitic steel with a microhardness of about 900 kg/mm^2.

In testing the adjustment of the PMT-3, considerable interference arose from vibrations of the indentor hanging freely in a soft suspension; under ordinary conditions, in which the

* G. N. Kalei, Dissertation, Institute of Science of Machines, Moscow (1967).

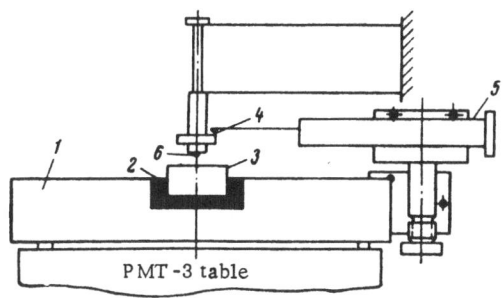

Fig. 20. Arrangement of the sensor of the profilograph for measuring vibrations, as used in [68].

apparatus is sited in a room subject to vibrations, these cause the impression sometimes to appear and sometimes to disappear under a load of 0.5 g. Figure 21 shows an oscillogram of the free vibrations of the indentor under the influence of random pulses only, with the apparatus sited in the basement of a building well-removed from severe vibration. We see that the vibrations are amplitude-modulated with a characteristic frequency of the order of 13 to 15 cps, the indentor vibrating all the time. A slight touch of the hand on the table carrying the apparatus leads to a sharp rise in the amplitude of the vibrations (Fig. 22). It was noted that, when the frequencies of the characteristic (free) vibrations and the vibrations of the perturbing forces coincided, a resonance phenomenon set in, leading to a four- or fivefold rise in the amplitude of vibrations of the indentor (Fig. 23).

However, the vibrations are quickly extinguished when the loaded indentor comes into contact with the microsection, since in this case the pyramid receives a second point of support. The amplitude of the vibration measured after magnification is extremely small (less than 0.02 μ) and in practice the vibrations may be neglected. It is found that low-frequency vibrations (on shutting a door, touching with the hand, and so on) affect the pyramid vibrations perceptibly. The influence of vibrations from the illuminating transformer was very slight, despite the fact that it was placed on the same table as the apparatus. Vibrations of large amplitude matching the intrinsic frequencies were particularly dangerous.

The applications of external vibrations from the motor with the indentor under load leads to an enlargement of the impression and to a false reduction in microhardness values. Thus Buckle [69] observed an 83% fall in hardness numbers for aluminum when using a small motor as vibrator, the fall being particularly sharp for small loads. In order to elucidate the influence of vibration tests were made on the two materials indicated with different hardness (using the ordinary method) for three dispositions of the SD-2 motor and loads between 5 and 200 g. The results appear in Figs. 23 and 24, from which we see clearly that the microhardness of all the materials tested falls sharply if the motor is placed on the loading mechanism. Vibrators from the SD-2 motor have less effect if it is placed on the support of the PMT-3, although

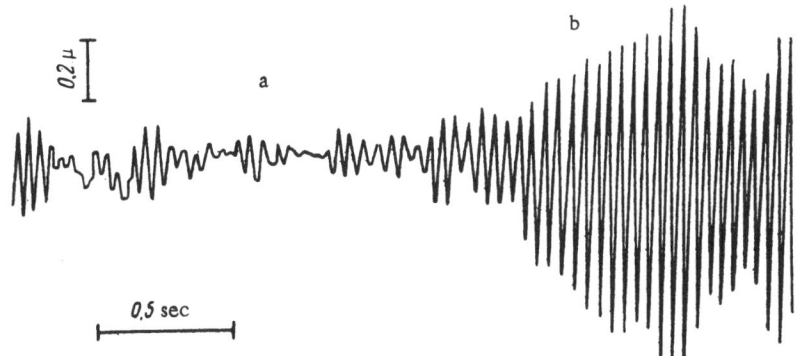

Fig. 21. Indentor vibrations in the absence of a load [68]: a) Free vibrations; b) resonance vibrations of the indentor.

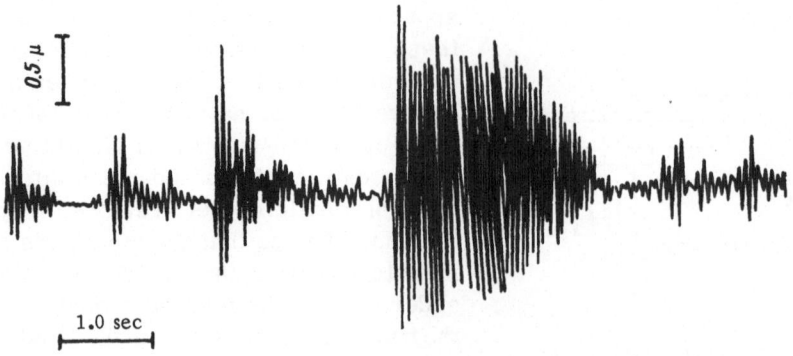

Fig. 22. Vibrations of an unloaded indentor on touching the table
with the hand [68].

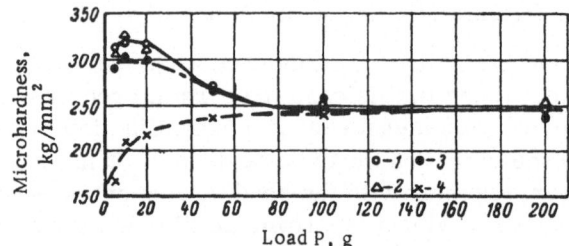

Fig. 23. Microhardness of 1Kh18N9 steel as a
function of load with external vibrations from
the SD-2 motor: 1) Without vibrations; 2)
SD-2 motor on the table; 3) on the support; 4)
on the loading mechanism.

the effect becomes greater on reducing the hardness of the material. The influence of a motor
placed outside the apparatus on the same table as the PMT-3 is almost imperceptible (Figs.
23 and 24).

In the experiments the time during which the motor was vibrating was kept as constant
as possible (of the order of 5 sec); however, there was still a scatter in the microhardness
values. This was partly due to the nonrhythmic operation of the reducer, i.e., the vibrations
created by the reducer were not always strictly identical in frequency and amplitude. The use
of another similar SD-2 motor as vibrator caused a fall in microhardness differing slightly
from the former. The perfection of manufacture of the individual SD-2 motors also clearly
had an effect.

All these factors must be considered when developing new versions of loading mechanism
and also studying microhardness with small loads.

Estimating Errors in Measuring Microhardness

The greatest error in measuring microhardness is usually associated with inaccuracies
committed when measuring the diagonals of the impressions. Many factors contribute to this:
deformation of the edges of the impression, nonuniformity of illumination, parallax, the method
of operating with the cross wires in the micrometer screw eyepiece, inaccuracy in calibrating
the loading mechanism (systematic errors), focusing, setting the cross wires of the micrometer eyepiece, and errors in taking the readings on the drum of the micrometer screw eye-

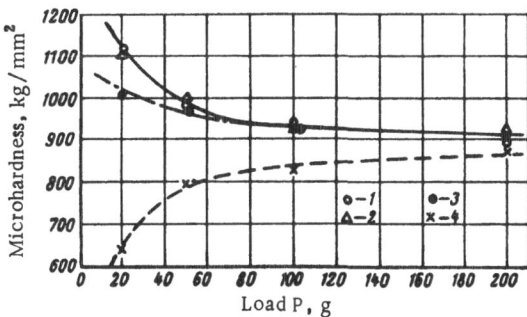

Fig. 24. Microhardness of a Johanson plate as a function of load under external vibrations from the SD-2 motor: 1) without vibrations; 2) SD-2 motor on the table; 3) on the support; 4) on the loading mechanism.

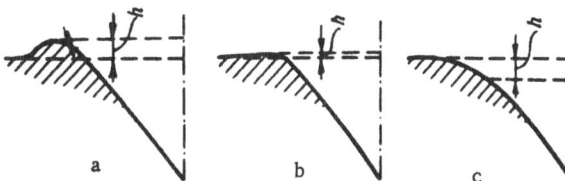

Fig. 25. Three types of characteristic deformations of the material at the edge of an impression when measuring microhardness.

piece (random errors). All these factors have a complex effect on the error in determining the diagonal of an impression.

Along the edges of the impression left by the pyramid in some materials a "protuberant" ridge projecting from the surface of the test object is formed (Fig. 25a); in others the edge of the impression remains at the level of the original surface (Fig. 25b), and in a third kind the surface near the edges of the impression lies below the orginal (Fig. 25c). All these types are encountered in materials of differing properties, depending on their tendency toward cold hardening. An impression of the type shown in Fig. 25c is observed in materials with a great inclination toward cold hardening.

The use of a pyramid with a square base is associated with the fact that the conditions of deformation in planes perpendicular to the direction of motion of the indentor are different in different directions and are repeated for the corresponding directions of the symmetrical parts of the cross section. For this reason the rise or fall in the surface of the microsection near the corner of the impression will be smaller than in the middle of its sides [70-72]. On inspecting in cross-section such an impression under the microscope, the sides appear either convex (Fig. 25a) or concave (Fig. 25b).

Up to the present it has not been made clear how one should allow for distortions in the shape of the sides of the impressions. It would be more accurate to measure the diagonal of an impression occurring at the level of the original surface of the undeformed material. However, this is difficult because of the dependence of the visible contour of the impression on the conditions of illumination.

It is found that when an impression with a protuberant ridge along the edges is so illuminated that the background is light and the impression itself dark, the contour of the impression appears much larger than when the illumination is in another direction so that the impression appears light and the background dark. Between the impression and background there is always a band of translational (intermediate) brightness, making the feature of the impression appear blurred. The error in determining the microhardness may accordingly reach 5 to 10%.

The accuracy of the measurement may be increased by raising the contrast of the image and the resolving power of the optical system. This can best be done by increasing the numerical aperture [60], since the use of large apertures tends to reduce the error committed when measuring the diagonal of an impression. The usual value of the error is from 0.5 to 1.0 μ, which for ordinary impressions lies between 1 and 5%.

The error arising may be further increased as a result of the parallax of the micrometer-eyepiece cross wires and the image of the object itself. In order to avoid this it is essential to adjust the eyepiece carefully so as to give a clear image of the wires and the microscope tube to give a clear image of the impression. The scale division of the microscope eyepiece should be determined for this setting, and then parallax will have a minimum effect.

Another possible error is associated with inaccurate setting of the eyepiece wires on the contour of the impression, which may be regarded as the beginning or end of the diagonal. This error may reach 0.5 μ for some observers using the PMT-2 and PMT-3; for any particular observer the error usually occurs in the same direction, i.e., either in the sense of too great or too small, and it therefore counts as a systematic error.

A random error arising when measuring the diagonal of an impression is associated with the fact that the reading on the drum of the micrometer screw eyepiece is carried out to an accuracy of ± 0.25 of a division. Such an accuracy corresponds to an accuracy of $\pm 0.07 \mu$ when working with a magnification of 500 times and falls on determining the average value of several measurements.

The micrometer screw eyepiece also involves other possible measuring errors. Thus for example one has to consider the dead travel ("backlash") of the screw, as well as inaccuracy and nonuniformity of the screw pitch, and one must accordingly always bring the cross wires up to the edge of the impression in the same sense and in the same part of the drum scale when measuring the diagonal of an impression.

In the foregoing paragraphs we have considered the most important questions relating to the technology of measuring microhardness, involving the choice of load, loading time, and time under load, and also the measurement of the diagonal of the impression. It is obvious from a consideration of these that the various possible deviations occurring in the load, loading speed, and time under load primarily affect the dimensions of the impression, and the microhardness value varies accordingly. On the other hand, there may also be deviations in the results of microhardness tests associated with measuring inaccuracies.

Since it is mainly the diagonal of the impression which is measured, it is mainly errors in measuring this diagonal which will have a decisive effect on the final result when using the microhardness method for physicochemical analysis. Hence, by dismissing possible errors of other origin not affecting the application of the microhardness method to physicochemical analysis, we may greatly simplify the problem of estimating the accuracy of the results of such measurements.

Let us present a classification of the errors which may occur when measuring microhardness and which are important when using the method for physicochemical analysis.

Errors Associated with Imperfections in the Loading Mechanism of the Form of Apparatus Employed.

1. Systematic errors (not vital for physicochemical analysis) arising as a result of:

a) a deviation from the norm in the mass of the load used;

b) incorrect calibration of the loading mechanism;

c) deviation from the norm in the shape of the diamond pyramid (incorrect angle between the faces, and presence of a "comb," unsatisfactory quality of the diamond).

2. Random errors (becoming systematic with automation of the loading mechanism) arising from:

a) imprecise reproducibility of the loading speed;

b) imprecise reproducibility of the ratio of the loading time to the time under load;

c) imprecise reproducibility of the time under load (influence of creep, shaking, and vibration);

d) imprecise reproducibility of the time to remove the load.

Errors Associated with Imperfection of the Optical System of the Microscope for a Specified Magnification of the Objective and Eyepiece.

3. Systematic errors (not vital for physicochemical analysis) associated with the method of bringing the cross wires up to the edges of the impression owing to:

a) the backlash of the micrometer-eyepiece screw;

b) nonuniformity of the pitch of the micrometer-eyepiece screw;

c) disagreement between the axis of the micrometer-eyepiece screw and the direction of the impression diagonal;

d) the thickness of the micrometer-eyepiece cross wires.

4. Random errors largely determined by the characteristics of the observer arising as a result of a failure to reset the following exactly:

a) aperture diaphragm;

b) focusing of the objective and eyepiece (parallax);

c) condenser and field diaphragm (direction and intensity of illumination);

d) micrometer-eyepiece wires on the edge of the impression (on both sides of the impression).

Thus if any physicochemical phenomena are studied with the same instrument, keeping the loads constant and the same arrangement of the loading mechanism, the first group of errors will have no effect on the result. By using a particular method of measuring the diagonal of the impression (method of bringing the crosswire up to the edge of the impression), the third group of errors may be reduced to a minimum. The use of automatic loading makes the second group unimportant for physicochemical research.

Hence the greatest attention should in this case be directed toward the fourth group of errors. If the main settings of the optical system are kept constant, the average error should be comparable with the scale division of the micrometer eyepiece drum. If one considers that during the work it is usual to change the focusing and the size of the diaphragm, to replace the light filters, and so on, then in this case the average error will increase and correspond to several divisions of the drum. These errors cannot be estimated theoretically; however, it has been established experimentally that the relative error in measuring the diagonal of an impression depends on the size of the diagonal, being 5, 10, and 20% for lengths of 20, 10, and 5 μ.

Analysis of the experimental results by the formulas of mathematical statistics may be recommended [73]. The average microhardness may be determined from the formula

$$H_\mu^{av} = \frac{1}{n} \sum_{i=1}^{i=n} H_\mu^i, \tag{1}$$

where n is the number of measurements; H_μ^i are the results of individual measurements.

The mean square deviation (dispersion) is determined from

$$\sigma^2 = \frac{1}{n} \sum_{i=1}^{i=n} \left(H_\mu^i - H_\mu^{av} \right). \tag{2}$$

The accuracy of determining the microhardness is given by

$$\Delta = \frac{\alpha^2}{n} \left(\frac{\sigma}{H_\mu^{av}} \right)^2, \tag{3}$$

where α is the parameter in the expression for the probability

$$W = \frac{2}{\sqrt{2\pi}} \int_0^\alpha \exp\left(-\frac{z^2}{2}\right) dz . \tag{4}$$

It is recommended that W should be taken as 0.9; then $\alpha = 1.65$ [74].

If necessary the results of the measurements may be expressed in the form of a table or graphically in the form of a frequency characteristic, showing the number of measurements associated with a particular value of the quantity measured. Experiments show that the curves or histograms of frequency characteristics obtained in measuring microhardness are very similar to a normal distribution.

CHAPTER 3

METHOD OF PREPARING SAMPLES FOR STUDY

In using microhardness as a method of physicochemical analysis it is desirable to concentrate our attention on three particular factors, on allowing for which the properties of alloys and other samples may be studied under identical conditions. The experimentally observed dependence of microhardness on composition, other parameters of state (such as temperature), or conditions of processing the alloy will then bear a regular character, in no way distorted by subsidiary phenomena which might seriously affect the results of absolute microhardness measurements.

The question as to the effect of the various factors on the microhardness measurements when studying phase diagrams was considered earlier [75].

Effect of Structure

The chemical inhomogeneity of the grains and the average grain size have a considerable effect on the results of an investigation. The structure of the samples must therefore be carefully prepared.

As a result of intracrystallite liquation the microhardness of a grain forming part of the solid solution of a cast alloy changes considerably on passing from the center to the grain boundaries. Hence the first fundamental requirement laid upon the structure of the samples used in studying phase diagrams is that of removing chemical inhomogeneity by way of a homogenizing anneal. A criterion indicating that a homogeneous grain composition has been achieved and dendritic liquation eliminated is that the microhardness assumes the same value in the center of the grain and at the periphery* (see Chapter 8).

In speaking of allowing for grain size when testing for microhardness we must remember the following. The force communicated to the grain by the pyramid is transmitted to the structural constituents in contact with the grain and to a certain extent is taken up in deforming these and the grain boundaries. This expresses itself in the influence of the so-called "backing" and grain boundaries. Depending on the nature of this influence, the structure of the sample may affect microhardness measurements carried out on its individual constituents in various ways. An experimental study of this question showed that on approaching the grain boundary its deformation resistance approached the deformation resistance of the neighboring grain [76-78]. It has been suggested [76] that the depth of the impression should not exceed 1/10 of

---

* We bear in mind the fact that the impression occurs at a certain distance from the boundaries, since in the immediate vicinity of the grain boundaries the microhardness will differ somewhat.

the thickness of the structural constituent being measured. If the depth of the impression equals the thickness of the measured phase, then the properties of this phase have little effect on the results of the measurement and the measurement will practically represent the hardness of the "backing."

The microhardness increases considerably on measuring near the grain boundaries, particularly if the impression lies near the join of several grains. This kind of join frequently occurs but is not always observed, since it may occur slightly below the surface of the microsection.

The grain boundaries have a particularly marked effect if the metal or alloy contains insoluble, highly dispersed particles concentrating at the grain boundaries [77-80].

Both effects may be eliminated by studying samples with a fairly coarse-grained structure; hence the second requirement imposed on the structure of samples used in studying phase diagrams by the microhardness method is that these should have a coarse-grained structure. In order to obtain such a structure the samples must be worked and annealed above the temperature corresponding to the onset of recrystallization. Usually this annealing coincides with the homogenizing anneal intended to eliminate the effects of liquation.

Effect of the Particular State of the Surface Layer

The state of the surface introduces the greatest changes into the results of microhardness measurements. The particular reasons for incorrect results in measuring microhardness arising from the particular state of the surface layer are as follows:

a) the phenomenon of hardening (strengthening) which takes place when the surface layers are deformed in cutting the samples and preparing the microsection for study;

b) the possible loosening of the surface layer in crystallization, deformation (mechanical working), surface oxidation, or recrystallization.

The hardening effect exerts a particularly strong influence even at a considerable depth (10μ and over); this appears differently in different metals and alloys and has been called surface cold hardening.

An experimental investigation of this question [52, 60, 76, 80] showed that surface cold hardening arising from mechanical working sometimes penetrates to a very considerable depth (Table 7). The plastic deformation of the surface layers of multiphased structures proceeds in a complex manner depending on the properties of the individual phases. This phenomenon may introduce considerable changes, not only into the absolute values of the results obtained but also into the general qualitative picture of the relative microhardnesses of the structural constituents.

The foregoing discussion relates to the influence of deformation in distorting the microhardness values obtained when the measurement only involves the penetration of the indentor into the outer surface layers of the test sample. However, when the surface of the metallographic specimen can be freed from the surface-hardened layer, the microhardness method may be successfully used to study plastic deformation and other associated phenomena in simple and complex alloys [81, 82].

In addition to the experimental investigation associated with the use of the microhardness method [82-84], the true character and intensity of local stress/strain configurations in metal samples has also been studied; this indicates that the results of microhardness measurements may be considerably distorted as a result of the severe, nonuniform deformation of the surface of metallographic specimens in the course of preparation.

TABLE 7. Depth of the Surface Layer Damaged as a Result of Mechanical Working of the Surface of the Section [60, 76, 80]

Material	Depth of penetration of surface hardening, μ			
	coarse grinding	fine grinding	polishing on cloth	polishing on a wax wheel
Copper	125	50	35-40	25
Brass	25	10-15	5	2.5
Aluminum	25	10-15	5	2.5
Quenched steel (martensite)	5	2.5	1.5	None

In order to obtain surfaces without surface cold hardening the following measures are taken:

1) casting on to a polished surface;

2) electropolishing after mechanical finishing;·

3) annealing of the prepared sections (recrystallization of the hardened surface);

4) chemical removal of the surface layer.

The practical application of these methods led to the conclusion that the first three, although to some extent removing the hardened surface layer, were only justified in special cases. Only the fourth was universal.

The first method is only applicable for studying materials in the cast state and not at all for alloys subject to appreciable oxidation on melting.

The electropolishing method offers the possibility of removing the undesirable surface layer completely, but it cannot always be used owing to the difficulty of choosing electrolytes and electropolishing conditions (particularly for multiphased alloys). In addition to this, electropolishing may introduce undesirable effects distorting the true picture (loosening of the surface layer, macroscopic relief, etc.).

The annealing of prepared samples before microhardness testing gives good results when using comparatively little-oxidized materials (such as aluminum alloys), when it causes no serious contamination of the sample material. However, this method cannot be used in cases in which the test material is very liable to oxidation, the evaporation of one of the components, and so forth.

The chemical removal of the surface hardening is a universal operation giving excellent results. The simplest and most convenient method, as well as the most reliable, for preparing samples to be subjected to microhardness tests in order to analyze phase diagrams is that of mechanical treatment (grinding and polishing) of the sample surface after appropriate heat treatment and subsequent chemical treatment.

However, in cases in which readily aging alloys are being studied, this method is unsuitable. In preparing specimens from quenched samples, intensive decomposition of the solid solution may develop, greatly distorting the true picture. In such cases it is better to anneal the prepared microsections and then remove the oxide film created by the annealing chemically. If in the course of prolonged annealing the microsections of readily aging alloys undergo changes (as a result of the volatility of the components and oxidation) to such an extent that they become unsuitable for analysis, a combination of microsection annealing with chemical removal of the

TABLE 8. Effect of Surface Hardening and Chemical Treatment on the Scatter in the Microhardness Data of Aluminum–Copper Alloys

Alloy	Microhardness of samples with a cold-hardened surface layer, kg/mm²	Average	Scatter (±)	Microhardness of samples with surface layer removed, kg/mm²	Average	Scatter (±)
Al	38, 32, 31, 35, 27, 26, 31, 29, 35, 37	32.1	6	20, 19, 19, 19, 20, 20, 18, 20, 19, 19	19.3	1
Al+0.5% Cu	39, 41, 35, 33, 31, 25, 31, 33, 31, 29	33.8	9	30, 31, 30, 29, 31, 30, 31, 31, 29	30.2	1
Al+1.0% Cu	45, 48, 39, 48, 54, 59, 51, 58, 62, 59	52.3	10	36, 37, 35, 36, 36, 37, 36, 36, 36	36.2	1
Al+1.5% Cu	72, 70, 65, 58, 61, 67, 60, 55, 52, 60	61.9	10	36, 36, 37, 36, 37, 35, 35, 36, 36	36.0	1
Al+2% Cu	59, 57, 49, 52, 47, 56, 61, 62, 54, 59	55.6	8	37, 37, 37, 36, 36, 37, 37, 35, 36	36.5	1
Al+3% Cu	64, 70, 69, 57, 59, 56, 70, 67, 58, 64	63.4	7	37, 38, 37, 36, 37, 36, 37, 36, 37, 36	36.7	1
Al+4% Cu	46, 54, 48, 59, 60, 63, 50, 57, 49, 55	53.1	8	45, 46, 44, 43, 45, 44, 44, 45, 44, 44	44.4	1.5
Al+6% Cu	61, 75, 71, 59, 72, 64, 58, 66, 63, 67	65.6	6	55, 54, 54, 54, 56, 54, 55, 53, 53, 54	54.2	1.5

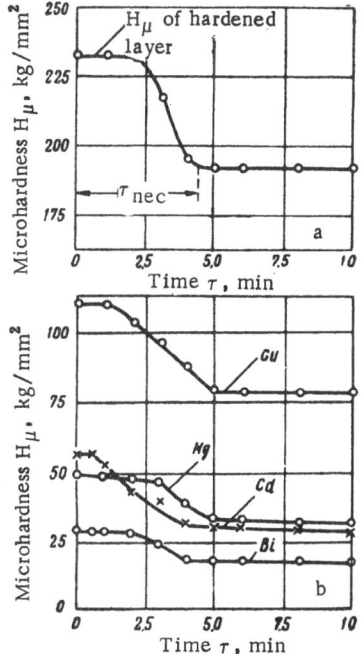

Fig. 26. Removal of surface hardening by the chemical treatment of the microsections: a) Nickel (etchant an 8% solution of CuCl$_2$ in ammonia); b) copper (etchant a 3% solution of FeCl$_3$ in 10% HCl), magnesium (etchant a 3% HNO$_3$ in alcohol); cadmium and bismuth (etchant a 5% HNO$_3$ in alcohol). The etching was carried out in all cases by forcible application of the reagent, with the removal of the etching products.

surface layer may be tried. In this case the samples, after prolonged heat treatment, are ground and polished, with subsequent removal of the hard surface layers by chemical means. During this operation natural aging of the alloys occurs, the results of which are eliminated by recovery treatment, consisting of a brief anneal to bring the alloys into a "freshly quenched" state. After this the surface of the microsections is treated chemically to remove the oxide film formed in the recovery treatment.

Practice shows that in every case chemical treatment of the surface of the microsections has a considerable effect on the results of the microhardness measurements.

Chemical Treatment of the Surface of Microsections

The chemical treatment of microsections prepared for studying microhardness (etching) may be carried out with the following aims:

1) the removal of the surface hardening due to grinding and polishing the sample;

2) loosening or removing the oxide film produced by the heat treatment of the polished sample;

3) revealing the microstructure of the sample immediately before the microhardness test.

In cases in which the microsections should not be subjected to high temperatures in order to remove surface hardening, this may be removed chemically. In carrying out this operation it is essential to choose a chemical reagent in such a way as to dissolve the base of the alloy preferentially and react more weakly with the secondary phases, the relative amounts of which (for example, when studying solubility) are usually small. The time for removing the surface hardening by etching has to be found experimentally for every sample. The time should be such that on further increasing it the microhardness remains constant. However, the dependence of the microhardness on etching time always has the same kind of character. Figure 26a shows a typical curve relating the microhardness to etching time for nickel. We see from this curve that in order to remove the hardened layer a specific etch time (τ_{nec}) is required, after which the microhardness remains constant (section of the curve with H$_\mu$ = const). Clearly in this case unhardened layers of metal are being subjected to the microhardness test.

Analogous curves were obtained for copper, magnesium, cadmium, and bismuth microsections (Fig. 26b). The etching operation for the removal of surface hardening may best be carried out in two or three stages, i.e., τ_{nec} is divided into two or three parts, in the intervals between which the etching products are removed. After the removal of the surface hardening by etching, the microsection should be carefully washed (best of all in warm distilled water) and dried, and then polished for a specific period (usually not more than 0.5 to 1 min) for the final removal of the etching products, washed in water again, and dried.

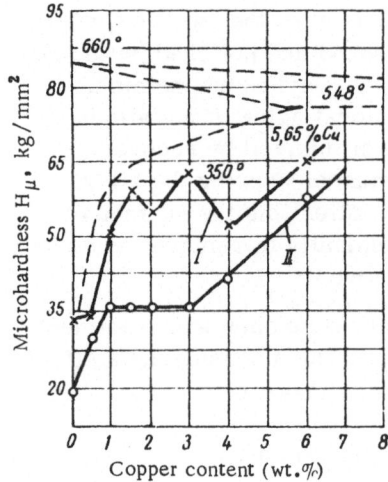

Fig. 27. Effect of surface hardening and chemical treatment on the character of the microhardness isotherm at 400°C for the aluminum−copper system: I) Curve taken from the surface of hardened samples; II) curve taken from the same samples after chemical removal of the surface hardening.

Practice shows that the careful carrying out of this operation gives reasonable results and sharply reduces the spread of the emergent data when compared with series of specimens not having their hardened surfaces removed (Table 8 and Fig. 27).

If we start from the data presented in the first half of Table 8 we obtain the picture reproduced in Fig. 27 (curve I). However, if we study the same samples after chemical treatment we obtain a sharp, regular picture (Fig. 27, curve II).

The results in the second half of the table represent the true microhardness of the solid-solution crystals and not that of the loose surface layer. A proof of this is the ordinary form of the microstructure and the fact that the results agree with those obtained on samples in which the surface hardening had been removed by annealing the prepared specimens.

After heat treatment carried out in order to remove the surface stresses arising as a result of grinding and polishing, the microsections have a tarnished oxidized surface not suitable for examination. The removal of the oxidized layer by polishing the surface of the specimen is inappropriate. It is far more convenient to use chemical etching. In choosing the reagent, due allowance must be made for the qualitative composition of the oxide film by considering the oxygen affinity of the alloy components, and also the interaction of the chemical reagent with the oxidized film and the base of the alloy. One must strive to obtain a reagent dissolving the oxide film efficiently but not interacting with the base. This is a difficult problem. In practice one has to use a reagent dissolving the oxide film but only slightly affecting the base of the alloy.

The etching time required to remove the oxide film is choosen for a particular series of alloys after two or three trial experiments with individual samples. After etching, the reaction products must be carefully removed by washing in distilled water. After this the microsection is "finished off" by brief polishing for a specific time (usually 30 to 40 sec), the surface being pressed uniformly and very lightly against the polishing wheel. This treatment leads to a sharp reduction in the scatter of the microhardness data when studying phase diagrams. In 30 to 40 sec the microhardness of the sample suffers no change, i.e., for this period the sample is unable to acquire any appreciable hardening; however, the etching products are entirely removed.

Etching to reveal the microstructure before a microhardness test is one of the most important operations in preparing the sample surface, and requires particular care. In choosing a reagent one must be guided by the principle that the reagent should interact only weakly with the base of the alloy (i.e., the crystals of the solid solution under consideration) but much more with other phases, so as to reveal the structure clearly in the shortest possible etching time. The time factor plays an extremely important part in etching the microsection before the test. The formation of a film of etching products often leads to a sharp fall in the apparent microhardness, which may be explained by the scheme depicted in Fig. 28. The etching time must thus be as short as possible.

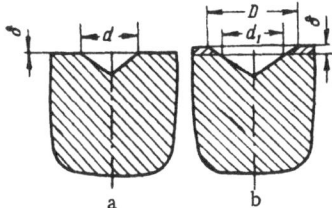

Fig. 28. Effect of a film of etching products on the apparent microhardness. a) Weak etching; b) strong etching; $D \gg d, d \approx d_1$; δ) film thickness.

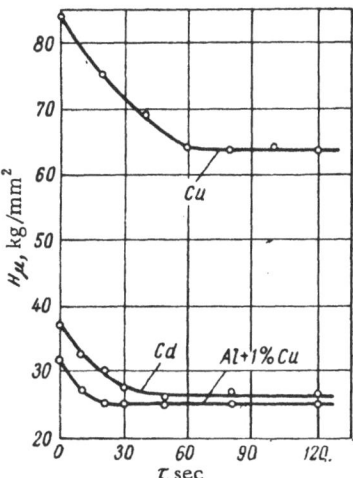

Fig. 29. Effect of etching time on the apparent microhardness of copper (etchant 3% solution of $FeCl_3$ in 10% HCl), cadmium (etchant 5% HNO_3 in alcohol), and an Al−1% Cu alloy (etchant 1% HF in water).

For illustration we present the results of a microhardness/etch time study relating to copper, cadmium, and an Al−1% Cu alloy (Fig. 29).

An experimental investigation [85] into the process of removing a surface layer existing in a stressed state showed that the etchant might react in various ways on the test surface. It was noticed, firstly, that the interaction between the reagent and the alloy was nonuniform in time; secondly, the character of the interaction varied in accordance with the qualitative and quantitative composition of the alloy. This must be taken into account when selecting a method of chemically preparing microsections for microhardness measurements.

Special attention was paid to the rate of etching the surface layer, the scatter in the microhardness values measured, and the relation between these two factors.

The rate of removing surface-hardened layer determines the reactivity of the particular etchant with respect to the alloy in question. In some cases this rate differs at different stages in the etching process. In addition to this, owing to differences in orientation, anisotropy of the properties, inhomogeneity of composition, and electrical effects intensified by these factors, the surface-hardened layer is etched away in a nonuniform manner, so that in some places the process is intensified and in others retarded. This primarily determined the scatter in microhardness values; however, the average microhardness obtained from a fairly large number of measurements (25 to 30) remains unaltered. Thus a simultaneous study of the variations in etching rate and the degree of nonuniformity of the etching process enables us to recreate the complete picture of the chemical removal of surface hardening. The etching rate is determined by the relation between the average microhardness value and the etching time, while the degree of nonuniformity of the etching process is determined by the scatter in the microhardness measurements, expressed in the form of an accuracy index.

The accuracy index is determined from the ratio of the mean deviation to the mean value and is expressed in percent. In considering the relation between the accuracy index and the etching time, special attention was paid to the general character of the variation, without any attempt at establishing a quantitative relationship (Table 9). For this reason, in the lower part of Fig. 30a, b, and c, the results are analyzed graphically so as to exhibit simply the level of the accuracy-index values and the general tendency of their variation.

After the removal of the surface layer, the microhardness method will not help in any further study of the kinetics of surface etching. However, the scatter in the results of microhardness measurements may enable us to draw certain conclusions as to the degree of uniformity in the further etching of the surface layer. The average microhardness will furthermore be affected by the growing amounts of oxidation products formed by the action of the etchant on the metal or alloy components and the products of exchange reactions, so that the

TABLE 9. Effect of the Etching Time Required to Remove the Surface-Hardened Layer on the Scatter in the Microhardness Measurements, %

Object of investigation	Spread of microhardness, % for etching time, sec															τ nec, min
	0	15	30	50	80	120	150	180	210	230	250	300	360	380	490	
Etchant, 3% solution of $FeCl_3$ in 10% HCl																
Copper Mo	18	—	10	6.0	2.0	2.0	—	1.0	—	0.8	—	0.8	—	·	0.7	5
Cu+5% Al	14	14	8.0	6.0	4.1	—	3.1	—	2.1	2.1	2.1	—	1.4	1.6	1.5	6
Cu+4% Ti	13	11	6.9	2.4	5.3	—	4.0	—	3.0	—	1.4	—	1.8	1.7	—	4.5
Cu+2% Al+2% Ti	10	13	7.0	4.3	5.8	—	4.7	—	2.5	—	1.6	—	1.3	1.6	—	4.5
Etchant, saturated solution of NH_4OH in H_2O_2																
Copper Mo	18	—	4.0	4.0	2.02	2.1	—	1.8	—	1.6	—	2.4	—	3.8	5.2	4
Cu+5% Al	14	19	18	22	6.4	—	6.1	—	4.6	—	11	—	9.2	7.8	14	—
Cu+4% Ti	13	11	11	9.1	9.6	—	12	—	10	—	9.6	—	17	15	—	—
Cu+2% Al+2% Ti	10	12	7.4	6.7	5.8	—	12	—	9.9	—	12	—	13	21	—	—
Etchant, $K_2Cr_2O_7$ + NaCl + H_2SO_4 + HF																
Copper Mo	18	—	5.9	3.2	2.0	1.8	—	0.9	—	0.4	—	0.3	—	0.8	0.8	5
Cu+5% Al	14	17	18	21	17	—	10	4.2	9.4	—	8.1	—	4.3	5.5	7.0	—
Cu+4% Ti	13	16	8.0	9.0	10	—	2.1	0.8	1.7	—	1.2	—	0.8	1.0	—	3
Cu+2% Al+2% Ti	10	11	12	10	4.8	—	10	9.6	9.8	—	6,8	—	9.7	8.0	—	—

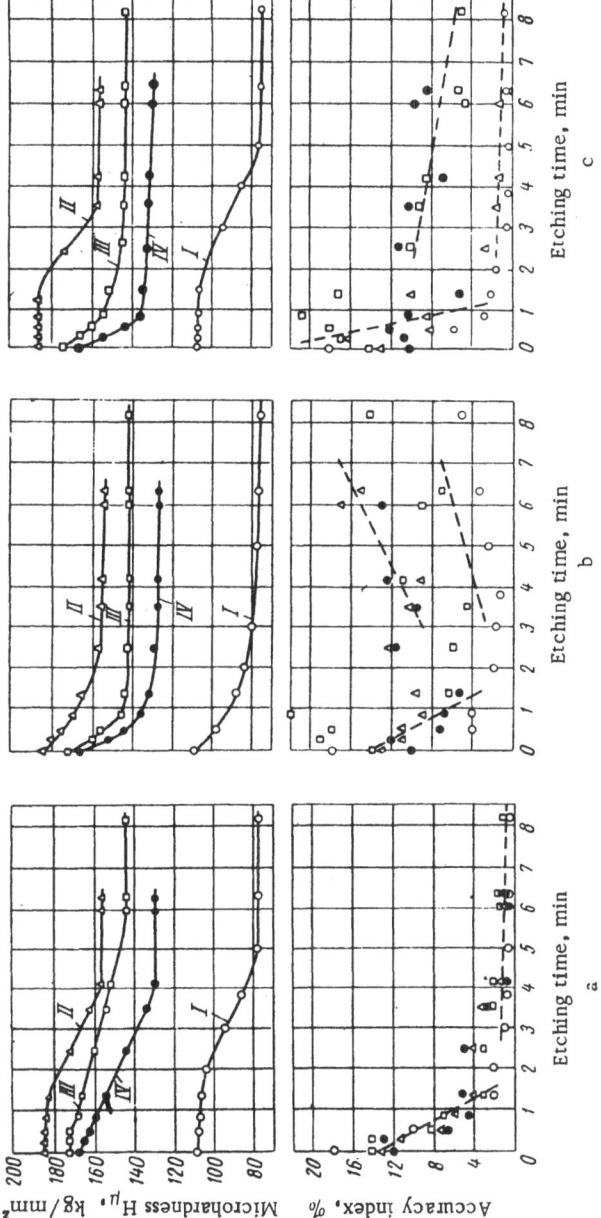

Fig. 30. Relation between microhardness and etching time: a) 3% solutions of $FeCl_3$ in 10% HCl; b) saturated solution of NH_4OH in H_2O_2; c) 2 g $K_2Cr_2O_7 + 4$ g NaCl + 8 cc H_2SO_4 in 100 cc of 5% HF; I) for Cu; II) for Cu–4% Ti alloy; III) for Cu–5% Al alloy; IV) for Cu–2% Al–2% Ti alloy.

microhardness will still vary in time after the removal of the surface layer, either increasing or diminishing.

In addition to the etch-time dependence of the microhardness (which determines the period τ_{nec}), another criterion to be considered when choosing an etchant for chemical preparation of a microsection is the spread or scatter in the microhardness values. This may suggest increasing τ_{nec}, or for certain purposes of physicochemical analysis may recommend the use of a lower value of τ_{nec}, giving a smaller scatter in the results of the microhardness measurements. It may also prove necessary to abandon the use of a particular etchant altogether.

The chemical method of preparing a surface in order to study copper-base alloys by the microhardness method has been used for a large number of alloys. The etchants recommended by various handbooks on microscope analysis were tested (Fig. 30). After grinding and polishing, the sample surface was carefully washed with warm water, then with ether, and then dried. During the etching the surface of the microsection was wiped with wadding dipped in the reagent. The wadding was pressed closely to the surface in order to remove etching products thus formed. Then the microsection was again washed and dried, after which the microhardness was measured immediately. A load of 20 g was employed, the loading time was 3 sec, and the time under load 5 sec. The results of the measurements were taken as the mean of 30 (for a large scatter) or 10 to 15 (for a small scatter).

It follows from Fig. 30 that in the majority of cases a somewhat passive process precedes the intensive etching away of the hardened layer. In this period the etching of the whole is, as it were, prepared; only insignificant proportions of the surface are removed, and the average microhardness values remain constant. The intensive process is evidently held up by oxidation products, which are not removed by washing in water and ether. It is also natural to suppose that the surface-hardened layer retains its mechanical characteristics down to a certain depth, and this is the reason for the unchanging values of microhardness in the initial stages of etching, despite the dissolution of the layer.

In alloys there may be a change in the surface properties, as a result of which the rate of dissolution increases.

An analogous case is observed, for example, on introducing aluminum into copper (particularly accompanied by titanium) for the etchants $FeCl_3 + HCl$ (Fig. 30a) and $K_2Cr_2O_7 + NaCl + H_2SO_4 + HF$ (Fig. 30b). When the oxidation products forming on the microsection are readily wetted and easily dissolved by the etchant employed, the preliminary period may vanish, the surface-hardened layer being etched away quickly (τ_{nec} small). However, in this case the surface etching is often nonuniform and this produces a scatter in the microhardness values, particularly after removing the distorted surface layer. Despite the fact that the etching speed gradually falls over the several stages, the results of nonuniform etching in the earlier stages still remain. The surface layer is loosened and its reflecting power is reduced, which leads to an increase in the spread of the microhardness values (etchant $NH_4OH + H_2O_2$, Fig. 30b) and a fall in the mean value.

If etching takes place at a lower rate, then there is a smaller scatter, even when the stability of the oxide film is reduced so that there is no preliminary period in the etching of the surface (etchants $FeCl_3 + HCl$ and $K_2Cr_2O_7 + NaCl + H_2SO_4 + HF$).

Thus it is easy to see the connection between the etch-time dependence of the microhardness and the scatter in the results obtained. The best of the available etchants will be the one characterized by the lowest rate of etching the surface layer. The surface layer will then be etched away more uniformly and the etching process may be more accurately monitored by microhardness measurements.

However, for obvious reasons τ_{nec} should not be increased too far.

Analyzing the resultant experimental data, we may readily conclude that the etchant $FeCl_3$ + HCl may be successfully used in preparing the surface of copper samples and Cu − Al, Cu − Ti, and Cu − Al − Ti alloys. The etching speed is greatest for Cu − Ti. The etching speed is the same for Cu − Al − Ti as for Cu − Al. The etchant NH_4OH + H_2O_2 is unsatisfactory.

The etchant $K_2Cr_2O_7$ + NaCl + H_2SO_4 + HF may be used successfully for preparing the surface of copper and Cu − Ti alloys. However, on introducing aluminum into the alloy the merits of this etchant no longer appear.

In the examples considered a change in the quantitative composition of the alloys had no effect on the qualitative picture and only appeared as a slight change in etching speed. Hence in applying the microhardness method to the study of metallic systems and choosing the chemical method of removing the surface hardening, it is sufficient to establish a qualitative picture for individual series of alloys (similar in structure) under the action of various etchants. The value of τ_{nec} may be determined at the same time.

CHAPTER 4

ASPECTS OF THE METHOD OF MEASURING
THE MICROHARDNESS OF SEMICONDUCTORS

Semiconducting materials are distinguished by strong covalent bonds, sharply directional in space. In measuring the microhardness of semiconductors one accordingly encounters complications associated with the anisotropy of the mechanical properties and plastic deformation, structural defects, and a great tendency toward crack formation and cleavage. It is important to remember these characteristics of semiconducting materials when measuring their microhardness and to adopt a corresponding test and measurement procedure [5, p. 206].

Anisotropy of Microhardness

The anisotropy phenomena liable to affect precision measurements of microhardness are quite well known. It is reasonable to expect that microhardness values obtained by measuring different faces of the same crystal will differ. So also there should be a difference between the microhardness values obtained with the diagonal of the impression set at different angles to the crystallographic axes of the crystal.

The effect of the relative orientation of the diamond point and the crystal on the results of microhardness measurements was studied by Schultz and Hanemann [56] on aluminum and antimony crystals. Two points were specially noted:

1) The variations in microhardness on a particular face of a crystal observed on rotating the pyramid around its own axis are much greater than the variations in microhardness on passing from one face to another;

2) in individual measurements one cannot establish the difference in the microhardness of different crystallographic planes, since the relative position of the pyramid (which is decisive) is arbitrary.

A study of the anisotropy of microhardness carried out by S. D. Dmitriev [4, p. 193] showed that Schultz and Hanemann's second point [56] required a certain amount of refinement.

Dmitriev [4] measured the microhardness of a number of crystals on various faces and also on one particular face, with rotation of the square diamond pyramid through a specific angle around its axis.

It was found that in beryl crystals the (1010) prism face was harder than the (001) pinacoid face; in zircon crystals the (100) prism face and the (111) dipyramid face were harder than the (001) pinacoid. In argonite the hardest of the three faces (010), (110), and (011) was the (011). The difference in microhardness of individual faces was considerable, the first Schultz and Hanemann [56] point was in general confirmed.

50

On the other hand, it was found that the elastic aftereffects in the crystal were often so large that the anisotropy of the microhardness on different faces of the crystal remained uncertain. It was recommended that a comparison between the microhardnesses of different faces should only be made when they belonged to the same zone and the impression diagonal was measured along the axis of the crystallographic zone. The anisotropy of the microhardness in the crystal could then properly be established.

N. Yu. Ikornikova [4] studied the anisotropy of the microhardness of synthetic corundum. The highest value of the hardness was found when the axis of the diamond pyramid was oriented at an angle of 60° to the principal axis of the corundum crystal. The lowest value of hardness was found when the direction of the axis of the diamond pyramid coincided with the crystal axis. The differences in the microhardness values obtained for directions parallel, perpendicular, and at an angle of 60° to the principal axis of the crystal corresponded to 1940, 2200, and 2315 kg/mm^2.

Existing experimental results confirm the importance of the anisotropy factors in microhardness measurements. This must be remembered when developing and testing methods of measuring microhardness for various purposes, as failure to do so might result in additional errors.

However, the anisotropy of the microhardness may also serve as a subject for independent study, since it may be used to secure the approximate orientation of the crystals.

R. I. Garber, S. Ya. Zalivadnyi, and F. S. Gorokhvatskii [86] studied the anisotropy of the microhardness of beryllium on spherical single crystals. The results led to the conclusion that the relation between the microhardness and the crystallographic direction could be represented graphically in the form of an ellipsoid of rotation, symmetrically disposed relative to the six-fold axis. On faces parallel to this axis the microhardness is 217 kg/mm^2. On varying the angle between the face studied and the six-fold axis from 0 to 90° the microhardness increases to a value of 350 kg/mm^2. Thus the axial ratio of the microhardness ellipsoid approximately equals 1.6.

M. S. Ablova studied the anisotropy of the microhardness of germanium [5].* It was found that the greatest microhardness occurred for the (111) plane ($\sim$1000 kg/mm^2) and the smallest for the (110) plane ($\sim$850 kg/mm^2).

In addition to this, some peculiarities in the microstructure of the impressions was observed [87]; these involved cleavages formed 30 to 40 sec after the removal of the load. The large, nonuniform elastic residual stresses in the impression plastically distort the latter. The greatest plastic distortion of the impressions after removing the load was observed in the (111) plane, for which the plastic deformation under load was the smallest.

It was also found that the anisotropy of germanium bore a different character in samples of the n and p types [88, 89]. For the n type the greatest microhardness corresponded to the (111) face. However, for the p type this plane was the "softest," i.e., it had the smallest microhardness. M. S. Ablova [86, 90] used the microhardness method to study the conditions for the existence of an electromechanical effect in germanium.

It was also noted in preliminary experiments relating to the microhardness of gallium antimonide [91] that the microhardness was smaller for n- than for p-type samples.

The anisotropy of the microhardness of silicon was studied by A. V. Sandulova and V. M. Rybak [92], taking account of the morphological characteristics of the crystals under examination.

---

* M. S. Ablova, Dissertation, Institute of Semiconductors, Academy of Sciences of the USSR, Leningrad (1956).

In order to discover the anisotropy in the microhardness of silicon single crystals, samples grown from the gas phase and having various shapes and a small number of dislocations were studied. A load of 200 g (giving the smallest scatter) was used with smooth loading in 25 to 30 sec. Only the impressions free from chipping (not more than 60% of the 100 to 150 taken) were considered. The surface was washed with alcohol. The most probable value was taken from the maxima of the frequency curves plotted with an interval of 50 kg/mm^2. There was no difference between the microhardness at the edges and in the center of the face. For the (111) face on single crystals in the form of regular octahedra or octahedra drawn out in the direction of the three-fold axis $H_\mu = 1650$ kg/mm^2, while on single crystals in the form of a cubooctahedron, in which this face was poorly developed, the lower value of $H_\mu = 1400$-1450 kg/mm^2 was obtained. For the (100) face on single crystals in the form of a regular cube and also needles 200 to 800 μ in diameter and cubooctahedra $H_\mu = 1100$-1150 kg/mm^2. It was found that for single crystals of all shapes the microhardness fell after 18 to 20 min etching in a solvent, and anisotropy ceased.

M. S. Ablova and N. N. Feoktistova [93] established anisotropy in the microhardness of indium antimonide, using n-type single crystals with a specific resistance of 0.006 $\Omega \cdot$ cm and a carrier concentration of $3 \cdot 10^{16}$ cm^{-3}. The measurements were made with the PMT-3 instrument at a load of 50 g. The results for the various faces (the surface being mechanically polished before measurement) may be arranged in the following order: (111) $-$ 233 kg/mm^2, (112) $-$ 231 kg/mm^2, (100) $-$ 224 kg/mm^2, and (110) $-$ 222 kg/mm^2. The accuracy of these measurements was estimated by the method of mathematical statistics: ± 1.4 kg/mm^2 with a probability of 0.9.

Some interesting investigations were also carried out in relation to the change in anisotropy with changing chemical composition of the material. Data were obtained for the tellurium−selenium system by the microhardness method in [94]. As a measure of anisotropy, the ratio of the larger diagonal of the impression to the smaller was taken, referred to the same ratio for tellurium (i.e., the anisotropy of tellurium was arbitrarily taken as unity). The following results were obtained:

Selenium concentration, % (at.)	0.00	0.10	0.20	0.50	1.00
Anisotropy	1.00	1.04	1.08	1.42	1.47

The rise in anisotropy with increasing amounts of selenium in the tellurium was explained as being due to the more intensive saturation of the covalent bonds between the atoms in the chain in the presence of selenium, since the Te−Se bond was rather stronger than the Te−Te. On further introduction of selenium (from 2 to 10 at.%) the anisotropy fell.

Apart from the differences in microhardness, detailed study of various crystallographic planes of semiconducting materials also revealed polarity of this property. In compounds of the $A^{III}B^V$ type the atoms of the group III and group V elements have different numbers of valence electrons and hence the surface states of these atoms on the $\{111\}$ and $\{\bar{1}\bar{1}\bar{1}\}$ planes are dissimilar. The physicochemical and mechanical properties of these surfaces should also differ [95].

G. V. Kukuladze and M. S. Mirgalovskaya [91] first established the appearance of polarity when studying the microhardness of semiconducting compounds. Experiments were carried out on samples of p-type gallium antimonide with a hole concentration of $1.5 \cdot 10^{17}$ cm^{-3}. The samples studied comprised dendrites grown in a hermetic apparatus with a graphite resistance heater in an atmosphere of purified helium. The choice of dendrites as subject for study ensured exact orientation of the $\{111\}$ and $\{\bar{1}\bar{1}\bar{1}\}$ surfaces; in the dendrites these faces were parallel and well developed ("principal faces"), and also, being clean and even, they needed no further grinding, polishing, or etching. The $\{111\}$ and $\{\bar{1}\bar{1}\bar{1}\}$ faces were etched in a mixture of acids after measuring the microhardness; etch pits only appeared in the $\{111\}$ face.

TABLE 10. Recommended Geometry of the Tips
for Studying the Anisotropy of Materials with
Face-Centered or Diamond Structure

Plane of measurements	Number of faces of regular pyramid	Angle between the base of the pyramid and the face	Angle between the faces of the vertex of the pyramid
(100)	4	54°46'	70°28'
(110)	4	36°14'	109°32'
(111)	3	70°28'	—

The measurements were made with the PMT-3 instrument, using loads of 10, 20, and 50 g with a loading time of 10 sec and a period of 5 to 7 sec under load. Only impressions with no traces of damage (60 to 80 impressions only yield 30% suitable for microhardness measurements) were taken into account.

The microhardness of the $\{111\}$ face was greater than that of the $\{\bar{1}\bar{1}\bar{1}\}$; the results obtained were approximately 485 and 370 kg/mm^2 respectively (taken from the maxima of the frequency curves).

The considerable difference between the microhardnesses of the polar crystallographic faces observed in the case of gallium antimonide presumably also occurs in other chemically complex materials and must be duly taken into account in precise microhardness measurements.

V. A. Kokoshkin [96] showed that spherical and conical tips were not very suitable for studying anisotropy; measurements were best made with a pyramidal indentor of special shape, corresponding to the crystal lattice. This recommendation was associated with the fact that the type and extent of the deformation arising in microhardness tests was largely determined by shear along the slip and cleavage planes of the crystal, so that it was important for the load to be oriented optimally with respect to these planes. This is particularly important in anisotropic crystals having few or no dislocations.

Several kinds of tips have been proposed for studying crystals with fcc lattices and lattices of the diamond type.

The first type of lattice is characteristic of many metals and the second of some important semiconducting materials. In all these structures the cleavage and slip planes belong to the $\{111\}$ family, grouped into four systems oriented at 70°28' to each other. Calculation shows that, for such cubic materials, tips of the kind indicated in Table 10 should be used.

Dislocation Density

Naturally, structural defects in the material under test must affect the results of microhardness measurements, since they resist the deformation of the material. Experimental data characterizing the relation between microhardness and dislocation density have primarily been secured for the following semiconductors: germanium, silicon, and indium antimonide.

Dale and Brice [97] studied the effect of dislocation density on the microhardness of germanium. Tests were made with a diamond pyramid having a square base with loads of 20 to 30 g applied at a velocity of 1 to 2 μ/sec and held for 15 sec. The surface of the samples was polished with aluminum oxide, after which etching was carried out in a mixture of hot acids for 3 to 4 min. The measurements were made on the (111) face and only those with no trace of damage were taken into account (not less than nine, constituting about 5% of the total number of impressions).

The results indicated a great difference in microhardness between dislocation-free germanium ($\sim$500 kg/mm^2) and germanium with various dislocation densities: 10^3 cm^{-2} ($\sim$600 kg/mm^2) and 10^6 cm^{-2} ($\sim$700 kg/mm^2).

These results correspond to the density of the edge dislocations revealed by means of "deep" etch pits (terminology of [97]). The density of the "shallow" (or "fine") etch pits was $1 \cdot 10^3$ cm^{-2} and was constant for all the samples studied. The influence of the stresses in the lattice, which appeared in the form of the "shallow" etch pits, was less substantial; when the density of these varied from 1 to 10^5 cm^{-2} the microhardness varied from 550 to 650 kg/mm^2. Meanwhile the density of the edge dislocations (based on the count of "deep" etch pits) remained at a level of $1 \cdot 10^3$ cm^{-2}.

M. G. Mil'vidskii and L. V. Lainer [98] established a relation between the microhardness of silicon and the dislocation density of the sample. The microhardness was measured with a PMT-3 and a load of 100 g. The surface of the samples was previously ground and polished chemically in an acid mixture for 2 to 3 min. For the (111) face the microhardness varied considerably (from 830 to 1250 kg/mm^2) on varying the dislocation density by one order (from $1 \cdot 10^3$ to $2 \cdot 10^4$ cm^{-2}). Outside this range of dislocation densities the microhardness varied little.

The method of preparing the surface for measurement in [98] may have altered the state of the surface layer considerably, as already indicated [92]. The result thus only qualitatively characterizes the true effect of the dislocations on the microhardness of the material.

Microhardness values for indium antimonide samples having a dislocation density varying between $2 \cdot 10^2$ and $2 \cdot 10^5$ cm^{-2} are presented in [5, p. 206]. This work was carried out with a PMT-3 using loads of 20, 50, and 100 g; loading was effected in 10 sec and the time under load equalled 5 sec. With increasing dislocation density the microhardness and brittleness of the indium antimonide samples both increased.

Microbrittleness

Many of the materials used in modern technology are expected to have a high resistance to mechanical or thermal shock. It is very difficult and indeed sometimes impossible to secure such properties in materials of high mechanical strength with a low capacity for internal-stress relaxation and an essential brittle nature.

In addition to this, quantitative methods of estimating the brittleness of materials have not yet been sufficiently developed to provide a satisfactory solution to such problems [99]. Furthermore, the laws governing the changes in brittleness on alloying have not been completely elucidated.

Much has been done in relation to the brittleness characteristics of refractory compounds. The characteristics of different compounds have been compared and correlated with other characteristics and with the position of the components in the Periodic Table. However, so far general laws relating brittleness to concentration have never been formulated; these would greatly extend the prospects of creating new materials with useful properties.

In order to study the manner in which metals and alloys (particularly steels) fracture in relation to their composition, structural characteristics, aging, and so forth, and also in order to solve various other metallographical problems, more and more use is now being made of the method of fractography, i.e., the study of structure in fractures. In recent years the fractographic study of steel has been carried out by means of electron microscopes. However, as a measure of brittle fracture this technique is only qualitative and hence unreliable.

The first quantitative characteristics of the brittleness of materials were proposed in connection with a study of the properties of minerals and refractory compounds. For comparing the brittleness of refractory compounds methods based on the compressibility of powders [100], the number of impacts on the sample leading to breakdown [101], and the work expended in the deformation of the material when producing an impression on the hardness tester [102] were proposed, as well as the electron-microfractographic method [103].

The method proposed by Palmquist [102] lies in determining the load (P_{cr}, in g) for which 50% of the impressions have cracks, and then calculating the work done by the force P_{cr} in impressing the pyramid (S_{cr}). The work S_{cr} was called "toughness" by Tsinzerling [104]. If we use an indentor having the form of a pyramid with a square base and a vertical angle of 136°, then, knowing the hardness (kg/mm^2) corresponding to the load P_{cr}, we may calculate the "toughness" from the formula

$$S_{cr} = 2 \cdot 10^{-4} \sqrt{\frac{P_{cr}^3}{H_\mu}}, \; g \cdot cm.$$ (1)

The so-called "microbrittleness" method has received quite widespread acceptance. This method enables us to characterize the brittleness of a material by the statistical curve relating the number of impressions cracking under the indentor of the microhardness meter to the value of the applied load. After reaching loads producing cracks in all the impressions, the curve expresses the relation between the number of cracks in one impression and the applied load. Clearly, for a specified load the more brittle materials will have a larger number of cracked impressions or a larger number of cracks in each.

The microbrittleness of corundum was studied by N. Yu. Ikornikova and L. A. Pikunova using this method [4].

I. N. Frantsevich and A. N. Pilyankevich [105] studied the brittleness of refractory compounds and showed that the curves so constructed might intersect each other; this tends to distort the problem of determining comparative brittleness.

Frantsevich and Pilyankevich came to the conclusion [105] that the reason for the intersection of the microbrittleness curves was a change in the stress field surrounding the test grain. In order to be able to compare different microbrittleness curves with each other, allowance must be made for the effect of sample stresses on their form. For this purpose the authors in question [105, 106] proposed calculating the relative derivative

$$\alpha = \frac{P}{n} \cdot \frac{dn}{dP},$$ (2)

along each microbrittleness curve, where P is the load at which the number of cracked impressions equals n (in parts or percents) and the derivative is calculated at the point characterized by the specified P and n values.

This method was used to study the comparative brittleness of silicon and a number of refractory compounds. One of the most important conclusions was that in cases in which the value of n = φ(P) was comparatively large and increased with increasing load, the grain size of the sample had a considerable effect on the brittleness; fine-grained samples were less brittle than coarse-grained. This may have been due to the fact that in a fine-grained sample the grain tested was subjected to a more evenly distributed all-round pressure than in a coarse-grained sample, so that greater loads could be sustained without cracking.

TABLE 11. Arbitrary Scale for Estimating the Brittleness Number
of Refractory Compounds

Brittleness number	Character of the impression
0	No visible cracks or chipping
1	One small crack
2	One crack not coinciding with the continuation of the diagonal of the impression. Two cracks in adjacent corners of the impression
3	Two cracks in opposite corners of the impression
4	More than three cracks. One or two chips at the sides of the impression
5	Shape of the impression broken up

In addition to this, it was found that comparative brittleness could not be characterized by comparing curves corresponding to loads for which all the impressions were cracked, since these quantities depended not so much on the physical nature of the test sample as the stress field surrounding each grain. The maximum loads under which no cracks were formed around the impressions differed very greatly for different materials, so that this quantity also could not be regarded as characteristic.

Hence microbrittleness tests may be carried out by reference to the form of the $n = \varphi(P)$ curves for medium loads giving roughly the same values of α. The samples should have the same grain characteristics.

However, an estimate of brittleness simply based on the rate of increasing damage with increasing load is also insufficiently precise. The nature of the developing cracks must also be taken into account.

When, in the intrusion of the diamond pyramid, the rupture strength of the material is exceeded and the sample is unable to dissipate the elastic stresses, the stresses concentrate at the boundaries of the impression and the sample ruptures. The greater the brittleness of the material, the more cracks occur and the more developed they become.

The brittleness factor is estimated in relation not only to the number of impressions with cracks or the number of cracks in an impression, but also the character of these. A so-called average brittleness number has been introduced; this is based on a five-point scale (Table 11, Fig. 31).

The overall brittleness number is determined from the formula

$$Z_P = 0 \cdot n_0 + 1 \cdot n_1 + 2 \cdot n_2 + 3 \cdot n_3 + 4 \cdot n_4 + 5 \cdot n_5, \qquad (3)$$

where n_0, n_1, n_2, n_3, n_4, and n_5 are the relative numbers of impressions out of the total number (usually 25 to 100) with the corresponding brittleness number.

Naturally the sum Z_P gives the fewer subjective results the greater the number of brittleness numbers obtained in the particular investigation.

In order to allow for the rate of growth of brittle failure of the material with increasing load P, one calculates the ratio of the increment in the overall brittleness number to the increment in the load $(\Delta Z / \Delta P) \approx (\partial Z / \partial P)_P$.

As a suitable brittleness index of the material, reflecting both the character of the brittle fracture and the rate of increase of brittleness with increasing load, we may use the product

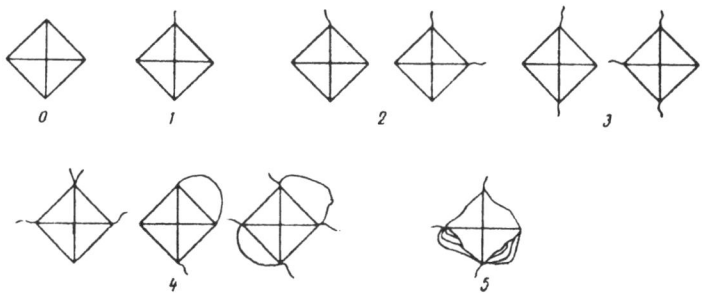

Fig. 31. Improved scale for comparing brittleness numbers (description given in Table 11).

of the overall brittleness number and the derivative of the latter with respect to load:

$$\gamma_P = Z_P \left(\frac{\partial Z}{\partial P} \right)_P. \qquad (4)$$

The brittleness index is determined in the PMT-3. Impressions are made on the crystals with a diamond pyramid under various loads. Impressions not corresponding to brittleness numbers 3, 4, or 5 may be used for an approximate determination of microhardness. It is better to determine the microhardness by reference to loads not producing any cracks near the impressions.

The instrument must be calibrated before the test.

As in the determination of microhardness, in determining the brittleness index it is important to prepare the surface of the sample for test very carefully and to make a correct choice of the regions to be tested on the surface of the microsection and in the field of view of the microscope. In addition to this, it is essential to regulate the loading time strictly as well as the time spent under load.

The method of measuring microbrittleness with a PMT-2 was developed in detail by N. Yu. Ikornikova for the case of corundum [4]. Two phenomena lie at the basis of the method of determining microbrittleness: the change in the number of impressions with cracks on increasing the load and the change in the number of cracks around one impression on increasing the load.

The corundum crystals were tested with seven to ten different loads (50 to 170 g) and 50 impressions were made at each load. The loading time was 15 sec and the time spent under load 5 sec. Special attention was devoted to preparing the specimen and orienting the sample on the PMT-3 table relative to the diagonal of the diamond pyramid. The various forms of brittle damage were analyzed in detail: cracks parallel to the impression diagonal, in some cases constituting continuations of the latter (vertical cracks), cracks formed under the surface of the sample (horizontal or sloping cracks), and finally cracks appearing as a result of the development of cracks under the sample surface. All forms of cracks were taken into account and analyzed.

The results were presented in the form of curves, similar graphs being plotted in order to compare the microbrittleness of different samples.

N. Yu. Ikornikova and L. A. Pikunova [4] established as a result of their preliminary study of microbrittleness that samples with a mosaic structure and traces of plastic deformation had a greater brittleness than homogeneous samples.

TABLE 12. Brittleness Indices of Metal-Like Compounds

Phase	Crystal structure	Overall brittleness $Z_{P=90}$	Rate of variation of brittleness $(dZ/dP)_{P=90}$	Brittleness indices	
				$\tau_{P=90}$ [100]	A [107]
WC	Hex.	20.5	0.200	4.09	18
TiC	Cub.	20.0	0.175	3.50	9
ZrC	»	32.5	0.100	3.25	2.67
VC	»	18.5	0.175	3.23	—
Cr_3C_2	Rhombohed.	16.0	0.150	2.40	—
Mo_2C	Hex.	3.0	0.025	0.75	—
W_2C	»	0	—	0	—
TaC	Cub.	0	—	0	—
TiN	Cub.	29.0	0.085	2.46	1.11
ZrN	»	—		2.47	—
W_2B_5	Hex.	21.0	0.125	2.62	—
TiB_2	»	13.5	0.100	1.35	2.66
Mo_2B_5	Rhombohed.	13.0	0.075	0.97	—
TaB_2	Hex.	9.5	0.06	0.57	—
NbB_2	»	8.0	0.065	0.52	—
CrB_2	»	2.0	0.05	0.10	—
ZrB_2	»	3.0	0.025 (?)	0.075	3.27
CeB_6	Cub.	28.5	0.15	4.10	—
BaB_6	»	19.0	0.175	3.33	—
LaB_6	»	9.5	0.05	0.475	—
CaB_6	»	5.5	0.045	0.248	—
$FeSi_2$	Tetrag.	37.0	0.10	3.70	—
$TaSi_2$	Hex.	10.0	0.25	2.50	—
$CrSi_2$	Hex.	26.5	0.075	1.99	—
$TiSi_2$	Rhombohed.	41.0	0.045	1.85	—
$ZrSi_2$	»	42.5	0.04	1.80	—
WSi_2	Tetrag.	12.0	0.06	0.72	—
$CoSi_2$	Cub.	15.0	0.045	0.68	—
$NiSi_2$	»	10.0	0.045	0.45	—
$MoSi_2$	Tetrag.	8.0	0.05	0.40	—

A detailed investigation into the microbrittleness of refractory compounds was carried out in the PMT-3 by G. V. Samsonov, V. S. Neshpor, and L. M. Khrenova [107]. The work was carried out with loads of 15 to 200 g . First the overall brittleness number was determined and then the brittleness index was evaluated. The results obtained in [107] are shown in Table 12, together with the results of some brittleness determinations based on the powder-compressibility method [100], which agree closely with the former. By comparing the overall brittleness numbers and indices for various compounds we find that brittleness generally increases in the order: silicide, boride, nitride, carbide.

I. N. Frantsevich and A. N. Pilyankevich [105] studied the microbrittleness of various refractory compounds on a titanium base and established a rise in brittleness in the order: silicide, carbide, boride, nitride.

An attempt was made in [107] and [105] to establish a correlation between the tendency of a material toward brittle failure and various other physical properties. It was shown first of all [107] that a reduction in the mean square displacement of the structural elements of the crystal lattice, i.e., a reduction in the mobility of these elements in the propagation of elastic waves, led to a rise in the brittleness of the substance. The reason for this was that, in loading, the elastic stresses were unable to propagate through the whole volume of the crystal, and were localized at the point of application of the load, subsequently engendering cracks.

In order to reduce the brittleness of metal-like compounds, these must be alloyed with components such as will tend to loosen the crystal lattice and reduce the forces of the interatomic bonds; this in turn increases the mobility of the structural elements of the lattice and hence should increase the relaxation capacity of these materials and lessen their brittleness.

However, as indicated in [105], in order to explain the changes in brittleness on passing from one compound to another it is insufficient simply to consider the forces of the interatomic bond; one must also allow for the type of bond and for different relaxation capacities.

CHAPTER 5

RELATION BETWEEN THE MICROHARDNESS OF SOLID-SOLUTION CRYSTALS AND THE COMPOSITION OF THE ALLOY IN TWO-COMPONENT SYSTEMS

Theoretical Microhardness/Composition Relation in Two-Component Systems

At the beginning of the twentieth century N. S. Kurnakov and his colleages began a systematic investigation of the physicochemical properties of metallic alloys in relation to their compositions. The year 1908 saw the appearance of the classical treatise by N. S. Kurnakov and S. F. Zhemchuzhnyi, "The Hardness of Metallic Solid Solutions and Certain Chemical Compounds" [108], in which were set out the general laws relating hardness to the composition of binary systems.

It was shown that the formation of metallic solid solutions was usually accompanied by an increase in hardness, while in the case of the formation of a continuous series of solid solutions the concentration/hardness relation was described by a smooth curve with a maximum at about 50 at.% (Fig. 32).*

This fact, first set out by Kurnakov and Zhemchuznyi, is of fundamental importance in connection with the possibility of using the microhardness method for physicochemical analysis, leading to the vital conclusion that the microhardness of a specific phase varies with composition.

Since the manner in which the composition of a particular phase varies with the composition of the alloy in an equilibrium system is determined by the phase diagram and hence obeys a specific law, we find that by studying the microhardness (constituting a function of composition) of a particular phase this law may be established.

From this we may draw conclusions as to how the microhardness of solid-solution crystals should vary with the composition of binary alloys when the phase equilibria in the latter are governed by phase diagrams of various types.

It is clear that when a continuous series of solid solutions is formed the general form of the concentration dependence of the microhardness will be practically the same as that of the macrohardness (Fig. 32).

---

* Subsequently this was repeatedly confirmed by experiment and justified from the metallo-physical point of view.

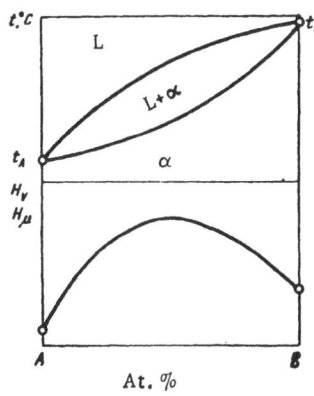

Fig. 32. Schematic hardness and microhardness relationship in a system with a continuous series of solid solutions.

When there is only limited solubility in the solid state, the microhardness of solid-solution crystals should increase in accordance with the alloy composition within the range of homogeneity up to the point of saturation at the temperature in question, subsequently remaining constant on passing into the two-phase region.

This constitutes the basis for using the microhardness method in order to plot the lines of phase transformations in phase diagrams.

Treating these principles as a basis and also considering N. S. Kurnakov's principles of continuity and correspondence [6, 7], S. A. Pogodin, L. M. Kefeli, and E. S. Berkovich [109] plotted the fundamental types of composition/microhardness diagrams of solid-solution crystals for the simplest cases of interaction between the two components; these diagrams enabled the solubility in the solid state to be determined (Fig. 33).

It was shown in [110, 111] that the microhardness method could be used in order to plot solidus lines and also lines of eutectoid and peritectoid transformations taking place in the solid state. In order to solve these problems, isotherms or polytherms of the microhardness of specific phase (for example, solid-solution crystals) are plotted.

Figure 34 presents some typical microhardness/temperature composition relationships on passing through the solidus line; these are based on Kurakov's principles. The same relationships hold when studying the corresponding transformations in the solid state.

The construction of phase-transformation lines by reference to microhardness isotherms is based on the same principle as the study of solubility in the solid state (Fig. 34a,b).

The use of the microhardness-polytherm method for this purpose, however, is based on the following. On studying single-phase alloys quenched from various temperatures, the micro-hardness of the solid-solution crystals remains constant on raising the quench temperature (or changes very little), right up to the line of the phase transformation; then, on passing into the two-phase region it rises or falls according to the nature of the interaction between the components (Fig. 34c, curve I, and Fig. 34d).

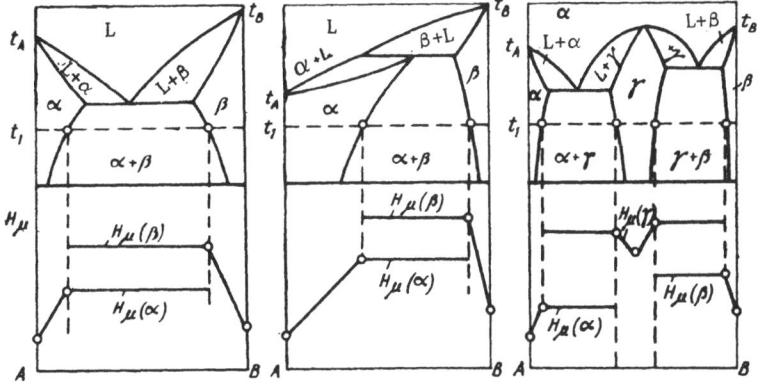

Fig. 33. Composition/microhardness diagrams for several cases of interaction between components.

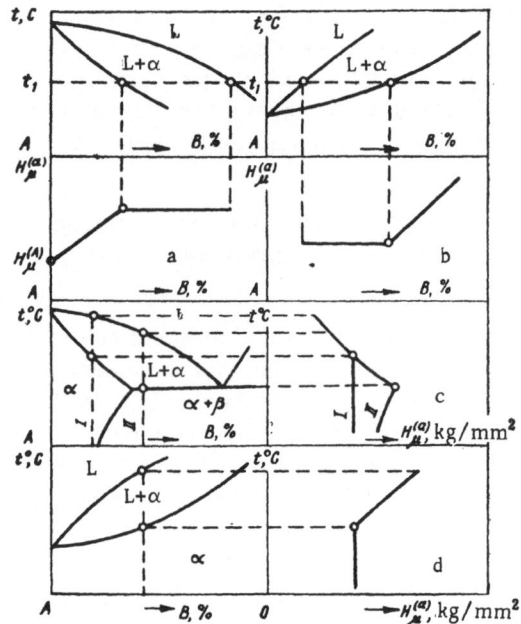

Fig. 34. Possible microhardness/composition and microhardness/temperature relationships on passing through the solidus line.

In plotting the microhardness polytherms of the crystals of a particular phase in a two-phase alloy, the character of the microhardness/temperature relation in one phase region is determined by the line representing the limiting saturation of the second component in this phase; on passing into another phase region it is determined by the solidus line (or the corresponding lines in the solid state); this enables us to mark a break on the microhardness polytherm corresponding to the line of the phase transformation [in the latter case the point lies on a line of nonvariant equilibrium: eutectic, peritectic, eutectoid, or peritectoid (Fig. 34c, curve II)].

We may obtain the theroetical microhardness/composition (or microhardness/temperature) relation (Figs. 33 and 34) after bringing the alloys into an equilibrium state by prolonged annealing at appropriate temperatures.

The conditions for obtaining these theoretical relationships when studying real systems are given in more detail in Chapters 6 and 7.

The Microhardness Method in Plotting Limited-Solubility Curves on Phase Diagrams of Two-Component Systems

Evaluation of Methods of Determining Solubility in the Solid State. In order to determine solubility in the solid state a number of methods of physicochemical analysis are currently employed. Apart from the microhardness method, the most widespread are the methods of microscope and x-ray analysis and also the method of determining the number of electric-charge carriers (when studying solubility in semiconductors).

In order to determine solubility in the solid state by the microscope method one usually prepares a series of alloys of different compositions, anneals them at an appropriate temperature, quenches them, and then studies the microstructure. The limiting solubility is determined by reference to the appearance of a second phase in the structure of the alloys. The method is extremely simple, convenient, and very reliable. However, it is impossible to determine the solubility boundary precisely in this way, since microscope analysis always requires a certain interpolation. In addition to this, in a number of cases microscope analysis may give an erroneous idea of the solubility value, since the second phase cannot always be clearly seen. When there are finely dispersed second-phase precipitates, hardly distinguishable in color from the general background, the microscope method may easily give much too high results.

The method of x-ray structural analysis is free from these disadvantages and enables the solubility limit to be exactly determined. When solid solutions are formed with increasing concentration of the dissolved element, the lattice constant of the solid solution varies over the single-phase region but remains constant on passing into the two-phase region. The break on the lattice spacing/composition curve determines the boundary of the solid solution. When using this method one usually constructs the curve relating the lattice spacing of the solid solution to its composition within the boundaries of the single-phase region (calibration curve) and then, by determining the lattice spacing of the solid solution in quenched, clearly two-phased alloys, previously annealed at an appropriate temperature, one finds the solubility limit in the solid state.

However, in a number of cases the prospects of using x-ray structural analysis are limited. For a low solubility the variation in lattice constant may be so slight that any determination of solubility by the method in question is very inaccurate, and sometimes even impossible.

When there is practically no variation in lattice spacing with varying concentration of the solution, x-ray analysis can no longer indicate whether any limiting solid solutions are formed in the system or not. An example of this case in the cadmium−bismuth system, which until recently was considered (on the basis of x-ray analysis) to constitute a classical eutectic system with no limited solubility.

However, use of the hardness [112] and microhardness [113] methods revealed the fact that at the eutectic temperature up to 5% of bismuth was dissolved in the cadmium. An analogous situation holds in the cadmium−tin system [112, 113]. (More details regarding solubility in cadmium-base alloys in the solid state are given in Chapter 7.) The lattice constant varies very slowly when various elements dissolved in substances having a crystal structure with a three-dimensional system of rigid, directional covalent bonds (for example, silicon or germanium). In these cases x-ray structural analysis gives extremely unreliable results.

The use of the x-ray structural method is also not justified when a solid solution is formed in a three-component system and each of the two additional elements changes the lattice constant in opposite senses.

The method of counting the number of electric-charge carriers has received wide acclaim when studying solubility in semiconductors, since the introduction of different additives leads to a sharp change in the electrical properties of the semiconductor. The carrier concentration is determined from the changes in the Hall constant R and the electrical conductivity σ. If these two quantities are known, the carrier mobility u may be determined and the number of carriers n calculated:

$$n = \frac{\sigma}{u \cdot e},$$

where e is the charge on the electron.

The chief failing of the method is the fact that the number of carriers thus determined is identified with the number of atoms of the dissolved element; in actual fact this is by no means always true. Existing dissolved impurities may also have a substantial effect; so may crystal-lattice defects (dislocations, vacancies, etc.).

In view of this, the method of calculating the number of carriers should only be regarded as an approximate way of determining solubility in the solid state; it only gives accurate results in individual cases. The experimental error in determining solubility by this method reaches 30 to 50%.

Apart from these methods, use is sometimes made of the electrical resistance, thermo-emf, and certain other properties; however, the use of these is very limited when compared with those just considered and we shall devote no further consideration to them.

TABLE 13. Heat Treatment of Al−Cu, Al−Mg, Al−Si, and Al−Zn
Alloys before Determining the Solubility by the Microhardness
Method*

Annealing temp., °C	Annealing time, h			
	Al—Cu	Al—Mg	Al—Si	Al—Zn
550	—	—	300	—
500	400	—	400	—
450	500	500	500	—
400	600	600	600	—
350	700	600	600	—
300	800	600	600	—
250	—	700	700	700
200	—	700	800	700

Note. After annealing the alloys were quenched from the temperatures indicated
in water.
*If the alloys were held at the temperatures given for shorter times than those in-
dicated, it was impossible to secure characteristics such as those shown in Fig. 35
because the solid-solution crystals of the two-phase alloys became heterogeneous
on the microscopic scale (Chapter 9).

The microhardness method compares favorably with those which we have been considering. Combining within itself a study of microstructure and the measurement of hardness
in individual structural components, this method is free from the disadvantages of the microscope and x-ray methods.

Being extremely sensitive, the microhardness method reacts to even the slightest variations in composition. However, the high sensitivity of the microhardness technique also has
disadvantageous aspects, since extraneous factors may exert an appreciable influence. One
must therefore pay special attention to careful preparation of the experiments (Chap. 3).

Determinations of solubility in the solid state by the microhardness method were first
undertaken by Buckle [114-118], who studied the solubility of beryllium, copper, titanium, and
thorium in aluminum at various temperatures. Buckle's microhardness data agreed closely
with the results of other authors using different methods.

We shall now give some more detailed consideration to the question of the reliability of
the results and the prospects of the microhardness method when studying solubility in the solid
state, using two groups of binary aluminum alloys of considerable practical importance as examples.

Solubility of Copper, Magnesium, Silicon, and Zinc in Aluminum.
Many papers have been devoted to a study of the solubilities of these elements in aluminum
[119, 120], using various methods of physicochemical analysis (chiefly x-ray and microscope).

In view of this it is interesting to compare the results of [119, 120] with those based on
microhardness measurements, so as to estimate their reliability.

The solubility of copper, magnesium, silicon, and zinc in aluminum was studied by the
authors, using the microhardness method. For this purpose, series of binary alloys were
prepared. The original material was aluminum of the AV0000 (99.998% Al) type, and also
magnesium, zinc, silicon, and copper containing no more than 0.005% of impurities. After
preparation and appropriate heat treatment (conditions indicated in Table 13), the composition of the alloys was established by chemical analysis.

The surface hardening was removed after grinding and polishing by brief annealing, the
samples being sealed in evacuated glass ampoules. After this the microhardness was measured under a load of 10 g. The results, averaged over seven to 10 measurements, were sum-

TABLE 14. Solubility of Copper, Magnesium, Silicon, and Zinc
in Aluminum at Various Temperatures

Temperature, °C	Solubility (wt.%)							
	Al—Cu		Al—Mg		Al—Si		Al—Zn	
	1*	2**	1*	2**	1*	2**	1*	2**
550	—	—	—	—	1.43	1.45	—	—
500	4.00	4.00	—	—	0.80	0.77	—	—
450	2.84	2.80	15.2	15.0	0.45	0.46	—	—
400	1.85	1.82	12.0	11.9	0.30	0.28	—	—
350	1.00	1.00	9.1	9.0	0.15	0.16	—	—
300	0.75	0.72	6.5	6.7	—	—	—	—
250	—	—	4.5	4.4	—	—	20.0	19.8
200	—	—	3.3	3.1	—	—	12.0	12.6

* Microhardness method.
** X-ray structural method.

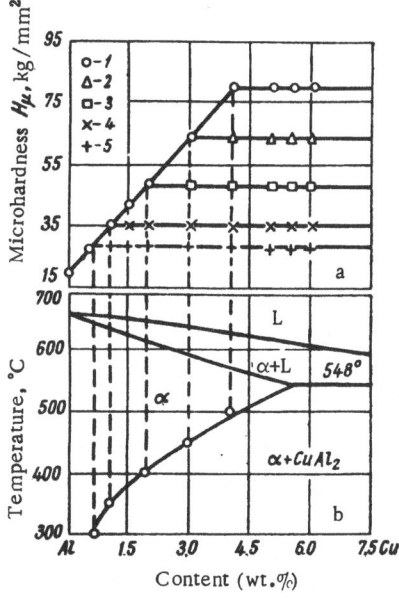

Fig. 35. Microhardness of solid-
solution crystals as a function of
composition for aluminum—copper alloys quenched after annealing at various temperatures (a),
as compared with the equilibrium
phase diagram of the aluminum—
copper system (b): 1) 500°; 2)
450°; 3) 400°; 4) 350°; 5) 300°C.

marized in the form of microhardness isotherms. By
way of example, Fig. 35 shows the microhardness/com-
position isotherms for the aluminum—copper isotherm
in comparison with the equilibrium phase diagram.

The solubilities of copper, silicon, and zinc in
aluminum at the corresponding temperatures are shown
in Table 14 and compared with the results obtained by
x-ray structural analysis [119, 120].

Analysis of the data given in Table 14 shows that
the solubility results obtained by the microhardness
method agree closely with the results of x-ray structural analysis. This convincingly indicates the reliability
of the microhardness method when used for determining
solubilities in the solid state.

Solubility of Some Transport Metals
in Aluminum. Let us give some more detailed consideration to the solubility of a number of transition
metals (Ti, Zr, Nb, Ta, Mo, and W) in aluminum, as determined by the microhardness method. This method
is of particular interest in the present case, since the
solubility of refractory metals in aluminum is as a rule
very small indeed and difficult to study by other techniques (the x-ray method in particular).

In addition to this, in recent years refractory transition metals have been more and more widely used in
the industrial production of aluminum alloys as modifying
and alloying additives [121].*

The solubility of these metals in aluminum in the solid state at various temperatures was
studied in [122-126] by the microhardness method.

* M. V. Mal'tsev, Dissertation, M. I. Kalinin Moscow Institute of Nonferrous Metals and Gold,
Moscow (1955).

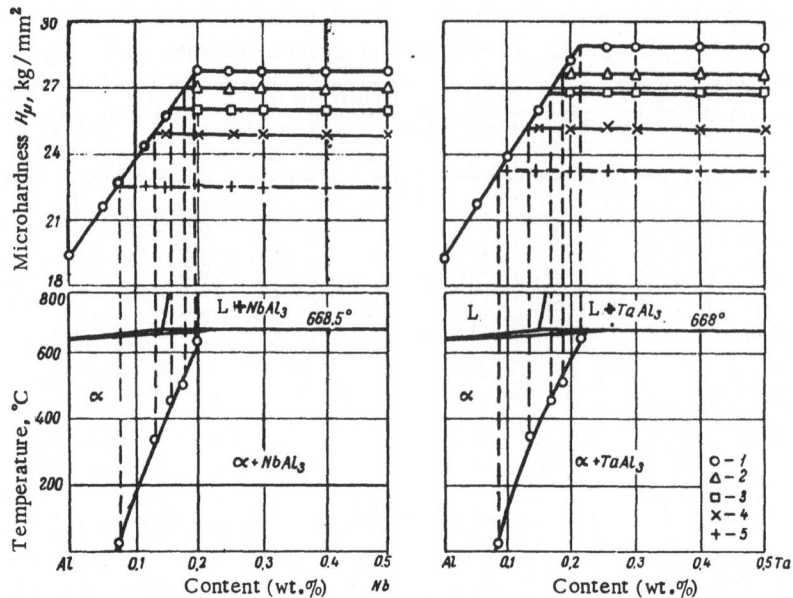

Fig. 36. Microhardness of solid-solution crystals as a function
of the composition of Al−Nb and Al−Ta alloys quenched after
annealing at various temperatures, and limited-solubility curves
of Nb and Ta in Al, °C: 1) 640; 2) 500; 3) 450; 4) 350; 5) 20.

By directly fusing the components (using aluminum of the AV000 type, 99.8% Ti, 99.9%
Zr, and also Nb, Ta, W, and Mo of purity no less than 99.8%), a series of binary alloys was
prepared.

In order to even out the chemical composition and bring the alloys into an equilibrium
state, the cast alloys were subjected to slight deformation (30%) and then homogenized for
200 h at 500°C and subsequently annealed in steps at 640, 550, 500, 450, 400, and 350°C for
50 h and quenched from the corresponding temperature in water.

After establishing the chemical composition of the alloys more exactly by chemical analysis and finishing the surface in an appropriate manner (Chap. 3), the microhardness was measured.

By way of example, Fig. 36 shows the microhardness of solid-solution crystals as a function of the composition of aluminum−niobium and aluminum−tantalum alloys.

An analogous situation holds for the other systems.

The results of the solubility determinations are given in Table 15.

We see from the graphs of Fig. 36 that, for each specific temperature, the solubility is
determined absolutely strictly by the intersection of the lines corresponding to 1) the micro-
hardness/concentration relation and 2) the microhardness of the solid solution completely saturated at the temperature in question, which has a constant value in two-phase alloys of different compositions.

The extrapolation of the limited-solubility curves plotted from the results of Table 15 up
to the temperatures of the nonvariant transformation yielded the limiting solubility of each of
the elements under consideration in aluminum, as shown in Table 16.

By considering the examples of the foregoing systems formed between aluminum and
transition metals, we see that the microhardness method is an exceedingly sensitive process

TABLE 15. Solubility of Titanium,
Zirconium, Niobium, Tantalum,
Tungsten, and Molybdenum in Aluminum

Temper-	Solubility, wt.%					
ature, °C	Ti	Zr	Nb	Ta	Mo	W
640	0.18	0.23	0.20	0.22	0.22	0.15
550	0.10	0.16	0.18	0.20	0.19	0.13
500	0.09	0.14	0.16	0.18	0.16	0.11
450	0.08	0.11	0.14	0.16	0.14	0.09
400	0.07	0.10	0.13	0.15	0.10	0.07
350	0.06	0.08	0.12	0.13	—	—
20	0.04	0.04	0.06	0.07	—	—

TABLE 16. Limiting Solubility of Titanium,
Zirconium, Niobium, Tantalum,
and Molybdenum in Aluminum

System	Temperature of the nonvariant transformation, °C	Limiting solubility (wt.%)
Al—Ti	665	0.28
Al—Zr	664	0.28
Al—Nb	668	0.22
Al—Ta	668	0.24
Al—Mo	660	0.25
Al—W	660	0.17

of physicochemical analysis, enabling the slightest trace of solubility to be determined. The solubility of titanium in aluminum obtained by the microhardness method in [123] agrees closely with Buckle's results [114-118] obtained by the same method.

The solubility values obtained for tantalum in aluminum by x-ray structural analysis [122] and the microhardness method [125] also agree closely.

These facts indicate the reliability of the results obtained by the microhardness technique.

Other Investigations into Solid-State Solubility Using the Micro-hardness Method. Apart from those already mentioned, there have been a number of other successful investigations into solid-state solubilities using the microhardness technique.

Studying the microhardness/composition relationship in the Mg—Sn and Mg—Pb systems, S. A. Pogodin, L. M. Kefeli, and E. S. Berkovich [109] observed a region of solid solutions based on the compound Mg_2Pb, although no such region occurred in the case of Mg_2Sn.

V. P. Elyutin and V. F. Funke [127] used the microhardness method to study the solubility of chromium in niobium and niobium in chromium.

In this way it was shown that the solubility of chromium in niobium at 1600, 1550, 1500, and 1400°C was respectively ~10.5, 6.2, 4.2, and 3.1 at.%. The solubility of niobium in chromium at 1600 and 1500°C was 20 and ~12.5 at.% respectively.

Apart from the microhardness method, the authors also used quantitative microstructural analysis for studying solubility. The percentage area of a microsection occupied by one phase or another was calculated (using A. A. Glagolev's method). The results obtained by the two methods agree closely with each other.

The solubility of titanium and copper was studied in [128] at various temperatures. The limited-solubility curve constructed agreed closely with results obtained by other methods.

Yan Van-Bok* studied the solubility of manganese and iron in aluminum at various temperatures, using x-ray structural analysis and microhardness measurements. The results agreed closely with each other.

E. M. Savitskii, V. F. Terekhova, and I. A. Novikova [129] successfully used the microhardness method to determine the solubility of neodymium in magnesium. At 530, 500, and 450°C and at room temperature magnesium dissolved 1.5, 1.2, 1.0, and 0.8 wt.% of neodymium respectively. The results obtained by the microhardness method agreed excellently with careful microscope analysis.

The microhardness method was used in [130] to study the solubility of calcium in lead and a limited-solubility curve was plotted. The microhardness data agreed very well with the results of Schumacher and Bonton, who used x-ray structural analysis [131].

E. M. Savitskii, V. F. Terekhova, and I. A. Markova [132] used microhardness to determine the solubility of yttrium in chromium at 1100, 1500, and 1700°C: 0.5, 0.6, and 1.0 wt.% respectively. An analogous result was obtained by microscope analysis.

The microhardness method was used in [133-135] for studying the solubility of aluminum, antimony, and phosphorus in germanium and silicon. The results obtained by the microhardness method agree closely with other results based on various other methods of measurement.

The data relating the solubility of antimony in germanium in [134] are extremely interesting, as they illustrate the great prospects of the microhardness technique. It was found that the maximum solubility of antimony in germanium at the eutectic temperature of 426°C was altogether 0.07 at.%.

The foregoing examples of solid-state studies based on microhardness measurements, and the comparison of the results so obtained with those obtained in other ways, show that the microhardness method is extremely sensitive and may successfully be employed in the solution of such problems.

The Microhardness Method and Plotting Solidus Lines
on Phase Diagrams of Two-Component Systems

The possibility of plotting solidus lines by the microhardness method was first mentioned in [137] when studying the ternary copper−chromium−zirconium system. Subsequently the possibility of using the method of plotting microhardness isotherms for this purpose was given a theoretical basis in [110, 111].

The region on the phase diagram lying between the liquidus and solidus lines is divided into two parts by the line corresponding to the onset of linear shrinkage: the solid−liquid state, in which the solid phase predominates and a framework of crystals is formed, and the liquid−solid state, in which the liquid phase predominates [138, 139].

Clearly, in heat-treating alloys, particularly prepared microsections, if one desires to preserve their shape, this factor must be taken into account; one must try not to pass into the region of the liquid−solid state.

_____

* Yan Van-Bok, Dissertation, M. I. Kalinin Moscow Institute of Nonferrous Metals and Gold, Moscow (1957).

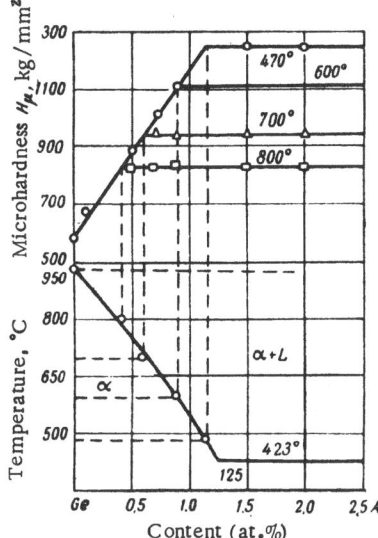

Fig. 37. Microhardness of solid-solution crystals as a function of the composition of Ge−Al alloys quenched after annealing at various temperatures, and part of the solidus curve in this system.

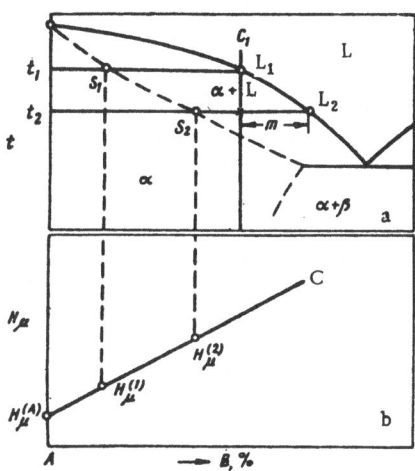

Fig. 38. Scheme illustrating the possibility of using the microhardness method for constructing the equilibrium solidus by studying alloys crystallized under nonequilibrium conditions.

Let us illustrate the possibility of using the microhardness method for plotting solidus lines by using the germanium−aluminum and silicon−aluminum systems, studied in [133, 134], as examples.

Two series of alloys were prepared from the purest original materials under conditions preventing oxidation (germanium and silicon with no more than $10^{-5}\%$ impurities and aluminum of the AV0000 type, 99.998% Al, were used).

The cast Ge−Al alloys were annealed at 470, 600, 700, 800, and 900°C for 1200, 1000, 850, 800, and 700 h respectively, for homogenization.

The Si−Al alloys were annealed at 1200, 1100, 1000, 850, 750, and 600°C for 170, 340, 720, 900, 1100, and 1500 h respectively. During their annealing the alloys were enclosed in sealed quartz ampoules filled with purified argon after evacuating to 10^{-4} mm Hg.

During the annealing the alloys were periodically quenched. The microstructure of the quenched samples and the microhardness of the solid-solution crystals were studied. Equilibrium was considered to have been reached when the microstructure and microhardness ceased varying with further annealing.

The microhardness of pure germanium and silicon was measured on single-crystal samples vacuum-annealed for three days at 900 and 1000°C respectively. Loads of 20 and 50 were used; in both cases similar microhardness values were obtained. From the mean of some 50 measurements the microhardness of germanium was found to be 580 ± 20 kg/mm² and that of silicon 630 ± 20 kg/mm².*

The silicon microhardness results lie close to the value obtained by Savitskii and Baron [140].

Figure 37 shows the results of microhardness measurements on solid-solution crystals for the Ge−Al system. We see from this graph that the microhardness isotherms clearly show sharp breaks indicating the solidus concentration at the specified temperature. The solidus line plotted from these results (Fig. 37) agrees closely with the results of [141, 142] which were based on other methods.

---

* These figures are of course not constants for the materials in question, since the microhardness is subject to the dislocation density. At the present time this point has not yet been pursued to its conclusion.

An analogous situation holds in the silicon−aluminum system; however, in this case the solidus is retrograde. The maximum extent of the solid-solution region along the concentration axis corresponds to 1000°C and equals about 1.1 at.% Al. At the eutectic temperature the solubility of aluminum in silicon is altogether 0.4 at.%.

Apart from the two cases mentioned, analogous solidus curves were plotted for the germanium−antimony, silicon−antimony, germanium−phosphorus, and silicon−phosphorus systems [133-135].

The solidus curve of the copper−titanium system was partly constructed by this method (in the region of the copper-base solid solution) in [128].

One of the authors showed in [143] that in certain cases the position of the solidus line could be approximately estimated by studying the microhardness of alloys crystallized under nonequilibrium conditions. It is well known that if diffusion in the solid phase is suppressed during crystallization the composition of the solid solution (in a section of the crystallite through the center) will vary in accordance with the equilibrium solidus line [144].

This is confirmed most clearly in experiments on the pulling of a crystallizing solid phase from the melt (Czochralski method [145]). Hence the composition of the solid solution in the center of crystallites formed by a nonequilibrium-crystallizing alloy (an alloy crystallizing with suppression of the diffusion in the solid phase) should be determined by the corresponding point on the equilibrium solidus line.

Clearly, if we know the composition of the initially separating portions of the solid solution (with diffusion in the solid state suppressed) and the temperature at which these portions are precipitated, we can determine the position of the point on the solidus line for the alloy in question. In order to determine the composition in the center of the crystallites we may (inter alia) use the microhardness method.* For this purpose we must know the relation between the microhardness and the concentration of the solid solution in the system under consideration. The temperature may easily be determined from the intersection of the ordinate corresponding to the alloy in question with the liquidus line, which is usually quite well known, since its construction by the method of thermal analysis ordinarily presents no difficulty.

Let us suppose that an alloy of composition C_1 (Fig. 38a) is crystallized under conditions ensuring suppression of diffusion in the solid phase, and in the center of the crystallites the microhardness equals $H_\mu^{(1)}$. Then the concentration of the solid solution corresponding to the equilibrium solidus is determined from the calibration curve relating the microhardness to composition ($H_\mu^{(A)} - C$ in Fig. 38b), while the temperature of the solidus t_1 is determined by the intersection of the ordinate C_1 with the equilibrium liquidus. In this way the coordinates of the point S_1 will be established.

Clearly, in order to determine the solidus by the method described, it is necessary that the equilibrium liquidus line should be known in advance, in addition to the calibration curve of $H_\mu^{(A)}$.

This method is distinguished by rapidity of preparing the alloys, since the conditions for the suppression of diffusion in the solid phase are easy to achieve by crystallizing the alloy in air or casting in metal or earthen molds.

------

* It was shown by I. I. Novikov, V. G. Lyutsau, and V. S. Zolotorevskii (Chapter 8) that, for a high degree of branching of the dendrites (the overall dimensions of these being revealed by a microstructural study), the composition in the center of a one-fold axis may differ considerably from that in the center of a higher-order axis, i.e., an arbitrarily selected micrograin.

However, it should be noted that the determination of the solidus line by this method is not always sufficiently precise, and in a number of cases it is quite impossible, since the dimensions of the impression left by the diamond pyramid may exceed the dimensions of the crystallization center having composition S_1.

In addition to this, not all the grains intersect the plane of the microsection in the middle. One must choose for study crystallites having average sizes most characteristic of the particular conditions of crystallization.

The concentration of the solid solution corresponding to the equilibrium solidus at the specified temperature may be determined more precisely by the microhardness technique in the following way. The alloy of composition C_1 (Fig. 38a) must be held at a temperature intermediate between the liquidus and the solidus (for example, at t_2, Fig. 38a). In this way a liquid of composition L_2 and crystals of the solid solution of composition S_1 corresponding to the equilibrium solidus should be in equilibrium. If, after appropriate holding at t_2, the alloy is crystallized in such a way that diffusion in the solid phase is suppressed, then, clearly, a region having a composition corresponding to the equilibrium solidus at the temperature in question should lie in the intersection of each solid-solution grain by the plane of the microsection. After measuring the microhardness of these regions (let us suppose that this is H_μ^2, see Fig. 38b), we may easily determine their composition from the calibration graph. In this way we shall establish the position of the point S_2 on the solidus line (Fig. 38a).

The dimensions of the regions of composition S_2 in the crystals of the solid solution of composition C_1 must be determined by the quantitative relation between the solid and liquid phases in equilibrium at temperature t_2. According to the lever rule, the relative number of solid-phase crystals of composition S_2 for the case in question is determined by the intercept m (Fig. 38a). The closer the temperature at which the alloy of specified composition is held in the two-phase region $L_2 + \alpha$ to the solidus temperature, the greater is the relative amount of solid-phase crystals and the more reliably will their composition be determined in the subsequent investigation.

In essence this second method reduces to the earlier-described method of plotting microhardness isotherms, or to be more precise it constitutes a modification of this.

The aluminum−copper, aluminum−silicon, and bismuth−antimony systems were studied in this way.

Aluminum alloys containing 8.5, 5.0, and 2.0 wt.% Si and 6.0, 4.5, and 3.0 wt.% Cu and bismuth alloys containing 20, 50, and 60 wt.% Sb were prepared.

After melting, the alloys were air cooled (thus ensuring the suppression of diffusion in the solid phase). In addition to this, all the alloys were cooled very slowly with the furnace from the moment at which they were entirely in the molten state to the temperatures indicated in Tables 17 and 18, and held there for 10 h, after which air cooling took place.

From the alloy samples thus prepared, microsections were obtained; after removal of the surface-hardened layer, the microhardness of these was measured in the PMT-3, furnished with an attachment for the automatic application of the load.

The results of the measurements are shown in Tables 17 and 18. On the basis of these results and the calibrating curves (Figs. 39 and 40) [113, 146, 147] the concentrations of the solidus points for the corresponding temperatures were determined. The solidus curves thus obtained for the aluminum−silicon, aluminum−copper, and bismuth−antimony systems are shown in Figs. 39 to 41.

TABLE 17. Results of Determining the Solidus in Aluminum Alloys
Using the Microhardness Method

Weight % of second component	t, °C	H_μ, kg/mm^2	C_S' (wt.%) *	C_S'' (wt.%) †
8.5 Si	600 ‡ 580	41.5 51.0	1.20 1.56	1.00 1.52
5.0 Si	630 ‡ 600 580	33.2 37.8 50.0	0.70 1.03 1.53	0.50 1.00 1.52
2.0 Si	654 ‡ 640 620 590	21.4 25.0 34.0 44.5	0.09 0.28 0.72 1.24	0.10 0.29 0.72 1.25
6.0 Cu	645 ‡ 630 610 590	34.0 37.6 49.7 70.0	0.85 1.08 1.83 3.12	0.75 0.99 1.77 3.00
4.5 Cu	650 ‡ 600 580 560	33.0 62.0 76.3 89.0	0.80 2.50 3.60 4.88	0.73 2.50 3.62 4.85
3.0 Cu	655 ‡ 640 620 600	23.2 35.5 44.0 62.3	0.18 0.92 1.50 2.54	0.13 0.90 1.50 2.50

* C_S' is the concentration of the solid solution corresponding to the equilibrium solidus line at the specified temperature, obtained from microhardness measurements.

† C_S'' is the same, obtained from electrical-resistance measurements and thermal analysis [119, 120].

‡ The liquidus temperatures for the corresponding alloys.

We see from the graphs in Fig. 39 that the resultant data, except for two points corresponding to aluminum alloys containing 8.5 and 5.0% Si cooled in air from the molten state, agree closely with the results obtained by studying equilibrium alloys on the basis of electrical-resistance measurements and thermal analysis [119, 120]. For comparison, Table 17 gives the concentrations of the solidus points C_S'' for the corresponding temperatures derived from [119, 120]. It is not difficult to see that the difference between these data and the results obtained by microhardness measurements (except for the two points mentioned) is very slight. The discrepancies observed in the determination of the solidus for the two alloys in question may in the present case be explained by the fact that the microhardness measurements are greatly affected by the microheterogenation of the grains of the solid solution, which may be strongly developed in these alloys under the cooling conditions envisaged.

The part of the solidus of the bismuth–antimony system constructed on the basis of microhardness measurements over the concentration range 50 to 90% Sb agrees very closely with the results of I. I. Novikov [112, 148], who studied the solidus in this system between 10 and 60 wt.% Sb,* and practically coincides with the theoretical solidus calculated by Ya. E.

---

*I. I. Novikov, Dissertation, Moscow Institute of Steels and Alloys (1964).

TABLE 18. Results of Determining the Solidus by the Microhardness
Method in Bismuth—Antimony Alloys

Weight %	t, °C	H_μ, kg/mm^2	C_S', wt.%
60	550*	71.5	86.8
	500	80.0	80.0
	450	87.5	68.4
50	525*	75.5	84.0
	475	81.3	75.6
	425	92.0	62.7
20	400*	93.2	58.0
	375	99.5	50.0

*The temperature of the liquidus for the corresponding alloys.

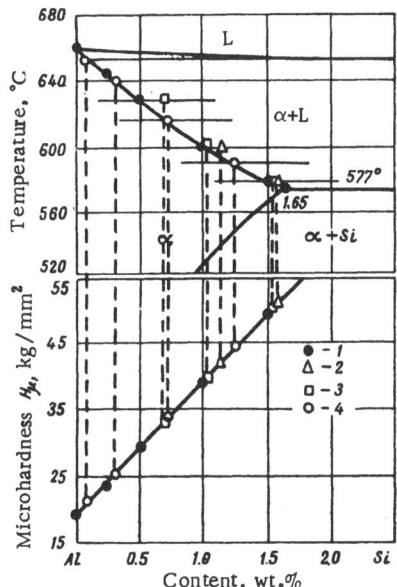

Fig. 39. Construction of the solidus
line in the aluminum—silicon system:
1) Results obtained on studying equilibrium solid solutions, with the construction of a calibrating curve and
the solidus line [124, 125]; 2-4) results obtained by studying alloys respectively containing 8.5, 5.0, and 2.0
wt.% Si.

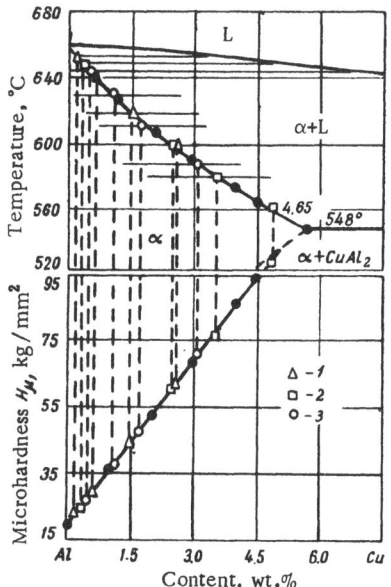

Fig. 40. Construction of the solidus
line in the aluminum—copper system: 1-3) Results obtained on studying alloys containing 3.0, 4.5, and
6.0 wt.% Cu (remaining notation as
in Fig. 39).

Geguzin and B. Ya. Pines on the basis of the energy of mixing [149]. The results obtained in
[143] may also serve as an additional proof of the mutual relation between the processes involved in the crystallization of solid solutions and the phase diagram.

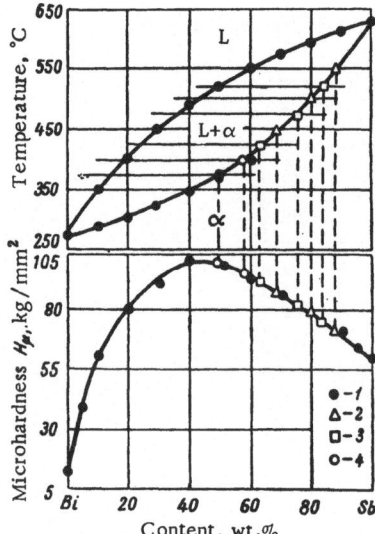

Fig. 41. Construction of the solidus line in the bismuth–antimony system: 1) On the H_μ graph, results taken from [154]; on the liquidus curve, data taken from various authors [117]; on the solidus curve, results taken from [117]; 2, 3, 4) results obtained on studying alloys with 60, 50, and 20 wt.% Sb, using the microhardness method.

Microhardness in Two-Component Semiconductor Systems

The concentration dependence of the microhardness of solid solutions in semiconductor systems has been repeatedly discussed. This has been due to the special nature of the structure and the character of the chemical bond in semiconductors. In particular the manner in which the hardness and microhardness vary with concentration on doping such semiconductors as germanium, silicon, and their structural analogs (compounds of the $A^{III}B^V$ type) has been considered. At the present time there are essentially two points of view. One of these reduces to the fact that the microhardness of the solid solution temperature may either rise or fall on increasing the concentration of the solution, depending on the particular manner in which the solvent and dissolved component interact. The other point of view amounts to the principle that the microhardness should always increase, since in the case of measurement at room temperature distortions of the crystal lattice, arising as a result of dissolution and leading to an increase in microhardness, should play a major part. Furthermore it is precisely in semiconductors that these distortions should appear most sharply, since the process of introducing foreign atoms into the lattice of the semiconductor is impeded by the rigidity of the covalent bonds characteristic of these. The change in the strength of the chemical bond on doping the semiconductor (the measurements being made at room temperature after annealing and quenching) should play a secondary role.

Before passing on to a brief description of experimental data relating to the microhardness/composition relation in semiconductor systems, it must be immediately stipulated that, in its correct sense, the term "semiconductor system" should be taken to means a system formed by two semiconductors (for example, germanium–silicon, selenium–tellurium, etc.) or a semiconductor and a doping element (often a metal). In the second case we consider that the semiconductor plays the part of solvent. By way of an example of such a system we may mention some such as germanium–aluminum, germanium–antimony, silicon–phosphorus, etc.

In the germanium–silicon system the microhardness of nonequilibrium alloys was studied as a function of composition by Liu Chen-Yuan [147].*

The original materials used for preparing the alloys were single-crystal germanium and silicon with carrier concentrations of ~10^{14} cm^{-3}, which corresponds to a purity of the order of 10^{-7}%. Alloys of different compositions were prepared by rapid cooling (in order to suppress liquation phenomena) with subsequent prolonged (six months) vacuum annealing at a temperature close to that of the solidus. In addition to this, some alloys were prepared

---

* Liu Chen-Yuan, Dissertation, A. A. Baikov Institute of Metallurgy, Academy of Sciences of the USSR, Moscow (1960).

by pulling from the melt with simultaneous feed maintenance (using the method proposed by D. A. Petrov). The alloys under consideration were also subjected to a long homogenizing anneal at high temperature.

All the alloys appeared single-phased under the microscope. The microhardness was measured with a load of 50 g, allowing for the methodical characteristics of its determination in semiconductors mentioned earlier. The results of the measurements and the plotting of a concentration/microhardness characteristic showed that the composition dependence of the microhardness in the germanium−silicon system constituted a curve passing through a maximum at about 45 to 50 at.% Si. Thus the manner in which the microhardness depends on concentration in a typical semiconductor system corresponds to the general scheme established for such systems (continuous series of solid solutions).

As regards systems of the second type (semiconductor-doping element), the manner in which the microhardness depends on composition within the range of homogeneity of the solid solution based on the semiconductor is in general similar to that found in metallic systems.

In the foregoing we have considered the dependence of the microhardness on the composition of solid-solution crystals in connection with the construction of the line of the solidus. These relationships have an analogous character when constructing parts of the limited-solubility curves in the same systems [133-135]. Similar relationships are obtained for a number of other germanium− and silicon−base systems [150-153]. We may accordingly conclude that when doping elements of the donor and acceptor type dissolve in germanium and silicon considerable distortions of the crystal structure develop, and the microhardness rises sharply. In this process the principal part is (according to our own view) played by the geometrical part [154]. Naturally on raising the temperature the manner in which the microhardness of the solid-solution crystals varies with concentration within the limits of the region of homogeneity based on the semiconductor may alter. On reaching a certain limit this relationship will be determined by the influence exerted by the particular doping element on the strength of the chemical bond.

However, in certain cases, when doping semiconductors of a specific structure, for example, that of the chain type (tellurium, selenium, etc.) with small traces of additives, there may be a slight fall in microhardness even at room temperature owing to the loosening of the structure and the weakening of the bond between the structural units. This was demonstrated by V. N. Lange and A. R. Regel' [94] by precision measurements of density, the density/concentration relationship showing a minimum on doping tellurium with selenium and sulfur.

The observed density minimum may be quantitatively explained on the principle of the truncation of the chains [155] if we consider that, for small quantities of additives, a "vacancy" of more than ten atomic volumes of tellurium is formed at each selenium or sulfur atom. On increasing the concentration of selenium or sulfur, the formation of "vacancies" ceases as a result of the combination of "groups of impurity atoms." It cannot, furthermore, be excluded that the effect observed may result from correlation between the formation of "microcracks" and the concentration of selenium or sulfur atoms in the tellurium [94].

In the authors' opinion, a change in the degree of defectiveness of the structure should lead to a change in mechanical properties. In order to verify this proposition, the microhardness of a number of Te−Se samples was measured in the PMT-3. The averaged results are presented below.

Since tellurium possesses a sharply expressed anisotropy of its mechanical properties, its microhardness in the direction of the C axis and perpendicular to the latter will be different: The impression of the diamond pyramid has a section of rhombic shape, the large diagonal being perpendicular to the C axis, in accordance with the stronger bond which exists between the atoms in the chain as compared with the interchain bond [94].

The hardness values calculated from the length of the smaller diagonal are given below, the hardness of tellurium itself being taken as unity:

Se content, at.%	0.00	0.10	0.20	0.50	1.00	2.00	3.00	10.0
Hardness	1.00	0.78	1.06	1.51	2.10	1.92	2.22	1.18

Thus for a concentration of 0.1 at.% Se there is a minimum of the hardness, indicating a "loosening" of the structure. A minimum of similar character may also be observed in the hardness curve based on the longer diagonal.

An analogous conclusion was reached by V. I. Veraks, V. N. Lange, and R. V. Sukhanova [156], who studied the effect of antimony doping on the microhardness of tellurium.

Determining Ranges of Homogeneity Based

on Semiconducting Compounds, Using the

Microhardness Method

Thermodynamic analysis of phase equilibria in binary systems shows that regions of solid solutions based on the components and chemical individuals always occur. It is particularly important to determine the extent of these regions or ranges for semiconducting compounds, since an excess of one component or another within the range of homogeneity will lead to extreme changes in electrophysical properties. It should be noted that in a number of cases solid solutions of the subtraction type may be formed; the effect of this will be that the range of existence of the homogeneous phase will no longer include the stoichiometric composition, which will lie altogether in the two-phase region. The deviation from stoichiometry may vary with increasing temperature and this may lead to substantial changes in electrophysical properties. In view of this it is of great interest to determine the position of the ranges of solid solutions based on semiconducting compounds. V. M. Glazov and colleagues [376, 377] considered the case of silver selenide and germanium and tin tellurides, and demonstrated the possibility of using microhardness measurements in order to solve this problem. We shall now briefly consider the results of these investigations, which in addition to their methodical value present special interest from the point of view of discovering the physicochemical nature of the semiconducting compounds Ag_2Se, GeTe, and SnTe.

It is well known that the silver chalcogenides are semiconductors with interesting physical properties. However, on crystallizing compositions corresponding to the stoichiometric ratio of the components pure silver is usually precipitated. It was desired to find whether this was a consequence of nonequilibrium crystallization or the result of incompleteness in the reaction of forming the compounds in the liquid phase.

Compounds of stoichiometric composition were taken for examination and crystallized under various conditions; they were also subjected to high- and low-temperature annealing and the microstructure of the resultant materials was analyzed. As original materials for preparing the samples, refined silver (99.99% Ag), tellurium of brand TA-1, twice vacuum-distilled, and selenium of brand V-3 were selected. Synthesis was effected by melting the components, taken in stoichiometric ratio and placed in evacuated (10^{-3} mm Hg) and sealed quartz ampoules. Crystallization was carried out after holding for many hours (up to 10) in the liquid phase and then cooling slowly with the furnace. The holding temperature was close to the melting point of the corresponding chalcogenides. A number of alloys were crystallized without holding at the melting point, and some by rapidly quenching the ampoules in water after prolonged holding at the temperature. However, silver precipitates appeared clearly in every case, even to the unaided eye, on the surface of the samples; under the microscope they were visible at a low magnification on unetched microsections.

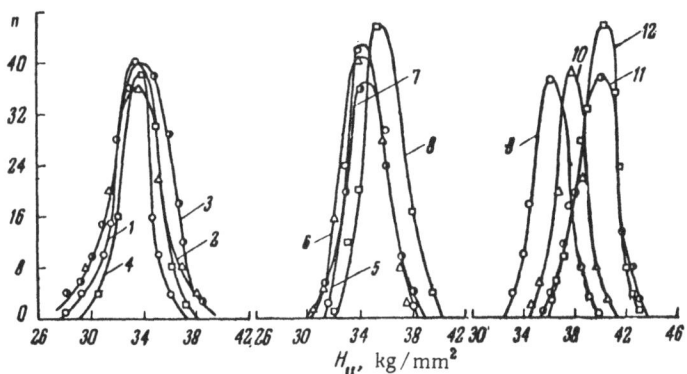

Fig. 42. Frequency curves of the microhardness
of the Ag_2Se-base phase in Ag−Se alloys of various compositions. Curves 1 to 12 correspond to
33; 33.2; 33.333; 33.4; 33.5; 33.6; 33.7; 33.8; 34;
34.2; 34.5; 35 at.% Se.

In addition to experiments on the crystallization of Ag_2Se and Ag_2Te, experiments were
carried out on the annealing of alloys obtained by slow cooling with the furnace after prolonged
holding in the liquid phase. The samples were annealed above and below the corresponding
polymorphic transformation temperatures (700 and 120°C respectively) occurring in both samples.

The samples were annealed in evacuated (10^{-3} mm Hg) and sealed quartz ampoules and
subsequently cooled in air. The annealing time was 200 h. However, not even the slightest
changes in the behavior of the alloys as regards the precipitation of silver were observed.

The results of these preliminary experiments suggest that the factors mentioned earlier,
i.e., nonequilibrium crystallization and failure to complete the reaction in the liquid phase,
could not play any great part in forming the final structures of Ag_2Se and Ag_2Te alloys of
stoichiometric composition; evidently the precipitation of the silver was unconnected with
these. This would suggest that the precipitation of the silver was due to the existence of equilibrium deviations from stoichiometry in these compounds.

In order to verify this proposition, two series of alloys containing 33, 33.2, 33.333, 33.4,
33.5, 33.6, 33.7, 33.8, 34, 34.2, 34.5, 35, and 37 at.% of selenium (or tellurium) were synthesized; after prolonged holding in the liquid state these were cooled slowly with the furnace
and then annealed at 120 (Ag−Se) and 130° (Ag−Te) for 200 h.

The microstructure and microhardness of the samples so prepared were studied.

In order to reveal the microstructure a mixture of 40% NH_4OH + 40% H_2O_2 + 20% H_2O
was used as etchant. For greater reliability the numerical values of the microhardness were
obtained after statistical analysis of the data, with the plotting of frequency curves. For this
purpose 100 impressions were made on microsections of alloys of each composition, and 200
impressions on the alloys of stoichiometric composition and certain others.

The microstructural analysis of the Ag−Se system showed that in alloys containing from
33.0 to 33.6 at.% Se silver precipitation occurred. Starting from 33.7 at.% Se these precipitates were practically absent, while alloys containing from 33.8 to 34.5 at.% Se were clearly
single-phased. Alloys with greater selenium contents contained Se precipitates.

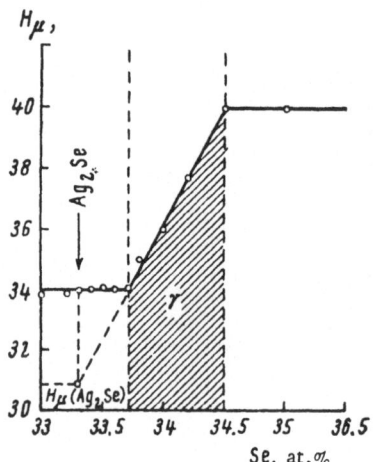

Fig. 43. Microhardness of Ag$_2$Se-base solid-solution crystals as a function of the composition of the alloys.

A qualitatively analogous picture occurred in the Ag−Te alloys. Here single-phased alloys were obtained over a concentration range of about 1 at.%, starting from approximately 34 at.% Te. It should nevertheless be noted that in this system the picture was less obvious than in Ag−Se.

On studying the microhardness the following results were obtained. For alloys of the Ag−Se system clear microhardness distribution curves were obtained; these appear in Fig. 42. We see from this figure that the maximum on the frequency curves corresponding to alloys with 33, 33.2, 33.333, 33.4, 33.5, 33.6, and 33.7 at.% Se represent roughly the same microhardness value, about 34 kg/mm^2 (curves 1 to 7). However, on further raising the Se content to 34.5 at.% the maximum on the frequency curves moves in the direction of higher values, and for the alloy containing 34.5 at.% Se it reaches 40 kg/mm^2.

The microhardness of the following alloy containing 35 at.% Se also equals 40 kg/mm^2. For greater reliability the frequency curve for this alloy was plotted from 200 measurements. Considering all the foregoing, it is not difficult to conclude that the range of concentrations in which the microhardness was varying corresponded to single-phased alloys. Figure 43 shows the concentration dependence of the microhardness of the Ag−Se alloys. From this we see that the microhardness of the phase based on silver selenide remains constant within the two-phase region Ag + Ag$_2$Se. The breaks in this curve clearly delineate the limits of existence of the single-phased region corresponding to an Ag$_2$Se-base solid solution. Thus the resultant data lead to the conclusion that the alloy corresponding to the stoichiometric ratio of the components in the compound Ag$_2$Se lies in the two-phase region.

In the case of the Ag−Te system no such sharp microhardness/composition relationship was achieved. The microhardness of the phase based on silver telluride varies less sharply as the composition changes (from 29 to 31 kg/mm^2), and this lies within the limits of sensitivity of the method for the load in question. Nevertheless it may be concluded from microstructural analysis that equilibrium deviations from stoichiometry also occur in the silver tellurides.

If the microhardness/composition relationship within the range of existence of the Ag$_2$Se-base-solid solution is extrapolated to the ordinate of the stoichiometric compound Ag$_2$Se, we find that the microhardness of this compound (if it existed) would be 31 kg/mm^2. Hence the introduction of about 10^{20} cm^{-3} of vacancies (corresponding to the silver sites in the Ag$_2$Se lattice), which corresponds to the right concentration boundary characterizing the existence of the phase in question, leads to a very slight change in microhardness. This justifies our considering that the distortion of the lattice on deviating from stoichiometry is insignificant. Presumably this is due to the substantial ionicity of the bond in the compounds considered, since the distortion of the lattice in substances with a sharply expressed covalent bond leads to incomparably greater changes in microhardness.

We may conclude from the foregoing that the microhardness method may successfully be used for determining the ranges of homogeneity of solid solutions based on semiconducting compounds. On studying SnTe and GeTe the problem was somewhat extended by virtue of the fact that alloys quenched after annealing at various temperatures were studied.

Let us now consider the presentation of the problem and the results in more detail.

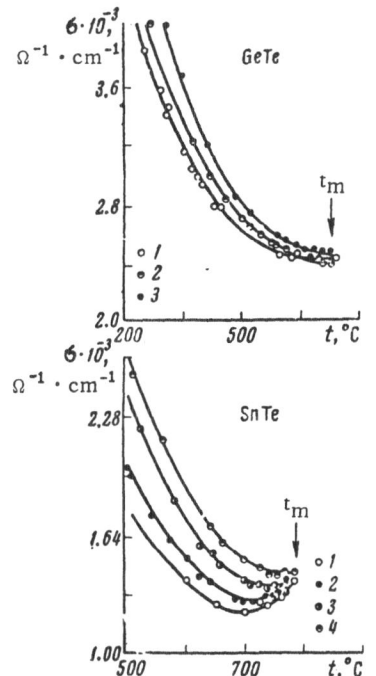

Fig. 44. Temperature dependence of the electrical conductivity of GeTe and SnTe (1 to 4 indicate the numbers of the samples).

It follows by considering the general character of the chemical interaction of germanium and tin with tellurium that compounds of the $A^{IV}Te$ (A^{IV} = Ge, Sn) [378] type are formed in the corresponding binary systems; however, more detailed investigations [379-382] showed that the alloys of stoichiometric composition (A^{IV}:Te = 1 : 1) lay beyond the limits of the single-phase region on the phase diagrams. According to the results of [379], the range of the solid solution based on SnTe at 600°C extended from 50.1 to 51.1 and at 750°C from 50.1 to 50.8 at.% Te. According to [380], at 400°C, the range of the solid solution based on SnTe extended between 50.1 and 50.9 at.% Te. According to [381], the range of the GeTe−base solid solutions at 250, 430, 500, and 600°C extended over the ranges 50.2-50.9, 51.3-51.5, 50.3-51, and 50.3-51.5 at.% Te. The range of homogeneity at 500 and 600°C given in [381] was based simply on microstructural analysis. It was shown from [382] that there was a reduction in the deviation of GeTe from stoichiometry on raising the temperature. It was noted that at 600°C the alloy with 50.00 at.% Te lay within the range of homogeneity, while the line representing the limited solubility of Ge in GeTe above this temperature lay to the left of the ordinate of stoichiometric composition.

The question as to the deviation of GeTe and SnTe from stoichiometry is particularly important, since the electrical properties vary sharply on forming solid solutions based on these compounds. In a number of papers the temperature dependences of the electrical conductivity, Hall effect, and thermo-emf have been studied [383-394] and the manner in which these properties vary with temperature changes at high temperatures has been noted. The temperature dependence of the electrical properties is treated in [383-391] from the point of view of the complex structure of the valence band. According to this model, the valence band of GeTe and SnTe comprises two subsidiary bands with different densities of states, associated with the presence of heavy and light holes. However, the character of the temperature dependence of the electrical properties at high temperatures lies outside the framework of this model [385]. The electrical conductivity, magnetic susceptibility, and thermo-emf were studied in [392-398] during the melting of GeTe and SnTe samples of stoichiometric composition and others containing an excess of tellurium; the authors interpreted the results as indicating that the deviation of GeTe and SnTe from stoichiometry diminished on raising the temperature.

In our own investigations we made a more careful study of the electrical conductivity of stoichiometric GeTe and SnTe right up to the melting points.

The results appear in Fig. 44. The electrical conductivity of all the samples studied fell sharply on heating. However, not far before the melting point the electrical conductivity of GeTe practically ceased varying with temperature, while that of SnTe even started rising. This in our view is associated with the development of intrinsic conductivity. It may be considered that the development of intrinsic conductivity arises as a result of the partial removal of degeneracy by virtue of a reduction in the deviations from stoichiometry.

In order to secure an experimental verification of this hypothesis, it was decided to measure the microhardness of samples of various compositions quenched after prolonged annealing

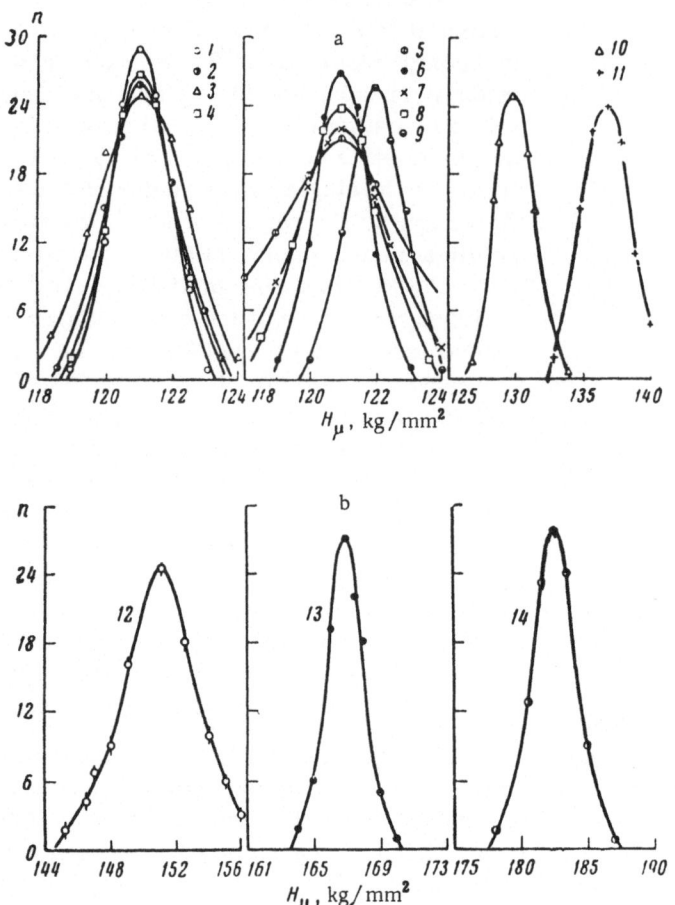

Fig. 45. Frequency curves of microhardness obtained for Sn−Te alloys. Compositions (at.% Te): 1) 49; 2) 49.5; 3) 49.7; 4) 49.8; 5) 49.9; 6) 49.95; 7) 50.0; 8) 50.05; 9) 50.1; 10) 50.2; 11) 50.3; 12) 50.5; 13) 50.7; 14) 51.0.

at high temperatures. Series of samples containing 49, 49.5, 49.7, 49.8, 49.9, 49.95, 50.0, 50.05, 50.1, 50.2, 50.3, 50.5, 50.7, and 51 at.% Te in the Ge−Te system and 49.5, 49.7, 49.9, 50.0, 50.1, 50.2, 50.3, 50.5, 50.7, and 51.0 at.% Te in the Sn−Te system were prepared. The original materials for preparing the alloys included germanium with a resistance close to the intrinsic value, tin of the OVCh 000 brand (containing about $5 \cdot 10^{-4}\%$ of impurities), and TA-1 tellurium, which after twice distilling in vacuum became spectrally pure. The alloys were prepared in evacuated (10^{-4} mm Hg) and sealed quartz ampoules. The cast alloys were subjected to a homogenizing anneal in sealed quartz ampoules for 50 h at 650°C (Ge−Te) and 30 h at 700°C (Sn−Te), after which the microhardness was measured on the PMT-3 with a load of 50 g. The loading time, time under load, and unloading time were each 10 sec. The microhardness was measured on unetched, freshly polished microsections. On each sample 100 impressions were made; then the frequency curves shown in Figs. 45 and 46 were plotted. As true microhardness, the values corresponding to the maxima on the frequency curves were taken. The maximum error in determining the microhardness was ±0.5 kg/mm². From these values the concentration/microhardness characteristics shown in Fig. 47 (curves 2 and 5) were plotted. Within the limits of the two-phase regions of Ge + GeTe and Sn + SnTe the microhardness of the solid-solution crystals based on GeTe and SnTe remained constant. On further raising the Te con-

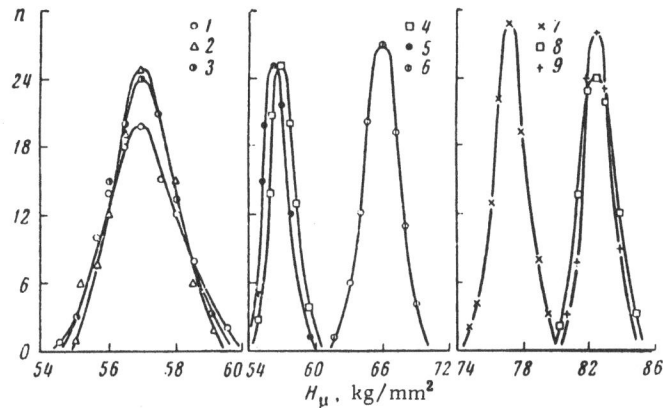

Fig. 46. Frequency curves of microhardness obtained for Sn−Te alloys. Compositions (at.% Te): 1) 49.7; 2) 49.9; 3) 50.0; 4) 50.2; 5) 50.1; 6) 50.3; 7) 50.5; 8) 50.7; 9) 51.0.

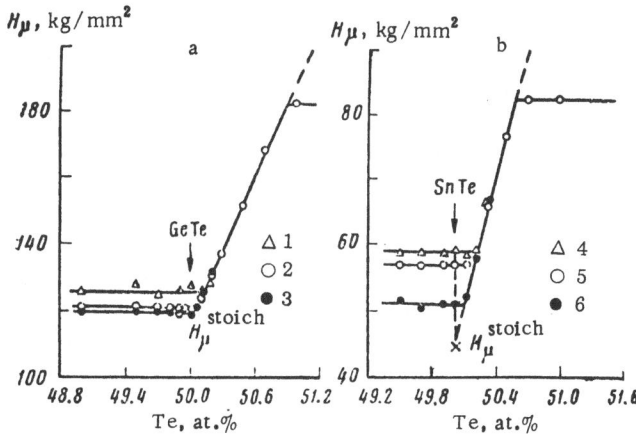

Fig. 47. Concentration dependence of the micro-hardness of (a) GeTe - and (b) SnTe -base solid-solution crystals. 1) 550; 2) 650; 3) 700; 4) 650; 5) 700; 6) 760°C.

tent the microhardness rose quite sharply in accordance with a linear law (the range of homogeneity for the method of investigation employed was established to an accuracy of 0.05 at.% Te). The range of concentrations within which the microhardness increased corresponded, according to microstructural analysis, to the range of homogeneity of the GeTe- and SnTe-base alloys. There was practically no deviation from stoichiometry in GeTe at 650°C. At any rate the deviation was under 0.05 at.% Te. The deviation of SnTe from stoichiometry at 700°K was slightly under 0.2 at.% Te, which was close to the value obtained for this temperature in [380].

Some of the samples studied at the temperatures in question were annealed in a vacuum of 10^{-4} mm Hg at 550 and 700°C (Ge−Te) and 650 and 760°C (Sn−Te) for 50 h. After annealing, the samples were water-quenched and then measured for microhardness. As in the previous cases, 100 impressions were made on each sample, and frequency curves were plotted. The microhardness values corresponding to the maxima on the frequency curves were taken as the true ones, and used to plot concentration/microhardness curves (Fig. 47, curves 1, 3, 4, and 6). It follows from these data that annealing at a higher temperature leads to a fall and at a lower temperature to a rise in the microhardness of the GeTe- and SnTe-base solid-solution crystals (as compared with curves 2 and 5). The microhardness of the single-phase alloys, however, remained almost constant, i.e., the concentration/microhardness curves were stable within the ranges of homogeneity of the samples based on the compounds in question, and their position was independent of annealing temperature within these limits. We see from the graphs represented in Fig. 47 that at 550°C the deviation of GeTe from stoichiometry is under 0.15 at.% Te, which is close to the value obtained in [380] for 500°C. At 700°C the deviation of GeTe from stoichiometry is practically zero. In the case of SnTe the picture is qualitatively analogous (Fig. 47b). At 760°C the deviation from stoichiometry for SnTe is under 0.1 at.% Te. These experimental results lead to the clear conclusion that, on raising the temperature above a certain limit, the deviations from stoichiometry diminish. This correlates with the results of the measurements of electrical properties at high temperatures, and also with the manner in which these properties vary when GeTe and SnTe melt [392, 393].

The linear character of the concentration dependence of the microhardness of SnTe- and GeTe-base samples within the ranges of homogeneity enables us to extrapolate the straight line until it intersects the ordinates corresponding to the stoichiometric compositions of the compounds in question, and thus to determine the microhardness which they would have in the absence of deviations from stoichiometry. The microhardness of GeTe and SnTe of stoichiometric composition equals 118 and 44 kg/mm^2.

Distribution of Alloying (Doping) Elements
in the Production of Single Crystals

In the last decade the development of semiconductor technology and the production of very pure substances have resulted in a wide extension of crystallization methods of purification and producing single-crystal samples of various elements and compounds. Among these methods the one most widely employed at the present time is that of pulling the crystallizing solid phase from the melt (Czochralski technique), together with the method of zone recrystallization.

In both cases the distribution of the doping elements or impurities in the resultant solids obeys a perfectly definite law [157]. Considering the relationship between the concentration of the solid solution and the microhardness, we may use this method in order to study the manner in which the doping elements or impurities are distributed along the length of the solidified bars obtained by the methods in question. The use of the microhardness method for this purpose is much more convenient than chemical analysis, since in using the latter the crystals have to be cut to provide samples, and this is often undesirable.

Using N. A. Sologub's attachment [14], samples up to 300 mm long may thus be studied by the microhardness method.

In addition to this, as we shall show later, the microhardness method may be used for a more refined study of the distribution of doping additives, which would be quite impossible by chemical analysis.

Figure 48 shows some graphs characterizing the variation in copper content (a) and microhardness (b) along crystals pulled at various speeds from an Al − 4.5% Cu melt (authors' data). We see from this figure that the microhardness curves agree qualitatively with the copper distribution curves derived from chemical analysis along the whole length of the crystal. Using the curves given in Fig. 48b and the calibration curve characterizing the relation between the

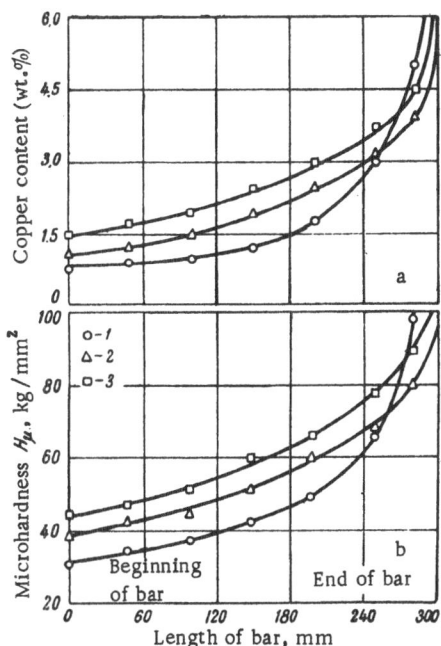

Fig. 48. Distribution of copper (a) and microhardness (b) along single crystals pulled from an Al−4.5% Cu melt at different velocities (mm/min): 1) 0.1; 2) 0.2; 3) 0.3.

microhardness and the composition of a solid solution of copper in aluminum (Fig. 39), we may calculate the copper concentration at the corresponding points of the bar and obtained the curves given in Fig. 48a.

Comparison between the copper-distribution data derived from the microhardness and those obtained by chemical analysis (Table 19) shows insignificant differences.

It follows from the foregoing that the microhardness method may successfully be used for studying the distribution of doping elements along crystals obtained by pulling from the melt and by zone recrystallization.

Use was made of this conclusion in [121, 129] in order to determine the character of the nonvariant transformation in the Al−Ti and Pb−Ca systems.

Some other authors [119, 120] proposed a eutectic or peritectic transformation in the Al−Ti system. In order to secure an unambiguous solution to this question, a sample of variable composition was pulled from a melt containing 0.1 to 0.12 wt.% Ti, and its microhardness was studied as a function of length. It was found that the upper part of the bar had a microhardness of ∼35 kg/mm^2, while the lower was some 10 kg/mm^2 smaller.

It was concluded from this that a peritectic transformation occurred in the Al−Ti system [123]. This was confirmed by differential thermal analysis.

In order to study the nature of the nonvariant equilibrium in the Pb−Ca system in [130], the microhardness was studied as a function of distance measured along a bar containing an average of 0.8 wt.% Ca, subjected to zone recrystallization. It was found that the microhardness was lower in the head of the bar than in the end.

The microhardness distribution obtained corresponded to calcium distribution consistent with a eutectic transformation in the system. Thermal analysis confirmed this conclusion.

When the crystals were pulled rapidly from the melt, under certain conditions a layer-like distribution of the doping components occurred. Using the microhardness method, D. A. Petrov and B. A. Kolachev [158, 159] determined the character of the copper distribution in crystals obtained from aluminum−copper alloys in the manner indicated.

Figure 49 shows the variation in microhardness along the samples obtained from aluminum−copper alloys at various pulling velocities. The curves clearly show the undulating variations in microhardness along the samples, corresponding to the layer-like copper distribution. On reducing the pulling speed the copper distribution becomes more uniform and the microhardness varies little along the crystal.

It was shown in [158, 159] on the basis of microhardness measurements carried out with the Al−Mg−Zn alloy AMTs that layer-like structures were also formed when producing bars under industrial conditions by the continuous-casting method.

TABLE 19. Copper Distribution
along Crystals Pulled from an
Al -4% Cu Melt at Various Speeds

Distance from beginning of bar, mm	Copper distribution (%) for a velocity (mm/min) of					
	0.10		0.20		0.30	
	1*	2**	1*	2**	1*	2**
0	0.75	0.70	1.15	1.10	1.50	1.50
50	0.85	0.82	1.30	1.30	1.70	1.75
100	1.00	1.00	1.50	1.46	2.00	2.08
150	1.25	1.20	2.00	2.00	2.50	2.50
200	1.75	1.78	2.50	2.40	3.05	3.00
250	3.00	3.10	3.15	3.20	3.75	3.80
280	5.00	5.00	4.00	3.85	4.50	4.40

*Chemical analysis.

**From microhardness measurements.

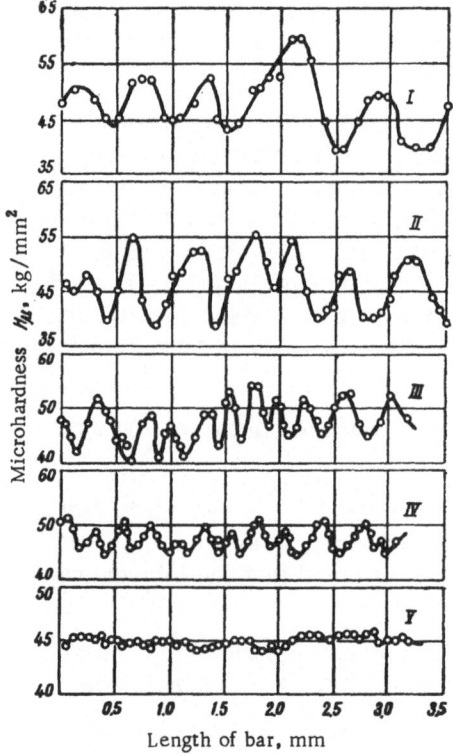

Fig. 49. Variation in microhardness along crystals pulled from
an Al -4% Cu melt at various speeds
(mm/min). I) 8.5; II) 2.6; III) 0.7;
IV) 0.25; V) 0.08.

Study of Diffusion Processes in Alloys

The possibility of using the microhardness method to study diffusion in metallic alloys arises, as before, from the specific relationship between the microhardness and the concentration of the alloying element in the solid solution.

The microhardness method was first used for studying diffusion velocity by Buckle [115], who studied the diffusion of copper into aluminum from the alloy; the copper concentration was determined from a previously plotted calibration curve similar to that shown in Fig. 40. The relation between the diffusion coefficient and concentration of the copper was studied at various temperatures. It was found that the diffusion coefficient increased for low copper contents. Increasing the amount of copper in the alloy had the effect that the diffusion coefficient became constant. Buckle's results [161] obtained by the microhardness method agree closely with other data obtained by the classical procedure [160].

A number of investigations have been made into the diffusion of the alloying elements of Duralumin into a cladding layer of aluminum, using the microhardness technique.

Buckle [161] showed that heating Duralumin to 500°C and holding at this temperature for 20 min caused the alloying elements to diffuse to a depth of about 50 μ.

Buckle and Keil [162, 163] made a detailed study of the diffusion of copper, magnesium, silicon, and manganese from Duralumin into a cladding layer of aluminum, using the microhardness method in conjunction with spectral analysis. In order to increase the surface studied, oblique cuts were made through the coating. The microhardness was measured at a load of 5 g. The rate of diffusion was studied at various temperatures with a holding period of 15 to 30 min. By way of example, Fig. 50 shows a graph characterizing the copper content and microhardness of the cladding layer and of the original Duralumin after heat treatment at 500°C. We see from this figure that the results based on the microhardness measurements and on the spectral analysis fall neatly on a single curve.

Diffusion from Duralumin with various Mg/Si ratios was also studied; analogous curves were obtained. Manganese had a low rate of diffusion.

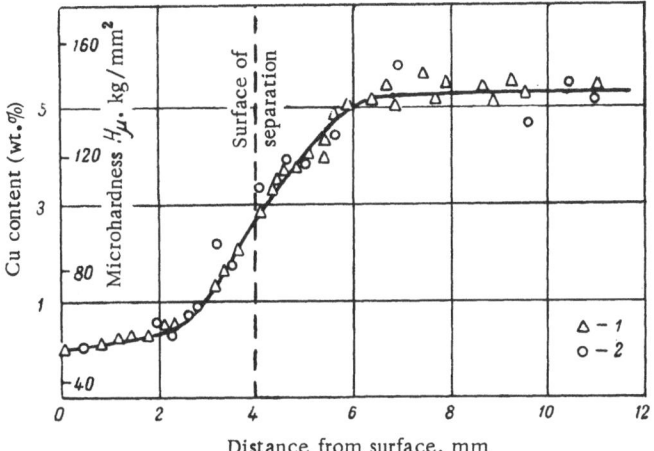

Fig. 50. Variation in the copper content and microhardness of cladded Duralumin samples after heat
treatment along an oblique cut through the cladding
layer: 1) From microhardness measurements; 2)
from spectral analysis.

An investigation similar to the foregoing was carried out in [164], using the microhardness method. The results of [162, 163, 164] agree closely with each other.

A. A. Bochvar and A. S. Titova [165] used the microhardness method for studying the
kinetics of the penetration of alloying elements from two industrial alloys, D1 (4.1% Cu; 0.6%
Mg; 0.5% Mn) and D16 (4.5% Cu; 1.5% Mg; 0.65% Mn) into the cladding layer at temperatures 503 ± 2, 506 ± 2, and $560 \pm 2°C$. The holding time at the specified maximum heating
temperature was varied from 5 min to 3 h, first in steps of 5 min and then after 1 h in steps
of 30 min. After completing a holding period at a specified temperature the samples were
water-quenched and subjected to full natural aging; then the microhardness was measured in
a PMT-3 with a 10 g load in a direction perpendicular to the separating line between the alloy
and the cladding layer and to the rolling direction.

By way of example, Fig. 51 shows typical curves representing the variation in microhardness over the cross section in the direction indicated for the D16 alloy, previously treated
at 506°C for 30 min and 2.5 h.

It was found as a result of this investigation that, despite the high degree of saturation
of D16 with copper and magnesium, the diffusing of the strengthening additives from the alloy
took place more slowly than it did from D1. For example, in 30 min at $503 \pm 2°C$ the depth of
penetration of the alloying elements into the cladding layer from D1 was 54 μ and from D16 only
40 μ. The authors explain this by chemical interaction between the alloying elements in the
aluminum-base solid solutions and the possible formation of the compound Al_2CuMg in D16.

Individual study of the diffusion of copper and magnesium from the binary alloys Al +
23.84% Cu and Al + 2.09% Mg showed that the rate of diffusion of these elements was approximately the same.

On the basis of the diffusion results obtained by the microhardness method, A. A. Bochvar
and A. S. Titova [165] concluded that cladded sheets of D1 and D16 alloys could sustain a maximum of two requenchings without any danger of loosing the cladding effect.

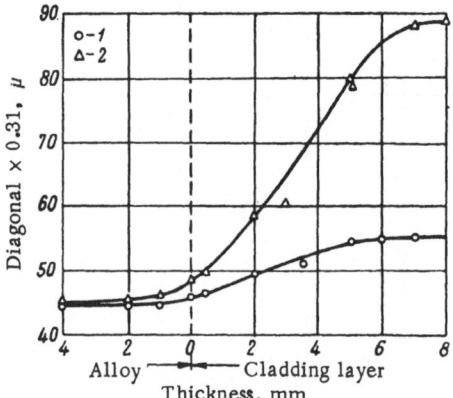

Fig. 51. Variation in the microhardness of cladded samples of D16 alloy in a direction perpendicular to the rolling direction and the line separating the alloy from the cladding layer after heat treatment at 506°C for 1) 2.5; 2) 0.5 h.

In a number of papers by Buckle and his colleagues [166, 116, 117, 167] the microhardness technique was employed in order to study the rate of diffusion of zinc into copper from copper alloys containing 10, 20, and 30% Zn, that of aluminum into copper from a Cu−16% Al alloy, that of beryllium into aluminum from an Al−1.06% Be alloy, that of tungsten into platinum, and conversely. In all these cases results agreeing closely with the results of other methods were secured.

When studying the mutual diffusion of zinc and copper at 380°C for 1 h, Buckle [168] used the microhardness method to establish the presence of layers corresponding to the ε, γ, and β phases. On further heating, the components were redistributed among the phases, the region of the ε phase on the zinc side concentrated, and the region of the γ phase widened.

M. I. Perfenova and N. A. Izgaryshev [169] used the microhardness method to reveal an intermediate phase between cadmium and silver, having a hardness of 278 kg/mm^2, far higher than the hardness of pure silver and cadmium themselves (44 and 20 kg/mm^2 respectively).

In addition to the cases mentioned, the microhardness technique has also been used for studying the diffusion of nickel into iron from a nickel coating [170] and also in a number of other investigations [171−173].

G. G. Maksimovich [174] studied the volatilization (diffusion and evaporation) of zinc from brass after holding in vacuum at high temperatures. In order to study this process the microhardness of the samples was measured and the changes in their absolute weight and specific gravity were also determined. It is well known that on heating an alloy in vacuum the alloy component with the greater vapor tension tends to evaporate. As a result of this process, the concentration of the volatilizing component of the sample surface diminished, a concentration gradient is created along the cross section of the sample, and a diffusion flow of the component arises from the inner layers of the sample to the outer.

In order to study the diffusion process, cylindrical samples 2 mm in diameter placed in special ceramic containers were heated in vacuum to a certain specified temperature and held there for a specified time. The rate of heating and cooling was varied over a wide range; the heating temperature was maintained automatically by means of a potentiometer, which simultaneously recorded the heating curve of the samples. In order to study the volatilization of zinc, the weight change in the samples was measured and the variation in microhardness across the cross section was also investigated. The samples were weighed before and after heating in a vacuum of about $1 \cdot 10^{-4}$ mm Hg. The microhardness was measured in the radial directions of the sample cross sections.

Figure 52 shows the curve representing the variation in microhardness over the cross section of an LS59-1 brass sample held in vacuum at 600°C for 8 h, and the external form of the sample after the impression of the diamond used for measuring microhardness. The figure illustrates the fall in zinc content on approaching the periphery.

Two cycles of tests were carried out, the microhardness being studied as a function of the holding period at constant temperature in vacuum and also as a function of temperature for

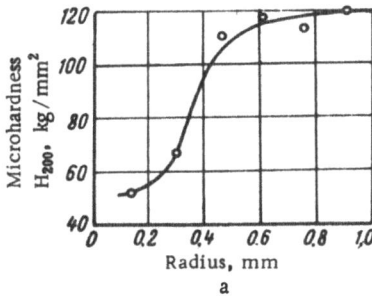
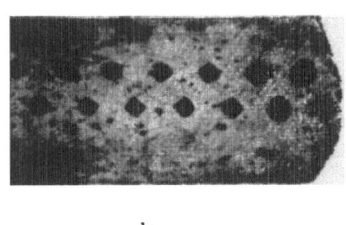

a

b

Fig. 52. Variation in microhardness over the radius of samples held at 600°C for 8 h: a) microhardness curve; b) external form of the sample with the impressions made by the diamond pyramid of the PMT-3.

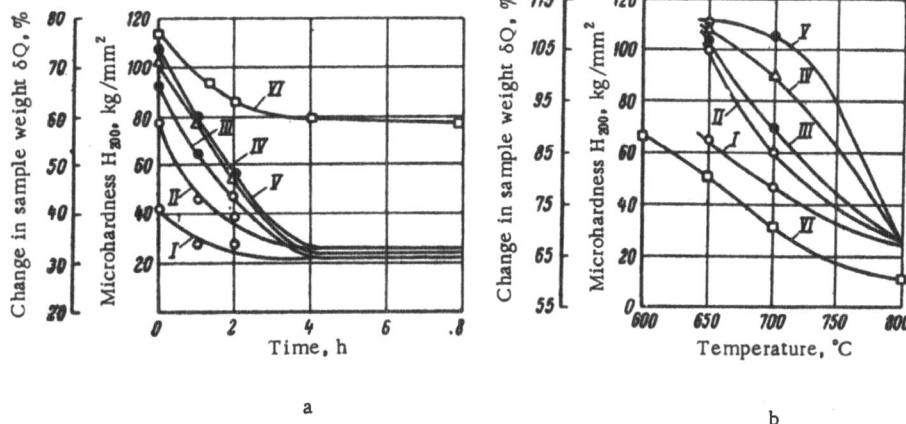

a

b

Fig. 53. Microhardness of the samples as a function of holding time and temperature in vacuum: a) At constant temperature; b) at constant holding time.

a fixed holding period. The results are shown in Fig. 53 (curves I to V in this figure correspond to microhardness measurements at a distance of 0.2, 0.4, 0.6, 0.8, and 1 mm from the sample surface). We see from this figure that for a holding temperature of 800°C and an insignificant holding time a high concentration gradient of the diffusing element is created. For a holding period of over 4 h the concentration gradient approaches zero and then (up to 8 h) the microhardness remains almost constant. This means that even after 4 h at 800°C nearly all the zinc has been lost from samples 2 mm in diameter.

During the heating of the samples to 800°C (without holding periods) and their subsequent cooling, a considerable amount of zinc evaporates from the layers close to the surface (curve IV, Fig. 53a), whereas in the deeper layers little diffusion occurs (curves IV and V). With increasing holding time the volatilization process gradually embraces all layers of the sample, and for a period of 4 h the zinc concentration evens out over the whole sample cross section.

The curves representing the change in zinc concentration for constant holding time as a function of temperature are given in Fig. 53a. For a holding time of 4 h and a holding temperature of 650°C the zinc evaporates intensively from the sample surface and a substantial diffusion flow of zinc arises from the layers close to the surface (curve I). At depths of 0.4 mm and more, even for a holding time of 4 h diffusion can hardly occur at all and the microhardness remains constant. With increasing volatilization temperature the diffusion gradually envelopes deeper and deeper layers of the sample. At a temperature of 700°C diffusion is appreciable from a depth of 0.4 to 0.6 mm but very slight at 0.8 mm and almost absent at 1 mm. On further raising the temperature, diffusion from the deeper layers of the sample becomes stronger and the concentration completely evens out at 800°C, as also follows from Fig. 53a.

The amount of evaporating zinc was determined as a function of time and heating temperature by weighing the sample before and after evaporation. Curve VI in Fig. 53a characterizes the change in sample weight δQ as a function of holding time at constant temperature (800°C). In addition to this we see that the evaporation of the zinc ceases almost entirely after a 4-h holding period. Further holding (up to 8 h) has no effect on sample weight. In Fig. 53b curve VI represents the amount of zinc evaporated as a function of heating temperature for a constant holding time.

The evaporation of the zinc is accompanied by the formation of vacancies, which after reaching saturation concentration unite into micropores. The material becomes porous. During the cooling of the sample from which evaporation has taken place the excess vacancy concentration in the "continuous" part of the material may lead to an increase in the dimensions of existing pores and the formation of new micropores. Some of the vacancies and micropores may, naturally, pass to the sample surface.

After holding the samples for 8 h at 800°C and cooling them, individual pores attain dimensions of several microns. The presence of vacancies and micropores in the depleted samples is confirmed by the change in the density of the samples during evaporation (the density of samples held in vacuum for 8 h at 800°C falls by an average of 17%). The large micropores may readily be inspected in the metallographic microscope.

The evaporation of the zinc from brass is naturally accompanied by the diffusion of micropores and vacancies to the sample surface. This is indicated by the considerable reduction in the size of the samples during heating in vacuum [174].

A. A. Bochvar and Z. A. Sviderskaya [175] used the microhardness method to study the softening of gold–copper alloys on heating, so as to estimate the effect of diffusion in ordering and disordering on resistance to plastic deformation. The microhardness was studied in a PMT-2 furnished with a furnace for heating the samples.* A load of 20 g was employed. The degree of softening was judged by reference to the difference in microhardness after short (30 sec) and long (1 h) periods under load, and also by reference to the slope of the temperature/microhardness curve.

Samples studied included an alloy of copper with 25 at.% gold corresponding to the compound Cu_3Au (temperature of the order–disorder transformation about 396°C), and also one with 50 at.% gold coresponding to CuAu (temperature of the order–disorder transformation about 424°C), as well as the pure components (copper and gold). The alloy with 25% Au was studied on passing out of the disordered state and vice versa. The disordered state of this alloy was reached by annealing at 650°C for 6 days and then water-quenching. The alloy with 50% Au was only tested in the course of disordering. Both alloys were brought into the ordered state by very slow cooling from 650°C after previously holding for a day. Then the samples were cooled for 7 h at a rate of 10 deg/h and then with the furnace to room temperature. The results of microhardness measurements on samples so treated as well as the pure components are given in Table 20. For the CuAu alloy the short- and long-term microhardness were also measured at 425°C; these were respectively 93.3 and 23.0 kg/mm².

---

* The construction of the attachment for the PMT-2 and the method of high-temperature testing for microhardness therein was developed by Z. A. Sviderskaya in 1948.

TABLE 20. Microhardness of Gold, Copper, and Gold − Copper
Alloys at Various Temperatures

Temperature, °C	Time, min	Microhardness, kg/mm²				
		Au	Cu	Cu₃ Au*	Cu₃ Au**	Cu Au
20	0.5	67.3	85.7	94.6	107.7	145.7
	60	50.1	80.9	90.6	102.4	133.7
300	0.5	37.4	54.5	97.3	132.9	171.7
	60	25.0	41.3	52.8	53.8	99.0
370	0.5	17.4	55.1	93.7	97.0	126.8
	60	13.7	—	35.3	40.9	43.5
380	0.5	16.6	—	87.3	94.2	122.4
	60	10.1	—	38.7	41.2	40.6
400	0.5	15.5	—	85.6	92.1	116.4
	60	9.1	—	37.2	32.4	39.0

*Disordered alloy.
**Ordered alloy.

The results presented in Table 20 show that in the course of ordering and disordering there is no such sharp intensification of creep processes such as occur, for example, in Al − Zn alloys.

In this connection, the authors consider that the absence of any intensification of creep processes in these alloys during ordering and disordering indicates that not every diffusion process associated with the motion of atoms is capable of producing a sharp increase in disordering and a corresponding rise in the ductility of the alloys at high temperatures. This is evidently associated with the fact that in the three-dimensional ordering and disordering of the solid solutions the transfer velocity of the particles is unable to rise to the level necessary for the healing of incipient fractures in the course of deformation, so that the alloys under consideration are not distinguished by a high degree of ductility and ready yield under stress [175].

Diffusion processes also include the decomposition of solid solutions. In order to study these processes once again the microhardness method has been used. Thus for example R. A. Akopyan [176, 177] demonstrated the possibility of using microhardness measurements for studying various stages of aging. Samples of Al − 8.4 wt.% Zn were quenched from between 200 and 450°C and subjected to subsequent natural aging at room temperature and artificial aging at 80 to 150°C. During the aging process the microhardness was measured and so were the Guinier − Preston (G − P) zones by the method based on small-angle x-ray scattering. Correlation between the results of the two measurements depended on the sample-quenching temperature. It was found that in the majority of cases the microhardness fell with increasing size of the G − P zones (the change was only slight on aging at 150°C, particularly after quenching from between 350 and 450°C). It was suggested that the increase in zone size taking place on quenching and aging had a greater effect on the microhardness than the simultaneous reduction in the number of zones and the increase in interzone distance (if one considers that the total volume of the zones formed remains constant).

The microhardness rose on natural aging. For example, after quenching from 200°C the microhardness increased from 40 to 54 kg/mm² in about 41 h and after quenching from 450°C it increased from 52 to 63 kg/mm² in about 20 min. Then the microhardness value stabilized. The results were used to calculate the activation energy of the process underlying the formation of G − P zones. For this purpose a graph was plotted between the logarithm of the aging time required to reach the maximum microhardness for the particular quench temperature and the reciprocal of the absolute quench temperature itself. For quench temperatures between 200 and 350°C this relationship was linear; the activation energy calculated from the slope was 15 kcal/mole. For higher quench temperatures (350 to 450°C) the activation energy was lower (10 kcal/mole), apparently because of the faster disappearance of the excess vacancies for a large number of quenching defects in the structure [177].

It is important to note that the process in question is due to a diffusion mechanism, as indicated by the similarity between the activation energy of the aging process and the activation energy of another diffusion process in aluminum, the formation of vacancies (17 kcal/mole) [178].

CHAPTER 6

USE OF THE MICROHARDNESS METHOD IN STUDYING
THE PHASE DIAGRAMS AND STRUCTURES OF
THREE-COMPONENT ALLOYS

Some Theoretical Premises

The possibility of using microhardness as a method of physicochemical analysis for studying ternary phase diagrams in the same way as binary systems arises as a consequence of the principles of correspondence and continuity enunciated by N. S. Kurnakov [6, 7]. It is true that the manner in which the composition of a particular interesting phase varies with the composition of the alloy on passing from one phase region to another is more complex in three-component systems. Nevertheless, since these laws are perfectly specific for every type of ternary diagram, the manner in which the microhardness of a particular phase varies in alloys of various composition should also be completely specific.

It was shown in the previous chapter that the microhardness of solid-solution crystals at a particular temperature* increased within the limits of the single-phase region, while on passing to the two-phase region (through the solidus line or the line of limited solubility) it remained constant. Clearly in the case of the quasibinary sections of the corresponding three-component phase diagrams the manner in which the microhardness of the solid-solution crystals varies should be the same, since the composition of the solid solution varies with the composition of the alloy along these sections in the same way as in binary systems.

The composition/microhardness isotherms should have also the same form when studying sections passing through conodes (tie-lines) characterizing the equilibrium of the phase in question with another phase in the particular two-phase region, since in this case also the composition of the phase in question remains constant on passing from the one-phase region into the two-phase region and equals that of the alloy completely saturated at the particular temperature under consideration.

However, when studying any new system one is often compelled to select sections disposed in a somewhat arbitrary fashion. Nevertheless, on the general basis of Kurnakov's physicochemical laws, only three types of possible relationship between the microhardness of the particular phase and the composition of the alloy can be envisaged on passing from the one-phase region of the phase diagram into the two-phase region,[†] that is, the microhardness of a particular phase may rise, fall, or remain constant in the two-phase region.

---

* The quench temperature is implied.

† In studying three-component phase diagrams one often has to examine the boundaries corresponding to a transition from a single-phase region into the two-phase region normally bordering the former.

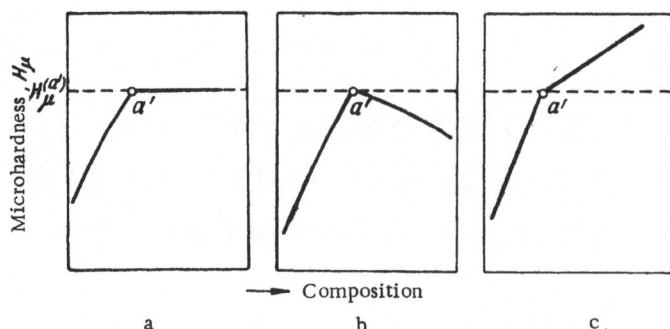

Fig. 54. Possible types of microhardness/composition
relationship for solid-solution crystals on passing from
a one-phase region of a three-component phase diagram
to a two-phase region.

Since the laws governing the changes in the composition of the alloy within the two-phase
region differ from those in the one-phase region, there will in any case be a break on the
microhardness isotherm on passing from one region to the other (point a', Fig. 54); from this
one may determine the boundary between the phase regions. The possible changes in microhardness on passing from a one-phase region of a ternary phase diagram to a two-phase region are illustrated schematically in Fig. 54a-c.

By studying the microhardness of any phase in relation to the composition of the alloy at
a particular temperature, we may also determine the boundaries separating the two- and three-
phase regions.

Thus on the basis of a preliminary consideration of the possible variations in the microhardness of a particular phase at a specified temperature with varying composition of the alloys in three-component systems we may conclude that the microhardness technique may be
applied to the construction of ternary phase diagrams also.

Construction of a Limited-Solubility Surface
and a Solidus Surface

The first indications as to the possibility of using the microhardness method for studying
ternary phase diagrams of metallic systems appeared in a paper by S. A. Pogodin and L. M.
Kefeli [179] on the $Mg-Sn-Pb$ system.

The microhardness method was first widely used in [137, 180] for plotting limited-solubility surfaces and (partly) solidus surfaces in the ternary $Cu-Cr-Zr$ system. For this purpose series of alloys were prepared along sections containing 0.25 and 0.5 wt.% Cr and 0.25
and 0.5 wt.% Zr, and also along the quasibinary section $Cu-Cr_2Zr$. The composition of the
alloys for microhardness analysis was chosen by reference to the approximate position of the
boundaries between the phase regions established under the microscope [181]. After 50% deformation at 800°C the alloys were annealed at 700, 800, 900, 940, and 1000°C for 20, 10, 5.5,
and 3 h respectively, after which they were quenched in water. The surface hardening of the
microsections was removed by etching for 0.5 min in a 3% solution of ferric chloride in 10%
HCl. After this the microhardness was measured. The results of the measurements (average
of 10 to 15 readings) were summarized in microhardness/composition curves.

By way of example Fig. 55 shows the microhardness of crystals constituting solid solutions of chromium and zirconium in copper as a function of the composition of alloys quenched

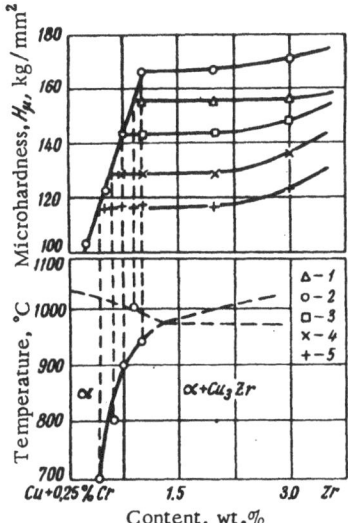

Fig. 55. Microhardness isotherms and polythermal section of the solubility surface for a constant (0.25 wt.%) Cr content in the Cu−Cr−Zr system at the following temperatures (°C): 1) 1000; 2) 940; 3) 900; 4) 800; 5) 700.

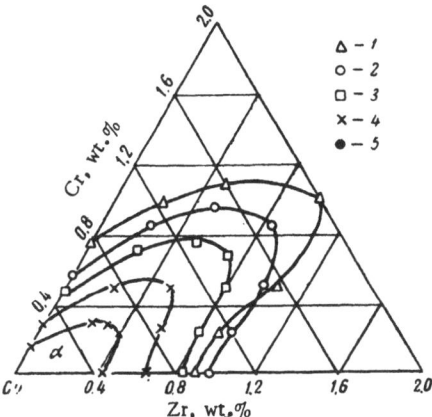

Fig. 56. Isotherms for the solubility of chromium and zirconium in copper at various temperatures (°C): 1) 1000; 2) 940; 3) 900; 4) 800; 5) 700.

after annealing at the temperatures indicated for 0.25 wt.% Cr. For the remaining sections the relations between the microhardness of the solid solution and the composition of the alloy are of a nature analogous to that illustrated in the figure.

After reaching a certain concentration corresponding to the saturation boundary at the temperature in question, the composition/microhardness isotherms of the α solid solution show sharp breaks. This enables us to mark the boundaries separating the region of the homogeneous solid solution from other phase regions on the phase diagram (Fig. 56).

The unusual form of the solubility isotherm at 1000°C (on the Cu−Zr side) is explained by the fact that at this temperature several alloys containing 0.25 and 0.5% Cr fused, since the melting point of the binary eutectic $\alpha + Cu_3Zr$ in the binary Cu−Zr system equalled 965°C. Hence over a certain range the isotherm for 1000°C is none other than the intersection of the plane of the isothermal section with the solidus surface of the Cu−Cr−Zr phase diagram. This is also indicated by the way in which the microhardness of the solid solution varies with the composition of the alloys along the sections with 0.25 and 0.5 wt.% Cr (see Fig. 55).

Comparison of the results obtained by the microhardness method with the results of microscope analysis [179, 180] indicates excellent agreement.

Thus the foregoing example (detailed study of alloys belonging to the Cu−Cr−Zr system) has demonstrated the possibility of applying the microhardness method in determining the surface of limited solubility and the solidus surface of three-component phase diagrams.

An investigation similar to that which has just been described was carried out by M. V. Mal'tsev and Yan Van-Bok [182] into alloys of the Al−Mn−Ti system. In addition to the microhardness method, x-ray structural analysis was also used in order to study the solubility in the solid state. Alloys using sections with 0.05, 0.1, and 0.3 wt.% Ti and also with 0.2 and 0.5 wt.% Mn were studied by both methods. The results are presented in Table 21.

We see from this table that the results obtained by the microhardness method agree closely with the results of x-ray structural analysis.

In [183, 184] the microhardness method was used to construct the solidus and limited-solubility surfaces in the Cu−Al−Ti system. The results agreed closely with microscope analysis.

TABLE 21. Limiting Solubility of Titanium and Manganese in Aluminum at 600 and 650°C

Constant content of component, %	Quench temperature, °C	Limiting solubility, %			
		x-ray method		microhardness method	
		Mn	Ti	Mn	Ti
0.1 Ti	650	1.08	0.10	1.09	0.10
0.3 Ti		0.95	0.30	—	—
0.2 Mn		0.20	0.21	0.20	0.22
0.5 Mn		0.50	0.16	0.50	0.12
0.05 Ti	600	0.96	0.05	1.00	0.05
0.10 Ti		—	—	0.65	0.10
0.20 Mn		—	—	0.20	0.16

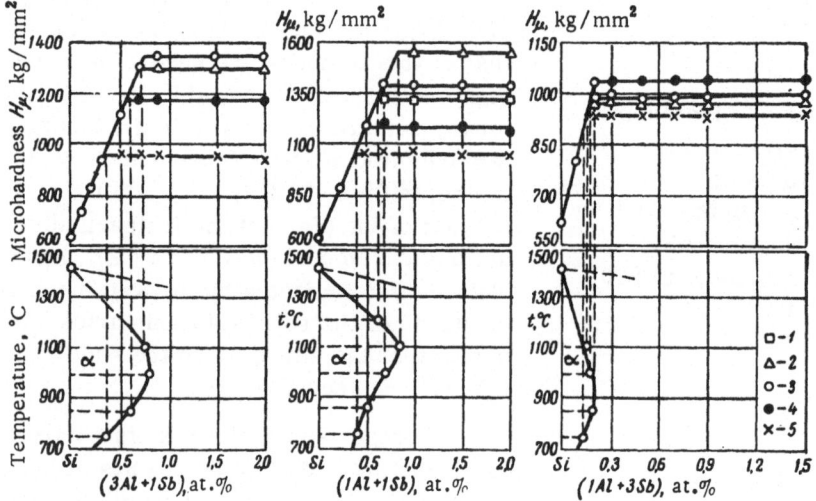

Fig. 57. Microhardness of the solid-solution crystals as a function of the composition of the alloys along the radial sections in the Si − Al − Sb system, together with parts of the solidus and limited-solubility curves based on these data for the following temperatures, °C: 1) 1200; 2) 1100; 3) 1000; 4) 850; 5) 750.

T. A. Badaeva and R. I. Kuznetsova [185] studied the simultaneous solubility of germanium and magnesium in aluminum along the Al − Mg_2Ge section by a combination of microhardness measurements and x-ray structural analysis. The results showed that the simultaneous solubility of magnesium and germanium in aluminum at 550°C was about 0.75 at.% Mg + Ge, falling rapidly to less than 0.2 at.% Mg + Ge at 450°C.

V. M. Glazov and M. V. Stepanova [186] studied the simultaneous solubility of phosphorus and iron in copper by the microhardness method. The results agree closely with those based on microscope analysis.

The microhardness method was also used* to study the simultaneous solubility of several elements of the third and fifth groups of the Periodic Table in germanium along the quasi-

*V. M. Glazov, Candidate's Dissertation, A. A. Baikov Institute of Metallurgy of the Academy of Sciences of the USSR (1959); Doctor's Dissertation, N. S. Kurnakov Institute of General and Inorganic Chemistry (1966).

binary sections Ge−AlSb, Ge−GaSb, and Ge−InSb. It was found that when ternary solid solutions of group III elements (Al, Ga, In) and group V elements (Sb) were formed in germanium the solubility of the elements in question was much higher than in the formation of the corresponding binary solutions. It was accordingly concluded that alloying elements of the donor and acceptor types exerted a mutual influence as a result of the chemical interaction between them, affecting their solubility in such semiconductors as germanium and silicon.

In view of this the individual and combined solubility of aluminum and antimony (and also aluminum and phosphorus) in germanium and silicon were studied at various temperatures in [133-135].

The results of the individual solubility determinations were mentioned in the previous chapter. The combined solubility of the elements in question in germanium and silicon was studied along sections corresponding to ratios of 3:1, 1:1, and 1:3, between the atomic proportions of the alloying elements. In order to bring them to a state of equilibrium the alloys were subjected to a prolonged homogenizing anneal in an atmosphere of purified argon. Figure 57 shows the microhardness of the solid-solution crystals as a function of the composition of the alloy, together with the boundaries of the single phase region along the radial sections indicated in the Si−Al−Sb system derived from these data. We see from this graph that the microhardness isotherms have a regular character. The transformation from the single-phase region of the diagram into the two-phase region is accompanied by a sharp break on the curve relating the microhardness of the solid-solution crystals to the composition of the alloy. In the other systems studied similar laws apply. Microscope analysis at high magnifications carried out on alloys quenched after prolonged homogenization at the corresponding temperatures yielded excellent agreement with the microhardness results in relation to the determination of the boundary of the one-phase region.

The simultaneous solubility of gallium and phosphorus and also indium and phosphorus in germanium was studied at various temperatures in [153]. The microhardness method was used in [152] to study the simultaneous solubility of gallium and arsenic and also indium and arsenic in germanium. The conclusions derived in these papers were analogous to those of [151].

The foregoing examples bear clear witness to the prospects of using the microhardness method for plotting the limited-solubility and solidus surfaces when studying three-component phase diagrams.

Effect of the Deviations of the Sections Studied from the Conodes on the Character of the Microhardness/Composition Relationships

As indicated in the foregoing, the sections to be examined when studying three-component phase diagrams have been chosen rather arbitrarily.

In so doing it may chance that the sections actually studied will deviate from the conodal sections (tie-lines), along which the manner in which a particular phase changes composition on passing from the one-phase region of the phase diagram to the two-phase region is exactly the same as in binary systems. When the chosen section (Fig. 58, I and III) deviates on one side or the other from the conode (section II), the concentration of the solid solution in the two-phase alloy may either increase or diminish with changing alloy composition, depending on the particular path of the solubility isotherm.

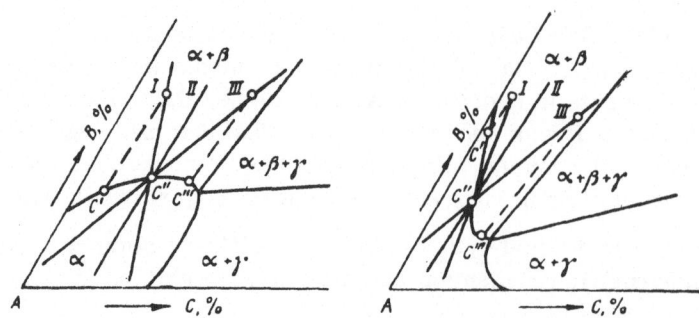

Fig. 58. Diagram illustrating possible changes in the composition of the solid solution in the two-phase region of the
phase diagram on deviating from the conode, depending on
the curvature of the solubility isotherm.

However, the influence of composition on the microhardness will not in this case be
particularly simple, since (depending on the behavior of the solubility isotherm), as the
action deviates from the conode, a fall (or rise) in the concentration of the solid solution as
a whole may be accompanied by a rise (or fall) in the concentration of one of the components.

At the same time the strengthening (hardening) effect of the alloying components in a
ternary solid solution may differ. In this connection it is important to know which of the components entering into the ternary solid solution will harden the latter to the greater extent
and how a change in the ratio of the alloying components in the solid solution of a two-phase
alloy will influence the character of the microhardness/composition isotherm when the sections deviate to one side or other of the conodes.

In order to solve these problems the concentration/microhardness relations of crystals
constituting solid solutions of magnesium and silicon in aluminum and zinc and tin in copper
were studied in [188], both along the conodal sections and along sections differing considerably from the conodal in one sense or another.

The disposition of these sections (I to VI) on the corresponding concentration triangles
and the composition of the alloys studied are indicated in Fig. 59a and b, which also show the
isothermal sections of the phase diagrams (Al−Mg−Si from [189] at 550°C and Cu−Zn−Sn
from [190] at 500°C).

Altogether 75 binary and ternary alloys were studied. The alloys were prepared in
graphite crucibles and cast into a cast-iron mould. The cast samples were deformed (worked)
by 20% on average and then annealed at 500°C (Cu−Zn−Sn) and 550°C (Al−Mg−Si) for 75 h.
The surface of the samples was prepared for the microhardness measurements by the method
described in earlier sections. The microhardness was measured at loads to 10 and 20 g. The
results of the measurements (mean of ten) appear in Figs. 60 and 61.

Figure 60a and b shows the microhardness/concentration relationship of the solid solution in the corresponding binary Al−Mg, Al−Si, Cu−Zn, and Cu−Sn systems. We see from
these figures that magnesium hardens the solid solution more than silicon and tin more than
zinc on dissolution in aluminum and copper respectively.

Figure 61a, b, and c shows the microhardness isotherms along the sections I to VI, the
positions of which are indicated in Fig. 59a and b. We see from these graphs that the microhardness of the solid solution varies quite regularly with the composition of the alloy. The
sharp breaks in these curves correspond to the points of saturation of the solid solution and
the transition from a single-phase region into a two-phase region. After passing from one

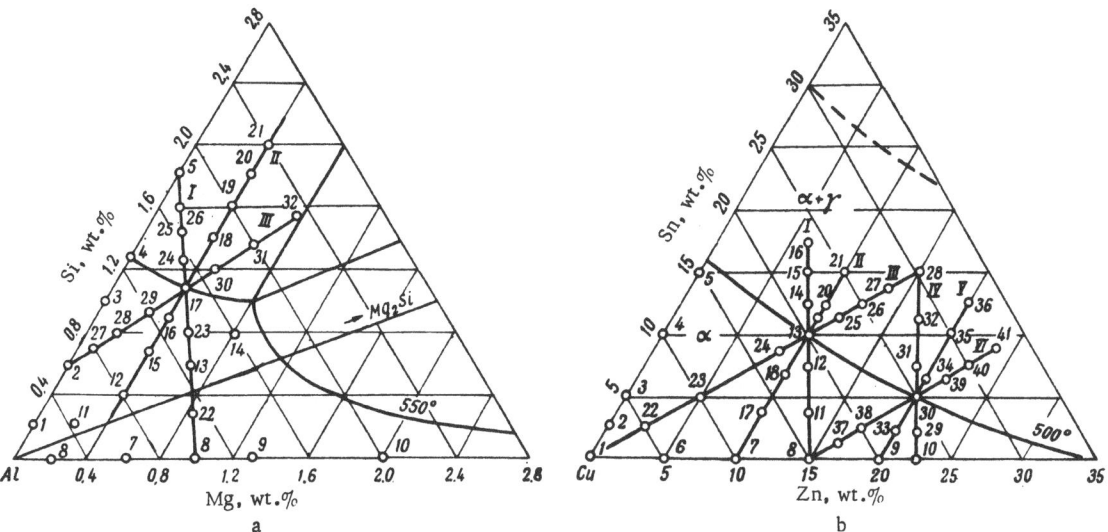

Fig. 59. Composition of the alloys studied and position of the sections relative to the conode:
a) at 550°C in the Al−Mg−Si system; b) at 500°C in the Cu−Zn−Sn system.

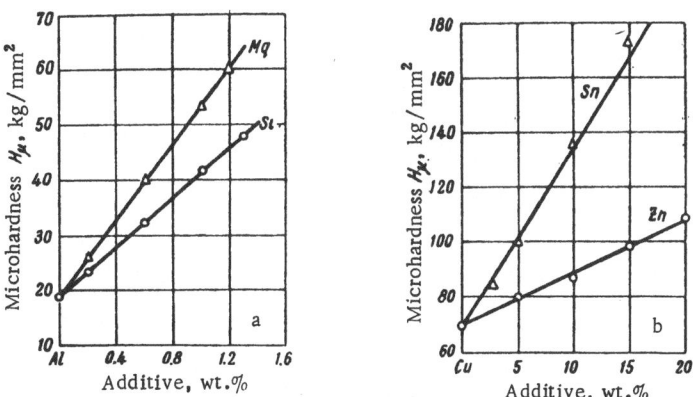

Fig. 60. Microhardness as a function of the composition of the solid solution: a) in the Al−Mg and Al−Si systems; b) in the Cu−Zn and Cu−Sn system.

to the other along section I (Fig. 59a) the microhardness of the solid-solution crystals in the Al−Mg−Si system falls. This is because the total concentration of the solid solution falls chiefly on account of the magnesium, which has a greater hardening effect than the silicon (Fig. 60a).

On passing from the one-phase region to the two-phase region along the conodal section II (Fig. 59) the microhardness remains constant, as it should do in accordance with the constant composition of the solid-solution crystals.

The behavior of the microhardness isotherms along section III in the two-phase region is determined by the increase in the amount of magnesium in the solution (Fig. 59a). Thus in accordance with the graphs shown in Fig. 61a the microhardness isotherms behave in different ways for different deviations from the conodes. However, desite this fact, the position of the point of limiting saturation at a specified temperature is established perfectly clearly in all

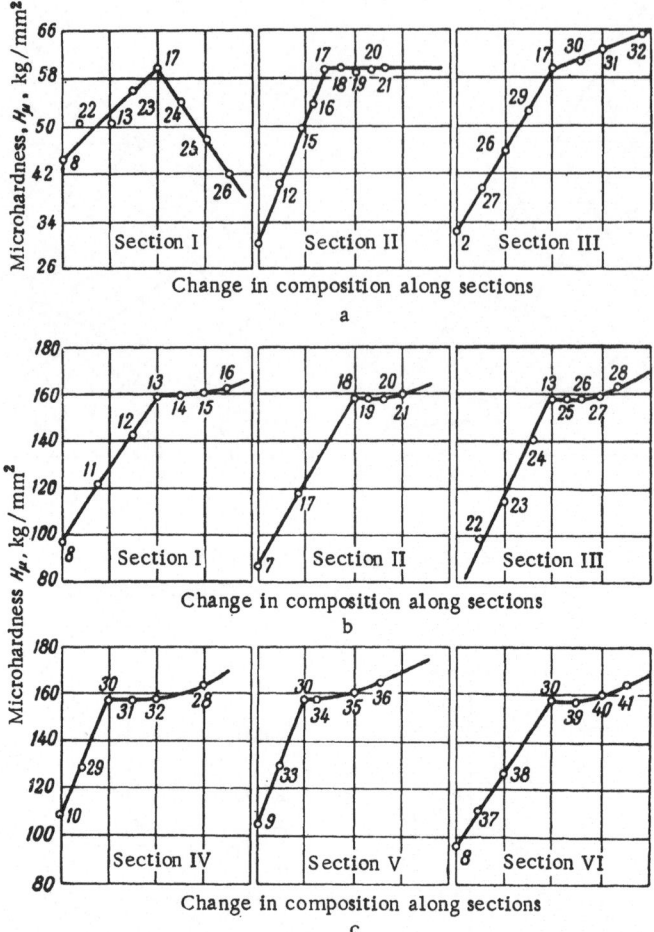

Fig. 61. Microhardness isotherms: a) at 550°C for
sections I to III in the Al−Mg−Si system; b) at 500°C
for sections I to III in the Cu−Zn−Sn system; c) at
500°C for sections IV to VI in the Cu−Zn−Sn system.

three cases. Hence even severe deviations of the sections chosen for study from the conodes
will not have any serious effect on determining the position of the point of limiting saturation
at the particular temperature. For section I (Fig. 59a) the position of this point is established
even more clearly than when the section passes through the conode.

It follows from the graphs presented in Fig. 61b and c that the microhardness isotherms
along sections I, III, IV, and VI (Fig. 59b) in the Cu−Zn−Sn system have the same form despite
the considerable deviations from the conodal sections II and V.

When the composition of the alloy varies along section I in the two-phase region of the
diagram $\alpha + \gamma$ the total concentration of the α solid solution diminishes. For example, the
total concentration of alloy 13, lying on the saturation boundary of the solid solution, equals
20% (by weight) of zinc and tin, while the concentration of the solid solution in alloy 15 (Fig.
59b) is 18.5%. However, in view of the fact that the section deviates in the direction of the more
hardening component, tin (see Fig. 60b), the concentration of which in the solid solution increases as the section approaches the Cu−Sn side, the microhardness of the crystals of the
α solid solution is almost unchanged, since the increment in microhardness associated with the
increasing amount of tin in the solution clearly compensates the reduction in microhardness

due to the fall in zinc content. On further varying the composition of the alloy along section I (Fig. 59b) the microhardness rises slightly, apparently as a result of the microheterogenization of the solid-solution crystals (see Chapter 9).

The microhardness/alloy composition relationship along the conodal section II needs no explanations. In the case of section III the zinc and tin content in solution changes in the opposite manner to that corresponding to section I, i.e., the total concentration of these elements in the solution increases, while the relative amount of the more hardening element (tin) falls. Hence the microhardness stays constant. Clearly the fall in microhardness as a result of the reduction in tin concentration in the solution is compensated by the increment on raising the zinc content. The manner in which the microhardness of the solid-solution crystals varies with the composition of the alloy is exactly the same for sections IV, V, and VI as for I, II, and III, and for the same reasons.

Thus, despite the extremely substantial deviations of the section from the conodes, the character of the microhardness isotherms in the $Cu-Zn-Sn$ system remains practically unaltered.

Both in the $Al-Mg-Si$ and in the $Cu-Zn-Sn$ systems, for any particular section the position of the point of limiting saturation at a specified temperature is established with complete clarity by the sharp break in the curve relating the microhardness of the solid-solution crystals to the composition of the alloy. Clearly in a number of cases when studying systems similar to the $Al-Mg-Si$ system the character of the microhardness/composition curves may be used to judge the orientation of the sections relative to the conodes and the position of the conodes themselves; this constitutes a useful supplement to thermal analysis when studying phase diagrams.

Additivity of the Microhardness Increment in the

Formation of Ternary and More Complex Solid Solutions

It was noted in [190] on the basis of an analysis of the relation between the microhardness and the composition of binary and corresponding ternary solid solutions that the microhardness was a linear function of the concentration.

When considering comparatively low concentrations the linearity is preserved on expressing the composition in either weight or atomic percent.* Essentially these relationships only constitute the initial sections of the microhardness/composition relation (lying close to the pure components) in a system with a continuous series of solid solutions (see Fig. 32). Hence, strictly speaking, these are not straight lines. However, within the range of 5 to 10% corresponding to the investigations in question (more often the concentrations are lower), the linearity of the microhardness/composition relationship holds fairly strictly.

Hence for binary systems $A-B$ and $A-C$ (Fig. 62) we may write

$$H'_\mu = H^0_\mu + C_1 K_B; \quad K_B = \frac{dH_\mu}{dC_B} = \tan\beta; \tag{1}$$

---

*V. Ya. Anosov [191, 192] showed that, if a particular property were represented by a straight line on the diagram in weight percent, then on expressing the composition in atomic percent it would be represented by a hyperbola. The same may be said of a corresponding transition from atomic to weight percent. However, in the initial section at fairly low concentrations linearity is preserved in both cases, since the part of the hyperbola associated with the corresponding transition is practically linear. Only the slope changes.

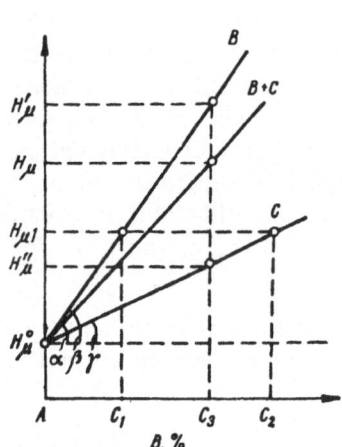

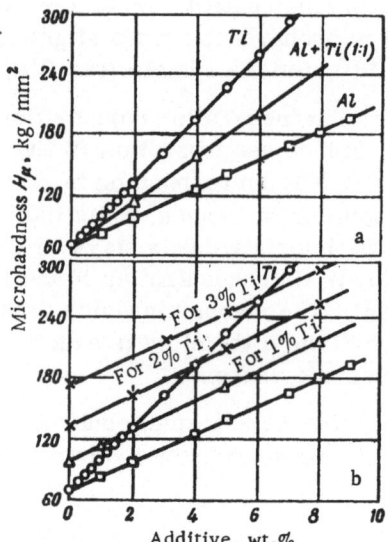

Fig. 62. Diagram illustrating the derivation of the equation representing the additivity of the microhardness increments associated with the formation of a ternary solid solution.

Fig. 63. Additivity of the microhardness increment in a solid solution of titanium and aluminum in copper relative to the microhardness increments of the corresponding two-component solutions.

$$H'_\mu = H^0_\mu + C_2 K_C; \quad K_C = \frac{dH_\mu}{dC_C} = \tan \gamma, \tag{2}$$

where H'_μ is the microhardness of the binary solid solution containing $C_1\%$ of component B in A or $C_2\%$ of component C in A; H^0_μ is the microhardness of the pure component A.

By comparing Eqs. (1) and (2) we see that

$$K_B \cdot C_1 = K_C \cdot C_2, \tag{3}$$

i.e., the concentration of $C_1\%$ of component B is equivalent in respect of its hardening effect on A to a concentration of $C_2\%$ of component C.

A quantitative comparison between the increment in the microhardness arising from the dissolution of the two alloying components taken in 1:1 ratio and the increment in microhardness arising from the corresponding binary systems showed in a number of cases that the increment associated with the change in the composition of the ternary solution was additive with respect to the increments associated with the binary solutions.

On this basis, considering as a very first approximation that the hardening effects of the components B and C on dissolution in A are independent, we may write (Fig. 62):

$$H_\mu = \frac{C_B}{C_B + C_C} H'_\mu + \frac{C_C}{C_B + C_C} H''_\mu; \quad C_B + C_C = C_3, \tag{4}$$

where H_μ is the microhardness of the ternary solid solution containing $C_B\%$ of component B and $C_C\%$ of component C; H'_μ is the microhardness of the binary solid solution of B and A containing $C_3\%$ of component B; H''_μ is the microhardness of the binary solid solution of C in A containing $C_3\%$ of component C.

After substituting the values of $H_\mu^!$ and $H_\mu^{!!}$ obtained from Eqs. (1) and (2) and making some appropriate transformations we have

$$H_\mu^{"} = H_\mu^0 + K_B C_B + K_C C_C. \tag{5}$$

Equation (5) is of an approximate nature, since in certain cases the microhardness/composition relationship may deviate from the linear law.

In addition to this (as we shall show later), deviations may arise from chemical interaction between the components in the solid solution. Clearly the relationship obtained should be satisfied best of all for low concentrations. Nevertheless, by using this relationship we may secure a reasonably good qualitative estimate of the way in which the microhardness/composition curve will behave on passing from the one-phase to the two-phase region. In this way we may furthermore discover the part played by each component exerting a specific hardening influence as a result of the formation of a solid solution in a ternary system.

A calculation of the relation between the microhardness of solid-solution crystals and the composition of the alloy along the corresponding sections in the Al − Mg − Si and Cu − Zn − Sn systems showed (see Figs. 59 and 60) that in the first of these systems the calculated and experimental curves resembled each other in general form. Quantitative differences were evidently associated with the appearance of chemical interaction between the magnesium and the silicon in aluminum-base solid solutions (more details will be given later), with possible experimental errors, and also with the fact that the assumptions made in deriving Eq. (5) were not absolutely rigorous.

In the second system (Cu − Zn − Sn), however, the experimental and theoretical curves were almost identical. Any differences occuring lay within the limits of experimental error.

The microhardness/composition relationship in binary alloys of the Au − Al and Cu − Ti systems was studied along a number of sections in [183, 184, 193].* It was found that the increment in microhardness was proportional to the quantity dissolved and that linearity was preserved up to the limiting concentration (Fig. 63a).

The analogous relationship also showed no appreciable deviations for the ternary solid solution. The microhardness increment in the ternary solid solution was additive with respect to the microhardness increments in the corresponding binary solid solutions.

In order to demonstrate the validity of the additivity rule in relation to microhardness increments in this system, Fig. 63b contains some geometrical constructions which emphasize that, for the cross sections of the Cu − Ti − Al phase diagram parallel to the Cu − Al side, the changes in microhardness with composition take place along parallel straight lines and are displaced from one another by a distance corresponding to the amount of titanium dissolved in the alloys.

Since modern alloys with advanced working characteristics contain three, four, and sometimes more alloying components, the principles and methods of studying multicomponent systems require further development. In particular, the general principles underlying the application of the microhardness method to the study of multicomponent alloys should be elucidated.

In view of this, the authors made an attempt at extending the microhardness-increment additivity rule to other systems comprising a large number of components [194]. In relation to multicomponent systems the additivity rule may be formulated thus: The numerical value of

*V. N. Vigdorovich, Candidate's Dissertation, M. I. Kalinin Moscow Institute of Nonferrous Metals and Gold (1958).

TABLE 22. Composition of the Alloys Studied

No. of alloy	Composition, at.%						
	Cu	Mg	Mn	Si	Cr	Ti	Al
1	0.8	0.19	0.39	0.15	0.13	0.03	Balance
2	1.1	0.32	0.40	0.16	0.14	0.03	
3	1.3	0.38	0.39	0.16	0.13	0.04	
4	1.4	0.40	0.40	0.20	0.15	0.05	

No. of alloy	Composition, at.%						
	Zn	Al	Sn	Cr	Ti	Zr	Cu
5	—	11.8	—	0.5	—	0.35	Balance
6	9.0	—	0.25	—	0.65	—	
7	9.0	—	0.14	—	—	0.50	
8	—	11.7	—	—	1.3	0.70	

the microhardness increment in a multicomponent alloy is equal to the sum of the microhardness increments associated with the formation of the corresponding binary solid solutions, i.e.,

$$\Delta H_\mu = \sum_{i=1}^{i=n} \Delta H_\mu^i, \tag{6}$$

where ΔH_μ is the microhardness of the multicomponent solid solution (additive property); ΔH_μ^i is the microhardness increment associated with the dissolution of component i.

To a very coarse approximation it is assumed that the hardening effects of the various alloying components are independent of each other.

Assuming the linearity of the concentration/microhardness relations in the corresponding binary systems we may write

$$\Delta H_\mu = \sum_{i=1}^{i=n} K_i C_i, \tag{7}$$

where K_i is the slope of the microhardness/composition (C_i) curve in the formation of a binary solution of component i.

In general the microhardness of a multicomponent solid solution may be determined from the equation

$$H_\mu = H_\mu^0 + \Delta H_\mu \tag{8}$$

or, allowing for expression (7):

$$H_\mu = H_\mu^0 + \sum_{i=1}^{i=n} K_i C_i, \tag{9}$$

where H_μ^0 is the microhardness of the solvent.

In order to verify these assertions, complex copper- and aluminum-base alloys were studied; the compositions are indicated in Table 22.

Table 23 shows the slopes corresponding to the linear parts of the microhardness/composition curves in the corresponding binary systems.

TABLE 23. Slopes (K_i, kg/mm^2 · at.%) of the Microhardness/Composition Relationships for Binary Aluminum- and Copper-Base Solid Solutions

System	K_i	System	K_i	System	K_i
Al—Mn	35	Al—Cr	56	Cu—Al	5.4
Al—Cu	46	Al—Ti	154	Cu—Ti	25,4
Al—Si	22	Cu—Zn	1,7	Cu—Cr	44.0
Al—Mg	34	Cu—Sn	10,5	Cu—Zr	114.0

TABLE 24. Calculated and Measured Microhardness of Complex Copper- and Aluminum-Base Solid Solutions

No. of alloy	Microhardness, kg/mm^2		No. of alloy	Microhardness, kg/mm^2	
	calculated	measured		calculated	measured
1	90	95	5	199	203
2	109	121	6	113	116
3	122	145	7	153	157
4	133	157	8	251	260

After preliminary deformation (working) by about 20% the aluminum alloys (Table 23) were subjected to prolonged homogenization at 500 ± 3°C for 200 h and then quenched in water. The copper-base alloys were subjected to 50% working and homogenized at 800°C for 60 h.

The microhardness was determined as the average of 15 to 20 measurements on the quenched alloys.

Table 24 shows the microhardness values determined experimentally and calculated by Eq. (9) on the basis of the data presented in Table 23.

On the basis of preliminary investigations, with the successive introduction of the various components into the solution, the microhardness appeared to increase in proportion to the change in concentration. This to a certain extent supports the idea that the individual components act independently on the microhardness of the solid-solution crystals, at any rate to a first approximation. It is evidently because of this that the experimental and theoretical results agree fairly well together.

There are nevertheless certain differences markedly exceeding those attributable to experimental error; these are evidently associated with the occurrence of chemical interaction between the components in the aluminum-base solid solutions.

As regards the copper-base alloys, in this case there is a considerably greater correspondence between the experimental and calculated data; this may well be associated with the fact that the copper is less active than aluminum and also with the greater limiting solubility of the elements introduced.

Thus the results obtained support the fundamental possibility of using the additivity rule in multicomponent systems.

Use of the Microhardness Method in Plotting Conodes in Two-Phase Regions of Ternary Phase Diagrams

In studying the three-component phase diagrams of metallic systems the positions of the conodes in the two-phase regions of the phase diagrams are usually not established, owing to the great difficulty in determining the chemical composition of the conjugate phases directly in the solid state, and even more so in the solid—liquid and liquid—solid states. Yet a knowledge of the position of the conodes in the two-phase regions enables the three-component phase diagram to be characterized far more completely, as it enables us to determine what composition a phase in equilibrium with a phase of a specified composition at a given temperature will have. In addition to this, knowing the position of the conodes, we may determine the compositions of the phases in equilibrium with one another when the composition of the alloy is varied within a particular phase region. As a result of this it may well become possible to predict the behavior of the alloy under service conditions from the point of view of the influence exerted by the quantitative relationship between phases of variable composition. This will enable us to make a reasonable decision in varying the composition of the alloy in any advantageous direction (within the phase region specified of course). This is particularly important for systems in which the compositions of the coexisting phases change sharply on changing the composition of the alloy within a specified phase region of the three-component phase diagram.

A knowledge of the position of the conodes in a two-phase region of the diagram in which one phase is liquid enables us to make a theoretical estimate of the redistribution of impurities taking place when crystallization is effected by pulling the materials from the melt or in zone recrystallization.

In view of the foregoing discussion it is essential that serious attention should be paid to the positions of the conodes when studying three-component phase diagrams.

In order to determine the desired position of the conode of a two-phase region in a three-component phase diagram, we must establish the composition of at least two alloys in which the compositions of the two coexisting phases are the same. Clearly it is sufficient to follow the composition of one of these phases only. For example, when constructing conodes in two-phase regions of a phase diagram in the solid or solid—liquid state, two figurative points must be found in the concentration triangle, corresponding to two alloys in which the crystals of the α solid solution will have identical concentration. It was shown in [193] that the laws relating the microhardness of the α solid solution crystals to the composition of the alloy in two- and three-component metallic systems could be used for this purpose.

Straight lines joining figurative points situated on the sides of the concentration triangle, symbolizing the compositions of binary solid solutions having identical values of microhardness, constitute "isosclers," or lines of equal microhardness. Schematically the position of the isoscler corresponding to the microhardness associated, for example, with a section of the ternary ABC system may be represented by a straight line KF (Fig. 64). The isosclers in the region of the ternary solid solution lie parallel to one another.

Not all points in the plane expressing the relation between the microhardness of a solid solution and the composition will correspond to realizable values. In this plane the region of microhardness values corresponding to a specific physical meaning for equilibrium solid solutions is limited by the limited-solubility or solidus surfaces, if we consider its projection on the plane of the corresponding isothermal cross section. If an isoscler corresponding to a specific value of the microhardness at the particular temperature intersects the solubility isotherm, then the point of intersection will correspond to the composition of the alloy (which is completely saturated at the temperature in question and has the specified value of microhardness. Knowing the microhardness of the solid-solution crystals in a two-phase alloy of a three-component system, we may determine its composition by reference to the point of intersection of the isoscler corresponding to this microhardness with the solubility or solidus isotherm. This offers the possibility of determining the fixed position of the conode.

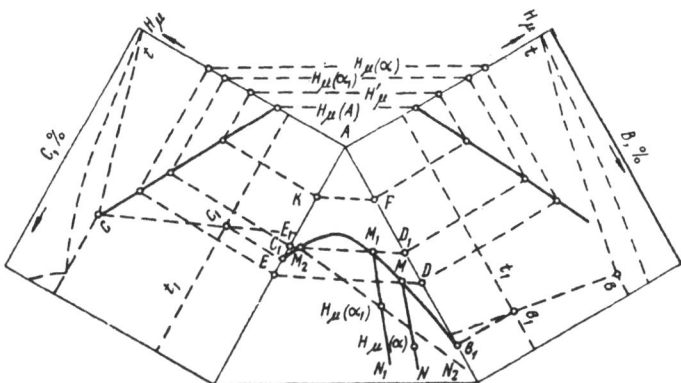

Fig. 64. Diagram to illustrate the method of constructing
conodes in the two-phase regions of the three-component
phase diagrams of metallic systems by the microhardness method.

Thus in order to discover the position of the conode we must determine the microhardness of the solid-solution crystals of an alloy lying in the two-phase region of the phase diagram (for example, $H_\mu(\alpha)$, Fig. 64), then draw the isoscler DE corresponding to this value of
the microhardness. After this, the point of intersection M of the isoscler DE with the solubility isotherm must be joined to the figurative point of an alloy containing solid-solution crystals
with microhardness $H_\mu(\alpha)$. In this way we secure the fixed position of the conode (Fig. 64).
All alloys for which the figurative points lie on the straight line MN contain in their structure
α solid solution crystals having an identical composition (determined by the point M) and identical microhardness $H_\mu(\alpha)$.

We may encounter a case in which an isoscler DE corresponding to a specific microhardness, for example, $H_\mu(\alpha_1)$, will intersect the isotherm at several points (for example, the points
M_1 and M_2, Fig. 64). In this case we shall have to decide which of the two possible lines (M_1N_1
or M_2N_2) constitutes the conode. This problem may be solved experimentally by studying the
microhardness of the solid-solution crystals as a function of composition along the two sections indicated. The particular section in which the microhardness of the α crystals is the
same will be the conode.

In determining which of the sections is the conode one may also be guided by other conodes
with unambiguous positions (for example, MN in Fig. 64), remembering that conodes must never
intersect.

In addition to this, information is sometimes available as to the possible position of the
second conjugate point of the conode, and this enables the problem to be solved without re-
course to experiment, since the general direction of the conode is then known. For example,
a case of this kind arises when the composition of the second coexisting phase is known and
constant.

Thus if we know the relations between the microhardness of the solid-solution crystal
and the composition of the alloy and also the position of the region corresponding to the α
solid solution in the ternary system, we may easily determine the position of the corresponding
conodes by using the method here described.

The possibility of using the microhardness method for determining the position of conodes
in the two-phase regions of three-component phase diagrams was confirmed experimentally

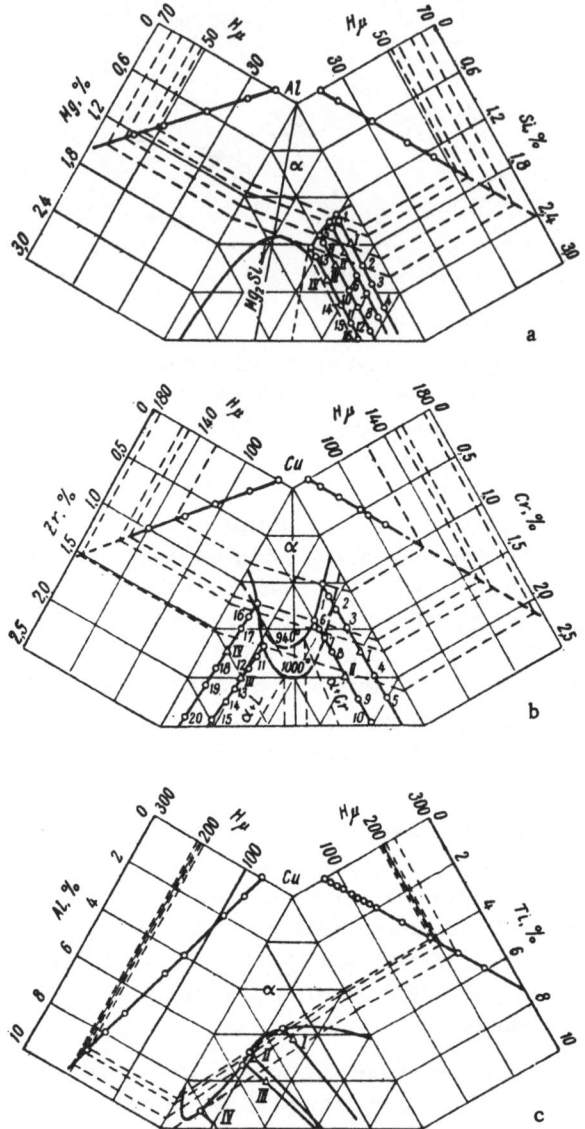

Fig. 65. Disposition of the conodes. a) In the
α + Si two-phase region of the three-component
phase diagram of the Al–Mg–Si system at 550°C;
b) in the α + Cr (sections I and II) and the α + L
(sections III and IV) two-phase regions of the three-
component phase diagram of the Cu–Cr–Zr system at 940 and 1000°C, respectively; c) in the two-
phase α + CuTi region of the three-component
phase diagram of the Cu–Al–Ti system at 850°C.

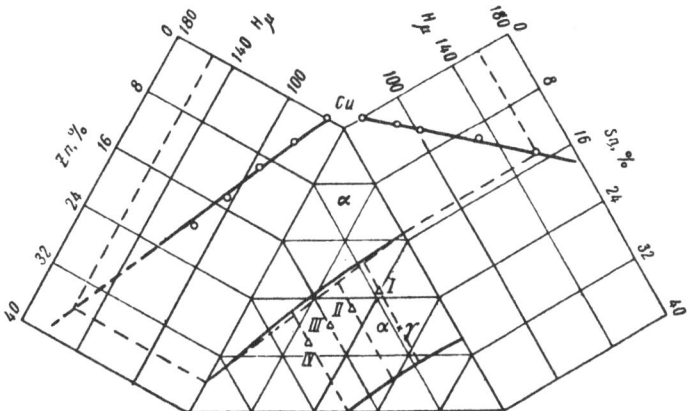

Fig. 66. Position of the isoscler corresponding to the microhardness of the α solid solution crystals of the two-phase alloys I, II, III, and IV in the three-component phase diagram of the Cu−Zn−Sn system at 500°C.

in [204] when studying the Al−Mg−Si, Cu−Cr−Zr, Cu−Zn−Sn, and Cu−Al−Ti systems. * These systems were chosen in view of the fact that the position of the region corresponding to α solid solution was reasonably accurately defined, and the relation between the microhardness of the solid-solution crystals and the composition of the alloy was also known for the corresponding binary systems.

In order to plot the conodes, alloys were prepared in several two-phase regions of the phase diagrams of the systems under consideration, four in each system. The Al−Mg−Si alloys were prepared in an electric induction furnace, using alundum crucibles, and cast into a graphite mold. The alloys of the copper-base system were prepared in a high-frequency furnace, using corundized crucibles placed in a deep graphite cylinder in order to create a reducing atmosphere. The alloys were cast into a cast-iron mold heated to between 200 and 300°C. The composition of the alloys (these are denoted by Roman figures in Figs. 65 and 66) derived from chemical analysis are given in Table 25. The cast samples were hot-rolled with a 50% reduction and appropriately heat-treated. The conditions of pre-roll heating and heat treatment for the alloys are indicated in Table 26.

The microstructure of the alloy samples thus obtained was studied and the microhardness of the solid-solution crystals determined. Results for the alloys whose compositions are shown in Table 25 as the average of 15 measurements are presented in Table 27.

The resultant data were taken together with earlier results relating the microhardness of the solid-solution crystals to the composition of the alloy, and used to plot conodes by the method described in several two-phase regions of the Al−Mg−Si, Cu−Cr−Zr, and Cu−Al−Ti phase diagrams (Fig. 65). In the Cu−Zn−Sn system (Fig. 66) the conode could not be plotted for reasons which we shall be considering shortly. In order to check whether the sections I to IV were in fact conodes in the systems under consideration, the microhardness of solid-solu-

---

* The construction of the conodes in these systems really requires the additivity rule of the increments in microhardness to be valid. We shall show later that in the Cu−Cr−Zr and Al−Mg−Si systems in the region of the ternary solid solutions corresponding to the quasibinary Cu−Cr$_2$Zr and Al−Mg$_2$Si sections there are considerable deviations from this rule. However, for the compositions studied in determining the positions of the conodes these deviations are slight.

TABLE 25. Chemical Composition of the Alloys* Intended for Determining the Position of the Conodes in the Two-Phase Regions

System	No. of alloy	Chemical composition, wt.%		
Al—Mg—Si	I II III IV	Mg 0.20 0.32 0.53 0.75	Si 1.50 1.50 1.50 1.50	Al Balance
Cu—Cr—Zr	I II III IV	Cr 1.50 1.48 0.5 0.26	Zn 0.25 0.50 1.50 1.48	Cu Balance
Cu—Al—Ti	I II III IV	Al 3.0 5.0 5.0 8.0	Ti 3.0 2.0 3.0 2.0	Cu Balance
Cu—Zn—Sn	I II III IV	Zn 23 26 28 30	Sn 23 14 12 10	Cu Balance

On expressing the composition in at.% the corresponding constructions lead to the same results as on expressing it in wt.%.

TABLE 26. Modes of Heat Treatment for Alloys to be Used in Determining the Position of the Conodes in the Two-Phase Regions

System	Temp. before rolling, °C	Holding time, h	Annealing temp., °C	Annealing time, h
Al—Mg—Si	450	3	550	75
Cu—Cr—Zn	900	1	940	5
			1000	3
Cu—Al—Ti	850	2	850	8
Cu—Zn—Sn	850	2	500	75

TABLE 27. Microhardness of Solid-Solution Crystals Formed by the Alloys in the Systems Studied

System	Microhardness of alloys, kg/mm²			
	I	II	III	IV
Al—Mg—Si	51	55	60	64
Cu—Cr—Zr	128	130	168	156
Cu—Al—Ti	196	198	203	224
Cu—Zn—Sn	158	157	158.5	156

tion crystals belonging to a number of alloys situated in these sections was studied (see Fig. 65a,b; alloys denoted by Arabic figures). The samples for study were prepared by the methods indicated. The results of the measurements are illustrated in Fig. 67. We see from these graphs that the microhardness of the solid-solution crystals belonging to alloys with compositions lying along sections I to IV remains constant everywhere, being equal to the microhardness of the alloy completely saturated at the corresponding temperature.

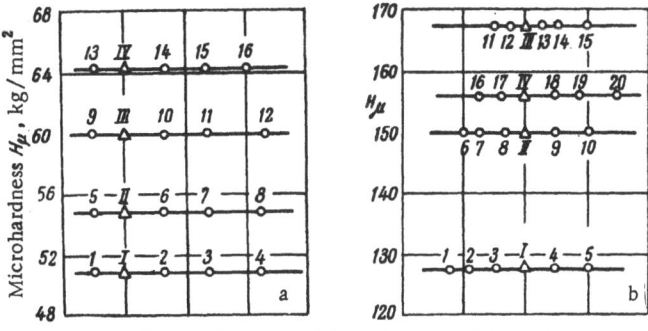

Change in composition along conodal sections

Fig. 67. Microhardness of solid-solution crystals
as a function of the composition of alloys on sections I to IV of (a) the Al − Mg − Si system at 550°C
and (b) the Cu − Cr − Zr system at 940 and 1000°C.

This indicates that the α solid solution crystals belonging to the alloys lying along these
sections have a constant composition and hence that these sections are in fact conodes. The
conodes in the Cu − Cr − Zr system plotted for a temperature of 1000°C lie in the (L + α) two-
phase region situated between the solidus and liquidus (Fig. 65b, sections III and IV). The test
samples in this region, heated to 1000°C, existed in the solid − liquid state, and this was subsequently fixed by quenching.

Taking the Cu − Al − Ti system as an example (Fig. 65c) we find that there may be cases
in which the position of the second conjugate point remains constant and all the conodes plotted
converge at this point. In the present case such a point corresponds to the compound Cu_3Ti in
which solubility of the components is almost nonexistent.

Special attention should be given to the results obtained in the case of the Cu − Zn − Sn
system. This system illustrates a certain limitation to the use of microhardness method for
plotting conodes in particular cases.

It follows from Table 27 that almost identical values of microhardness were obtained for
all four alloys. This is due to the fact that the variation in the microhardness with the composition of the alloy associated with a change in solubility is compensated by the varying ratio
of the components alloying the solid solution. As a result of this the microhardness remains
constant. The isoscler therefore intersects the isotherm at a very small angle and practically
merges with it. The exact position of the conodes cannot be determined by the microhardness
method in this system.

The experimental results obtained confirm the excellent prospects of using the microhardness method for determining the positions of the conodes in the two-component regions of
three-component phase diagrams.

Chemical Interaction between Alloying Components

in Ternary Solid Solutions Based on Metals

and Semiconductors

The question as to the particular state in which alloying components dissolved in the base
of an alloy find themselves and the manner in which the chemical interaction taking place between them affects the properties of the alloys is of fundamental importance at the present time

in connection with the doping of semiconductors and also various aspects of the theory of heat resistance [196, 197].

In many cases chemical interaction may occur between the dissolved elements in ternary solid solutions and also between the dissolved elements and the solvent, culminating in the formation of the corresponding chemical compounds.

When binary solid solutions are formed in systems containing chemical compounds, it has been established in a number of cases by thermal analysis that the heat of formation of the primary solid solution referred to a gram-atom of the dissolved substance is (as a rule) no smaller than the heat of formation of the chemical compound closest to this solid solution (i.e., the compound formed by the solvent and dissolved substance), also referred to one gram-atom of the dissolved substance. However, when solid solutions are formed in systems having no chemical compounds, the thermal effects are relatively slight. On this basis it has been suggested [198] that the solid solutions formed to the accompainment of thermal effects similar to the heats of formation of the corresponding chemical compounds may be considered as solutions of these compounds in the parent material.

In addition to the foregoing case, the formation of ternary solutions may be accompanied by chemical interaction between the alloying components. This interaction should clearly appear in the most obvious manner when the binary system formed by the alloying components contains a strong chemical compound, while a quasibinary section occurs in the ternary system.

It was shown earlier that the increment in microhardness associated with the formation of a ternary solid solution was additive with respect to the increments associated with the formation of the corresponding binary solutions. It follows from this that the relation between the microhardness and the concentration of the solution in the ternary system may be expressed graphically by a plane passing through the straight lines expressing the concentration dependence of the microhardness in the corresponding two-component systems.

If there is a strong chemical interaction between the alloying components in the ternary solid solution (this would plainly be expected to manifest itself most sharply if there is a simple relation between their concentrations, i.e., if we are considering the corresponding quasibinary section), this cannot fail to produce deviations from additivity, as in this case the hardening effects of the alloying components on dissolution cannot be considered independent.

An important conclusion to be drawn from this is that deviations from additivity in the microhardness relationship may be used in order to judge whether there is any chemical interaction between the alloying components in ternary solid solutions.

Certain papers describing work on the Al−Mg−Si system carried out by methods such as the measurement of electrical resistance, temporary rupture resistance (tensile strength), hardness, and Young's modulus as well as x-ray structural analysis have indicated that a chemical compound of the Mg_2Si type may be formed in a ternary aluminum-base solid solution.

A minimum corresponding to a quasibinary section was observed in [199] on the curves characterizing the mechanical properties of alloys corresponding to sections intersecting the quasibinary section Al−Mg_2Si in the region of the ternary solid solution. The authors accordingly concluded that molecular formations of the Mg_2Si type existed in a ternary solid solution of magnesium and silicon in aluminum.

X-ray structural analysis of the same alloys carried out in [200] revealed a difference between the experimental lattice-constant curves and curves calculated for the case of an atomic structure of the solid solution. It was therefore concluded that the solution of Mg_2Si in aluminum was molecular.

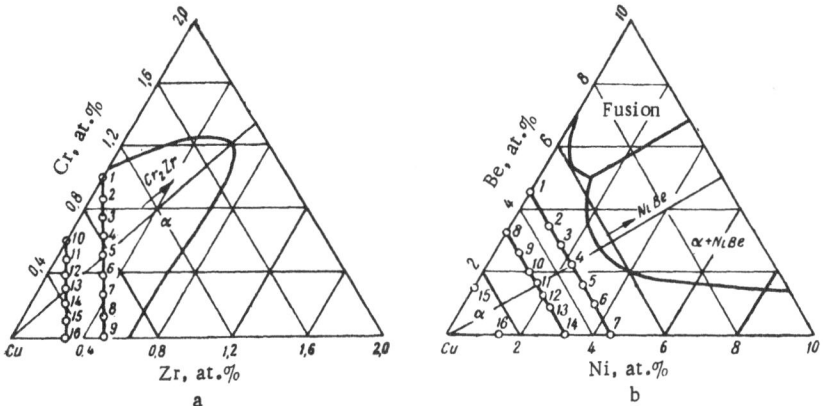

Fig. 68. Disposition of the test sections and solubility isotherms at 1000°C: a) in the Cu−Cr−Zr system; b) in the Cu−Ni−Be system.

The electrical resistance of Al−Mg−Si alloys was studied in [198] along a section with 99 at.% Al between 480 and 500°C. When this section intersected the quasibinary section the resistance/composition isotherms showed a sharp minimum. T. A. Badaeva explained this by the fact that as a result of the formation of Mg_2Si molecules within the primary solid solution the number of electron-wave scattering centers diminished, so that the electrical resistance did likewise.

The conclusions drawn in [198-200] were later confirmed by x-ray diffraction [201] and also by measurements of Young's modulus [202].

All these investigations into the Al−Mg−Si system indicate that the chemical interaction between the alloying elements in ternary solid solutions may be so intensive as to have a substantial effect on the physicochemical properties.

At the same time it should be noted that there is little experimental material regarding this question. In view of this the problem of studying the interaction of alloying elements is very important.

It is not quite clear why chemical interaction appears in some cases and not in others. No relation has been established between the properties of the chemical compounds and their behavior in the ternary solid solutions in equilibrium with them. In order to solve all these problems there must be a further accumulation of experimental data relating to the physico-chemical properties of solid solutions and the plotting of chemical composition/property diagrams.

A study of these problems is particularly vital in connection with the doping of semi-conducting materials.

In a number of investigations it has been noted that there is a considerable increase in the solubility of certain alloying elements in the presence of others and vice versa.

The observed increase in solubility is associated with the appearance of chemical interaction between the alloying elements, although the nature of this interaction is as yet uncertain.

It has simply been noted in a number of cases that the character of the interaction between alloying elements in ternary solid solutions, particularly those based on germanium and silicon, is closely connected with the thermal stability of the compounds with which these so-

TABLE 28. Chemical Composition of the Copper-Base Alloys Studied

No. of alloy	Composition, at.%		No. of alloy	Composition, at.%	
	Cr	Zr		Ni	Be
1	1.12	—	1	—	4.53
2	0.88	0.06	2	1.00	3.55
3	0.75	0.14	3	1.78	2.77
4	0.60	0.20	4	2.25	2.20
5	0.51	0.25	5	2.73	1.81
6	0.38	0.33	6	3.62	0.96
7	0.25	0.39	7	4.50	—
8	0.12	0.46	8	—	3.21
9	—	0.52	9	0.50	2.94
10	0.60	—	10	1.12	2.20
11	0.49	0.07	11	1.63	1.60
12	0.38	0.13	12	2.25	1.10
13	0.32	0.17	13	2.87	0.47
14	0.24	0.20	14	3.25	—
15	0.11	0.24	15	—	1.54
16	—	0.49	16	1.50	—

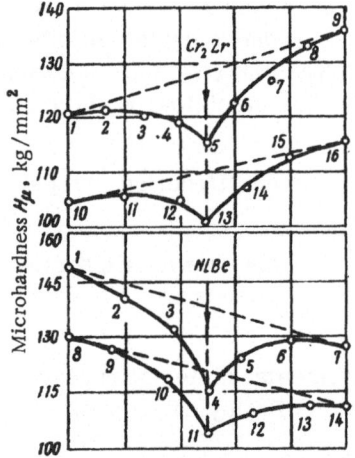

Change in composition along sections

Fig. 69. Relation between the microhardness and composition of a solid solution of chromium and zirconium and also nickel and beryllium in copper along the sections indicated in Fig. 68.

lutions are in equilibrium. Clearly the use of the microhardness method for solving this problem should lead to a more rapid solution.

The microhardness of ternary solid solutions of chromium and zirconium and also nickel and beryllium in copper was studied in [203] with due allowance for the possibility of there being an interconnection between the interaction of the alloying elements in the solid solution and deviations from the microhardness-increment additivity rule. In both systems there was a quasibinary section ($Cu-Cr_2Zr$ and $Cu-NiBe$ respectively). It was considered that on dissolution of these compounds in copper molecular solutions were formed. Apart from purely theoretical interest, the study of these systems was of practical value in connection with the preparation of heat-resistant alloys with a high electrical conductivity. The range of solid solutions in these systems was studied in considerable detail by microscope analysis, microhardness measurements, and x-ray structural analysis in [204].

The maximum solubility of chromium and zirconium and also nickel and beryllium in copper is reached at about 1000°C. On being quenched from high temperatures, alloys of both systems are quite stable at room temperature. Aging is only observed on heating above 300 to 400°C.

Series of ternary (and also binary) alloys were prepared along sections intersecting the quasibinary systems $Cu-Cr_2Zr$ and $Cu-NiBe$ (Fig. 68a,b).

The disposition of the test sections on the corresponding concentration triangles was governed by the nature of the solubility isotherms on expressing the concentration in at. and wt.%.

The composition of the alloys derived from twice-repeated chemical analysis is given in Table 28.

The cast alloys were hot-rolled (after heating to 900°C) with a 50% reduction, annealed at 1000°C for 2 h, and quenched in cold water. Microsections for the microhardness measurements were prepared from these samples by the method described in the foregoing.

A preliminary study of the microstructure of the alloys showed that all were single-phased and had a uniform polyhedral structure.

On the basis of preliminary investigations and also earlier considerations in the present monograph we may fairly conclude that the concentration/microhardness relationships representing the binary Cu—Cr, Cu—Zr, Cu—Ni, and Cu—Be systems are strictly linear within the ranges studied. The increments in microhardness for 1 at.% of chromium, zirconium, nickel, and beryllium are respectively 44, 114, 12, and 20 kg/mm².

The microhardness was measured in a PMT-3 furnished with a mechanism for the automatic application of the load.

The results of the measurements are shown in Fig. 69 in the form of a relation between the microhardness of the solid solution and the composition of the alloys along the sections indicated in Fig. 68.

We see from these graphs that the microhardness/composition relationships for the solid solutions have sharp minima at the intersections of the test sections with the quasibinary $Cu-Cr_2Zr$ and $Cu-NiBe$ lines.

Hence the microhardness/composition relationship corresponding to the systems in question is by no means characterized by a plane of general position, as suggested by the additivity rule, but by a surface with a sharp singular fold corresponding to the quasibinary $Cu-Cr_2Zr$ and $Cu-NiBe$ sections.

From general considerations of physicochemical analysis this fact might be associated with the formation of chemical complexes of Cr_2Zr and $NiBe$ in the lattice of the Cu base solid solution. From the point of view of physical nature, the phenomenon may be explained by the fact that the crystal lattice of the corresponding solid solutions is distorted to a lesser degree than would be the case for a disordered arrangement of the dissolved atoms, since the distortions are localized around the Cr_2Zr and $NiBe$ molecules so that additional slip planes are freed.

The microhardness of the solid-solution crystals thus becomes much lower than would be the case if we were simply dealing with a disordered distribution of the atoms of the alloying components in the copper lattice.

Thus on simultaneously alloying copper with chromium and zirconium or with nickel and beryllium, introduced in a ratio corresponding to the formation of the respective chemical compounds, within the limits of the solid solution these have a less effective hardening influence than they would have if introduced separately.

For example, if 2 at.% Be is dissolved in the copper, the microhardness rises by about 40 kg/mm^2. However, if the same amount of nickel is also added to this alloy, the microhardness of the latter will not increase, as might be expected, but will fall by about 4 kg/mm^2. At the same time, if 2 at.% Ni is dissolved in the copper on its own, the microhardness will rise by 24 kg/mm^2. On introducing 0.2 at.% Zr into copper the microhardness rises by 34 kg/mm^2, while if 0.4 at.% Cr is added to this alloy (the Zr/Cr ratio here corresponding to the compound Cr_2Zr) the microhardness of the alloy falls by 5 kg/mm^2, whereas 0.4 at.% Cr by itself, on dissolution in copper, increases the microhardness of the latter by some 18 kg/mm^2.

Thus within the range of the solid solution the alloys with the lowest degree of hardness are those lying on the quasibinary $Cu-Cr_2Zr$ and $Cu-NiBe$ sections. At the same time it is well known [204, 205] that the heat resistance reaches a maximum for alloys belonging to these sections on passing from the one-phase into the two-phase regions, when the alloys contain finely divided inclusions of the second phase within their structure.

This may serve as an indirect indication of the fact that, in the present case, it is precisely the excess phase which has the decisive effect on the heat resistance of these alloys.

The conclusions drawn in [203] regarding the possibility of chemical interaction taking place between the alloying elements in copper-base solid solutions were later supported by thermo-emf measurements in [29].

A comparative study of the microhardness and electrical resistance of aluminum- and copper-base solid solution in the $Al-Mg-Si$, $Al-Mg-Ge$, $Cu-Cr-Zr$, and $Cu-Ni-Be$ systems was carried out in [206].

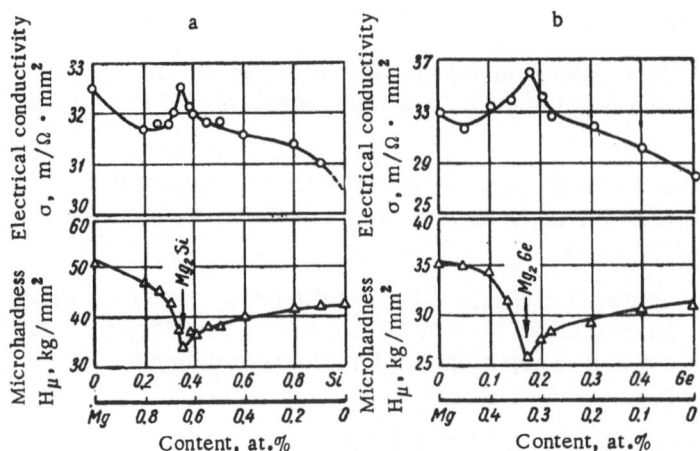

Fig. 70. Microhardness and electrical conductivity as
functions of the composition of the solid solution in (a)
Al — Mg — Si (section with 99% Al) and (b) Al — Mg — Ge (section with 99.5% Al) systems.

A series of alloys lying (as regards composition) on sections intersecting the corresponding quasibinary sections in the region of the aluminum- or copper-base solid solutions
were prepared for study.

In the Al — Mg — Si and Al — Mg — Ge systems sections with 99 and 99.5 at.% Al were studied
respectively. In the Cu — Ni — Be system the section with 95.5 at.% Cu and in the Cu — Cr — Zr
system the section with 1.0 at.% Cr and 0.6 at.% Zr were studied.

As original materials for preparing the alloys, aluminum of the AV0000 type (99.998% Al),
vacuum-remelted MO copper, germanium and silicon at least 99.9999% pure, and magnesium
of the MG1 type (99.92% Mg) were used, as well as Cu — Cr, Cu — Zr, Cu — Ni, and Cu — Be alloys. After preparation the samples were analyzed chemically.

The cast samples were worked by 30 to 50% and annealed at 550 (Al — Mg — Si), 450 (Al —
Mg — Ge, and 1000°C (Cu — Cr — Zr, Cu — Ni — Be) respectively for 50, 100, and 2 h, and then quenched
in cold water. The alloys thus obtained were used to produce microsections for measuring the
microhardness and samples for measuring the electrical conductivity. The microsections and
samples so prepared were further annealed in evacuated and sealed quartz ampoules at corresponding temperatures and water-quenched. The measurements were made immediately after
quenching. The microhardness was measured in a PMT-3 furnished with a device for automatic loading. The microhardness of the aluminum alloys was measured with a load of 10 g
and that of the copper alloys with one of 50 g. The averages of 10 to 12 measurements were
taken. The electrical conductivity was measured with an IÉ-1 inductive conductivity gage.
The averages of eight to ten measurements were taken.

The results of all the measurements are shown in Figs. 70 and 71. We see from these
diagrams that in all systems considered the intersection of the test sections with the corresponding quasibinary sections was marked by sharp singular microhardness minima and electrical-conductivity maxima.

This indicates a chemical interaction in the ternary copper- and aluminum-base solid
solutions, revealing itself most sharply for a ratio of the structural components corresponding
to stoichiometry of the chemical compound.

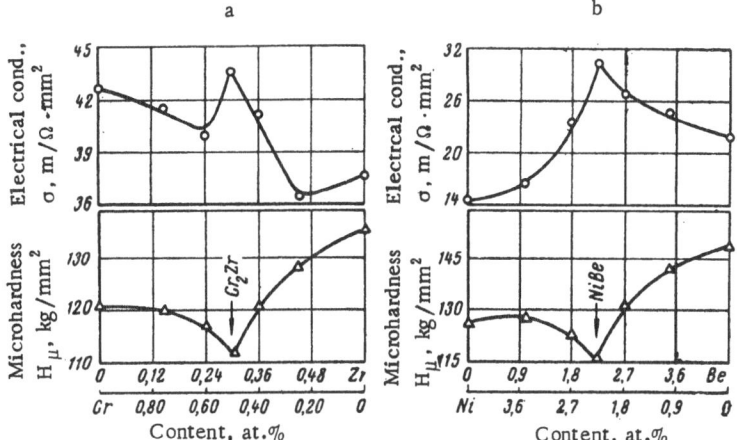

Fig. 71. Microhardness and electrical conductivity as functions of the composition of the solid solution in (a) Cu−Cr−Zr (section with 1% Cr and 0.06% Zr) and (b) Cu−Ni−Be (section with 95.5% Cu) systems.

These values of the microhardness and electrical conductivity of Al−Mg−Si alloys are in good qualitative agreement with those described earlier.

It should be noted that the electrical-conductivity maximum observed when studying the quenched alloys is much more sharply expressed than it is in high-temperature measurements [198]. This may be due to the fact that at high temperatures the scattering of the carriers at thermal lattice vibrations plays a major part, so that the effect of the increase in electrical conductivity associated with chemical interaction between the alloying components in the ternary solid solution is partly obscured.

It is evidently for the same reason that the effect in question disappears altogether in dilute Al−Mg−Ge solutions on measuring the electrical conductivity at high temperatures [185]. At the same time, on measuring quenched alloys under conditions precluding aging (Fig. 71), the electrical conductivity increases substantially at the point of intersection between the test section and the Al−Mg_2Ge quasibinary. The predominant chemical interaction between the magnesium and germanium in the aluminum-base solid solution also appears very sharply when studying the microhardness of the quenched alloys.

Thus the Al−Mg−Ge system is in this respect qualitatively completely analogous to the Al−Mg−Si system.

The experimental data obtained agree closely with the character of the liquidus surfaces studied in [185] and confirm the predictions regarding the possibility of there being Mg_2Ge molecular formations in aluminum-base solid solutions.

The observed electrical-conductivity maximum in alloys lying (as regards composition) on the quasibinary sections Cu−Cr_2Zr and Cu−NiBe (Figs. 70 and 71) indicates that the solid solution in heat-resistant copper alloys based on these systems possesses a high electrical conductivity. This finds a specific physicochemical explanation in the light of our own data.

The character of the composition/property diagrams obtained (Figs. 70 and 71) may be explained on the assumption that, as a result of chemical interaction between the alloying elements, the distortions of the crystal lattice are localized at specific points and the crystal lattice as a whole is less distorted than would be the case if we were simply dealing with a disordered distribution of the atoms of the dissolved elements in the lattice of the solid solu-

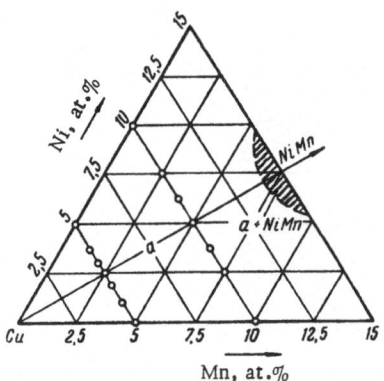

Fig. 72. Composition of the alloys and disposition of the sections under consideration on the Cu−Ni−Mn concentration triangle.

tion. The observed substantial deviations from the additivity rule in relation to the increment in the microhardness of these alloys may also be associated with this.

Evidently this causes the rise in electrical conductivity, since the foregoing characteristics of the crystal lattice belonging to ternary aluminum- and copper-base solid solutions should promote an increase in the mean free path of the electrons.

It should be noted nevertheless that in all the systems in question there is a stable, congruently melting chemical compound in equilibrium with the solid solution (for example, Mg_2Si in the Al−Mg−Si system or Cr_2Zr in the Cu−Cr−Zr system). Hence we can only judge the presence or absence of chemical interaction between the dissolved components in the corresponding ternary solid solutions by analyzing the character of the chemical composition/property diagrams; we cannot control the interaction process, since the dissolved compound is stable right up to melting and also in the liquid state.

In order to observe the process of chemical interaction between the dissolved components in the ternary solid solution it is essential that the melting point of this solution should be above the temperature of thermal stability of the dissolved compound.

It is precisely this situation which holds in the copper−nickel−manganese system studied in [207] by the microhardness method. This system is composed of three binary systems: Cu−Ni, Cu−Mn, and Mn−Ni; the phase equilibrium in these is described by phase diagrams with unlimited mutual solubility of the components at high temperatures. However, at lower temperatures a compound NiMn (Kurnakov phase) is formed in the Ni−Mn system in the solid state, and this is stable as far as about 900°C. The character of the phase equilibrium in the Cu−Ni−Mn system was studied by Chao Pao-Ch'ang, according to which the Cu−NiMn section is quasibinary, while the ordered θ' phase (NiMn) with a face-centered tetragonal crystal lattice observed in the binary NiMn system at low temperatures also occurs in the structure of ternary alloys containing up to some 86 wt.% of copper.

In order to study the character of the chemical interaction between nickel and manganese in the ternary alloys at various temperatures, alloys lying within the region of the copper-base solid solution along sections intersecting the quasibinary Cu−NiMn section at 95 and 90 at.% Cu were prepared. As original materials, electrolytic copper, nickel, and manganese were used, these being vacuum-remelted at $\sim 1 \cdot 10^{-4}$ mm Hg for refining purposes. The alloys were prepared in evacuated and sealed quartz ampoules.

After preparation the alloys were worked with a 30 to 50% degree of deformation and homogenized for 2 h at 900°C. After this all the alloys were analyzed chemically. Figure 72 shows the composition of the alloys and the disposition of the test sections on the Cu−Ni−Mn concentration triangle. It was shown earlier that there were considerable deviations from the microhardness-increment additivity rule when interaction occurred between the dissolved components, the maximum value being reached when the sections under consideration intersected the quasibinary sections.

─────

*Chao Pao-Ch'ang, Dissertation, M. I. Kalinin Moscow Institute of Nonferrous Metals and Gold (1957).

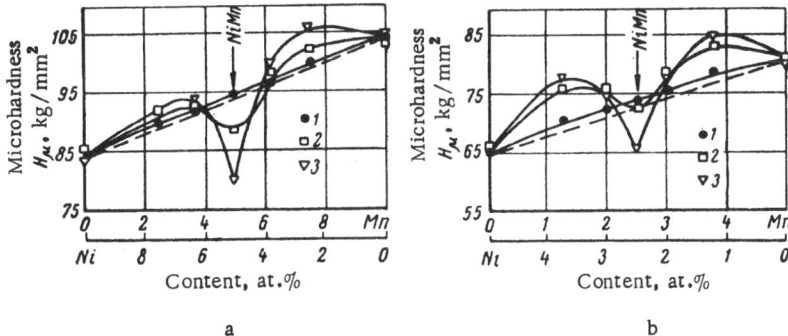

Fig. 73. Relation between the microhardness and composition of the alloys along sections at 90 and 95 at.% copper (1, 2, 3 significantly annealing at 900, 700, and 500°C, respectively).

For measuring the microhardness three microsections of each composition were prepared. The finished microsections were sealed in quartz ampoules evacuated to $1 \cdot 10^{-4}$ mm Hg and annealed at 900°C (first batch of samples), 700°C (second batch), and 500°C (third batch). The alloys annealed at 900°C were water-quenched and those annealed at 700 and 500°C were cooled in air. After this heat treatment the microsections were etched in a 3% solution of ferric chloride in 10% HCl, using the method described in Chapter 3.

Then the microhardness was measured at a load of 50 g. The average values of 10 to 15 measurements were taken. The results of the measurements are shown in the form of chemical composition/property diagrams in Fig. 73. We see from these diagrams that after annealing at 900°C the microhardness/composition relationship along the sections indicated in Fig. 72 was characterized by curves very close to the additive straight line (shown as a broken line in Fig. 73). This may indicate a disordered arrangement of the nickel and manganese atoms in the lattice of the copper-base solid solution.

The relationship obtained for the alloys annealed at 700°C is characterized by a tortuous curve with a smooth minimum in the middle and two smooth maxima at the sides. The relationship obtained for the alloys annealed at 500°C, however, is distinguished by a sharp singular minimum corresponding to the intersection of the sections under consideration with the quasibinary Cu−NiMn section.

Furthermore, the side maxima found on the curves obtained by annealing at 700°C are more sharply visible in the present case. The sharp microhardness minimum indicates that there has been considerable chemical interaction between the dissolved components.

Comparison of the graphs presented in Fig. 73 thus shows the following. Additional annealing at 500, 700, and 900°C carried out for 40 h in order to investigate the stability of the state of the alloys caused no appreciable changes.

At 900°C there is no preferential interaction between the nickel and the manganese; at any rate it has no effect on the character of the chemical composition/microhardness diagram. At 500°C the effect is quite obvious, as indicated by the singular minimum of the microhardness at the intersection of the sections under consideration with the quasibinary Cu−NiMn. At 700°C there is an intermediate state.

It is natural to suppose that the observed change in the character of the chemical composition/microhardness diagrams on raising the annealing temperature is due to the dissociation of the compound NiMn formed in the copper-base solid solution. At 900°C this compound is completely dissociated, at 500°C it is probably stable, while at 700°C the dissociation is partial.

The composition/microhardness curves here obtained are analogous in general form to the curves relating the composition to the mechanical properties obtained by G. G. Urazov and T. I. Šhushpanova [199] and also N. N. Sirota and colleagues [202].

The side maxima in these curves are hard to explain; their appearance is in no way associated with the occurrence of aging, since in the present case we are concerned with solid solutions lying in that part of the concentration triangle for which they are stable over the whole temperature range.

It may be concluded from our own experimental data that chemical reactions may occur between the dissolved components of ternary solid solutions in a manner never hitherto investigated, just as in the case of liquid (e.g., aqueous) solutions.

The direction of these reactions depends on the temperature and is clearly governed by the thermal stability of the corresponding chemical compound.

The microhardness/composition relationship was studied in [133] along sections intersecting the quasibinary for silicon- and germanium-base solid solutions in the Ge−Al−Sb, Si−Al−Sb, Ge−Al−P, and Si−Al−P systems.

In all the systems studied except Si−Al−Sb the relationship between microhardness and composition along the sections indicated is distinguished by a sharp minimum corresponding to the quasibinary sections Si−AlP, Ge−AlSb, and Ge−AlP. In the case of the Si−Al−Sb system, however, although there are deviations from additivity, the minimum on the corresponding concentration/microhardness curves is very diffuse. As in the previous case, the minimum on the microhardness curves of the systems considered is clearly associated with chemical interaction between the alloying elements and the formation of chemical compounds AlSb or AlP in the germanium- and silicon-base solid solutions. The character of the microhardness/composition relation in the Si−Al−Sb system is due to the fact that on dissolution in silicon aluminum antimonide is largely dissociated into its components.

On dissolution in germanium, however, the dissociation is much weaker. As regards aluminum phosphide, this is evidently thermally more stable and thus dissociates very little on dissolving in germanium or silicon. The conclusions as to the behavior of aluminum antimonide on dissolution in silicon and germanium are confirmed by the character of the solubility isotherm and also by an analysis of the liquidus curves in the corresponding phase diagrams by means of the Schroder−Van't Hoff equation [208, 209]. This is also indicated by the behavior of pure aluminum antimonide at the corresponding temperatures in the molten state and also during its formation from the components, either independently or in a medium of molten germanium or silicon [210-215].

In view of the foregoing considerations regarding the nature of solid solutions of aluminum and antimony in silicon it appeared probable that aluminum antimonide might be formed by the prolonged annealing of the solid solutions in question. In order to verify this, a corresponding series of alloys were enclosed in quartz ampoules (previously evacuated to $1 \cdot 10^{-4}$ mm Hg and filled with purified argon) and subjected to annealing at 800°C for three months. Measurements carried out on solid-solution crystals with compositions corresponding to the section at 99.8% Si showed that the microhardness of all the alloys except that lying on the quasibinary section Si−AlSb remained practically constant, while the microhardness of the alloy indicated fell by more than 60 kg/mm^2. After this further annealing at 800°C for three months was carried out. After this the microhardness of the alloy lying on the quasibinary section fell by another 70 kg/mm^2. The character of the microhardness/concentration relationship along the corresponding sections thereupon became the same as in the other systems studied.

On the basis of the foregoing results we may conclude that chemical interaction takes place between the alloying components in solid solutions of aluminum and antimony in silicon (with an equiatomic ratio) as a result of prolonged annealing, and that this is accompanied by a unidirectional change in the microhardness of the solid solution. This latter circumstance opens the possibility of developing forms of heat treatment for silicon containing large amounts of aluminum and antimony so as to obtain materials with more advanced properties.

CHAPTER 7

USE OF THE MICROHARDNESS METHOD IN STUDYING THE PHASE DIAGRAMS OF QUATERNARY SYSTEMS

It has been shown that the microhardness method may be successfully used for studying the phase diagrams and structure of alloys belonging to two- and three-component systems. In recent years A. M. Zakharov [216, 217] has shown that the method may also successfully be used for studying the phase diagrams of four-component systems.

Plotting the Boundaries between Phase Regions

in Quarternary Systems

The use of the microhardness method for plotting the boundaries between phase regions in binary and ternary systems is based on finding breaks on the curves relating the microhardness of the α solid solution to the composition of the alloys, since on passing from the one- to the two- or three-phase region of the phase diagram the laws governing the composition and hence the microhardness of the α solid solution change.

In the case of quaternary phase diagrams the position is somewhat complicated in view of the more complex way in which the composition of the α solid solution changes on passing from one phase region into another. Nevertheless, it follows from the principles of continuity and correspondence formulated by N. S. Kurnakov [6, 7] that the fundamental laws governing changes in the microhardness of a quaternary α solid solution should not differ in principle from the existing types of composition/microhardness diagram characterizing the α solid solutions of binary and ternary systems.

Taking account of the foregoing laws governing the changes in the microhardness of the α solid solution on passing from one phase region into another in binary and ternary systems, the possible types of composition/microhardness diagram characterizing the α solid solutions of various secondary sections of quaternary phase diagrams* may be expressed in the manner shown in Fig. 74.

In the practical analysis of quaternary phase diagrams we most frequently desire to plot the boundary between the one-phase and two-phase ($\alpha + \beta$) regions (Fig. 74a). A horizontal microhardness characteristic may occur beyond the bend if the secondary section selected is quasibinary or coincides with the direction of the conodes in the two-phase $\alpha + \beta$ region. If

---

* The diagrams are plotted on the assumption that the alloys under consideration are brought into an equilibrium state; for simplicity the microhardness curves are shown as straight lines.

120

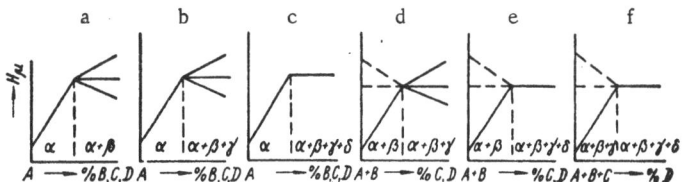

Fig. 74. Possible types of composition/microhardness diagram of a quaternary α solid solution for various secondary sections of the quaternary system A – B – C – D.

the secondary section is chosen arbitrarily (with respect to the conodes in the two-phase $\alpha + \beta$ region) the microhardness of the quaternary α solid solution beyond the bend may either rise or fall.

The transition from a single-phase α region into a three-phase $\alpha + \beta + \gamma$ region (Fig. 74b) may occur through a point lying on the limited-solubility curve, in the present case that representing the solubility of components B and C in A. In the particular case in which the secondary section coincides with the plane of the conodal triangle characterizing the equilibrium of the α, β, and γ solid solutions, the composition of the quaternary α solid solution in alloys of the ($\alpha + \beta + \gamma$) region remain constant; it follows from this that the microhardness isotherm beyond the bend will be a straight line parallel to the horizontal axis.

The case shown in Fig. 74c is rare, since the secondary section has to pass through the point of maximum solubility, in the present case that of the components B, C, and D in A (i.e., through the vertex of the conodal tetrahedron characterizing the equilibrium of the α, β, γ, and δ solid solutions). The horizontal form of the microhardness/composition relationship beyond the bend is due to the fact that, within the limits of the four-phase volume $\alpha + \beta + \gamma + \delta$ the composition of the α solid solution (as in the remaining solid solutions) always remains constant.

It is not difficult to explain the remaining types of diagram relating the composition to the microhardness of the α solid solution on the basis of the foregoing laws governing the changes in the microhardness of a quaternary α solid solution on passing from a one- into a two-, three-, or four-phase region of the phase diagram.

Since in the two-phase $\alpha + \beta$ and three-phase $\alpha + \beta + \gamma$ regions the microhardness of the α solid solution may either rise or fall, the number of possible versions of the microhardness curves which may arise on passing from the two- into the three-phase region is nine (Fig. 74d). In a particular case, the secondary section in the two-phase $\alpha + \beta$ region may be "conodal," and in the three-phase $\alpha + \beta + \gamma$ region it may be disposed in the plane of the corresponding conodal triangle; then the microhardness isotherm of the quaternary α solid solution will be represented by a single horizontal straight line in both regions.

The microhardness isotherms may occur in three forms on passing from the two-phase $\alpha + \beta$ or three-phase $\alpha + \beta + \gamma$ regions into the four-phase $\alpha + \beta + \gamma + \delta$. The first of these cases (Fig. 74e) is possible when the secondary section "intersects" an edge, and the second case (Fig. 74f) when it "intersects" the face of the conodal tetrahedron characterizing the equilibrium of the α, β, γ, and δ solid solutions.

Since the laws governing the changes in the microhardness of the quaternary α solid solution differ in different phase regions, it is clear that a break of greater or lesser clarity should appear on the microhardness isotherms on passing from one phase region into another. It is by the breaks in the microhardness curves that one may judge the position of the boundaries between the phase regions on the secondary sections of the quaternary phase diagrams.

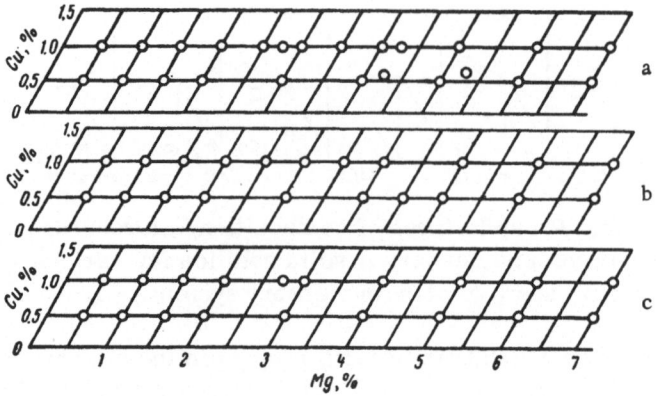

Fig. 75. Position of the alloys under consideration on
the projections of the secondary sections of the quaternary Al − Zn − Mg − Cu system with the following zinc
contents (%): a) 4; b) 6; c) 8.

The clearest illustration of the possibility of using the microhardness method for plotting
the boundaries of the α solid solution was given by A. M. Zakharova [216]. In addition to microhardness, x-ray structural analysis was employed in conjunction with measurements of electrical conductivity and microscope analysis. A number of secondary sections of the quaternary
Al − Zn − Mg − Cu system were plotted on the basis of microscope analysis in [218] and the combined solubility of zinc, copper, and magnesium in solid aluminum was studied at 430 and 460°C.
In plotting the secondary sections the solidus temperature of the alloys was determined by differential thermal analysis, using a Kurnakov pyrometer.

Three groups of alloys with 4.6 and 8% Zn containing 0.5 and 1.0% Cu and from 0.5 to
0.7% Mg, belonging to six secondary sections (Fig. 75), were studied. Alloys 200 g in weight
were prepared from aluminum (purity 99.95%), magnesium (99.945%), zinc (99.95%), and an alloy of aluminum with 50% copper; melting was carried out in an electric furnace in corundum
crucibles under a layer of carnallite. Chemical analysis of the alloys showed that these deviated little from the calculated compositions. Bars turned (machined) to dimensions of d =
35 and h = 50 mm were annealed for 48 h at 400 ± 5°C and then upset under the press with a
75 to 80% degree of reduction, after which they were used to prepare samples 15 × 15 × 10
mm in size.

Microscope study of the alloys were carried out at 460, 430, and 200°C and the microhardness, lattice constant, and electrical conductivity were measured at 460 and 430°C, using
samples quenched from these temperatures after prolonged annealing to bring them into an
equilibrium state. The annealing was carried out in the following way: In evacuated quartz
ampoules the samples were heated slowly (at 30 deg/h) to a temperature of 460 ± 5°C and
held for seven days; then some of the samples were quenched and the rest cooled to 430°C.
After holding for ten days some more samples were quenched, and the rest were cooled in
stages (holding for 15 days at 350 and 18 days at 300°C) to 200°C, thereafter being quenched
in cold water after a further 28 days. Thus the holding periods for quenching from 460, 430,
and 200°C were 12, 20, and 78 days, respectively.

In order to identify the various phases under the microscope the following etchants were
used: a 10% NaOH solution at 20 and 60 to 80°C, etching time 30 to 60 and 10 to 15 sec, respectively, Keller's reagent (0.5% HF + 1.5% HCl + 2.5% HNO_3 + 95.5% H_2O), 20 to 30 sec, 0.5% HF
solution 15 to 30 sec, concentrated HNO_3 5 to 7 sec, 2% solution of HNO_3 (in alcohol) 15 to 20
sec, and concentrated HNO_3 vapor 7 to 10 sec.

In order to measure the microhardness microsections were prepared from the alloy samples by the method described earlier. The etchant was a 10% solution of NaOH. The microhardness was determined in a PMT-3 under a load of 20 g with a loading time of 2 to 3 sec and a time of 7 to 8 sec under load. In order to measure the lattice constant, back-reflection x-ray diffraction patterns were obtained on a flat film. The photographs were taken in copper radiation, the lattice constants being calculated from the rear lines of the diffraction pattern obtained from the (333) and (511) planes. The electrical conductivity of the alloys was measured by the eddy-current method.

The secondary sections thus plotted as well as the measured values of microhardness (H_μ), lattice constant (a), and electrical conductivity (σ) of the alloys belonging to these sections are shown in Fig. 76. We see from the graphs that, depending on the temperature and composition of the alloys, the following may be in equilibrium with the quaternary aluminum solid solution: in alloys with 4% Zn the θ (CuAl$_2$), S (Al$_2$CuMg), and T phases (the latter being a solution between the ternary compounds $Al_2Zn_3Mg_3$ and Al_6CuMg_4 of the Al$-$Zn$-$Mg and Al$-$Cu$-$Mg systems); in alloys with 6 and 8% Zn, in addition to the foregoing, an M phase (MgZn$_2$) or a solution between the compounds $MgZn_2$ and AlCuMg of the Mg$-$Zn and Al$-$Cu$-$Mg systems.

These sections confirm the view that the principal part in the hardening of alloys of the V95 type by heat treatment is played by the M (or MgZn$_2$) phase.

The sections thus plotted agree with the results obtained by studying the phases precipitated during the aging of the solid solution in ternary Al$-$Zn$-$Mg and quaternary Al$-$Zn$-$Mg$-$Cu alloys.

We also see from Fig. 76 that the results obtained by measuring the microhardness, lattice constants, and electrical conductivity of the alloys at 430 and 460°C agree with each other and with the results of microscope analysis. In particular, the microhardness and lattice constant of the α solid solution rise in the one-phase region, while the electrical conductivity falls as the magnesium content increases. In the two-phase (α + T) region the microhardness and lattice constant continue to increase, while the electrical conductivity falls, although to a lesser extent than in the α region. The boundary between the α and α + T regions is established particularly sharply on the microhardness and electrical conductivity isotherms.

The increase in the microhardness and lattice constant and the fall in the electrical conductivity in the two-phase α + T region are associated with the fact that, within this region, the magnesium concentration in the aluminum solid solution increases as a result of the failure of the test sections to coincide with the direction of the conodes in the Al$-$Zn$-$Mg$-$Cu tetrahedron. Analogous phenomena were noted in the previous chapter for the case of three-component systems.

Thus the foregoing data relating to the comparative study of the solubility of alloying elements in aluminum (in the Al$-$Zn$-$Mg$-$Cu system) verify the possibility of applying the microhardness method to this problem of great practical importance.

The microhardness method was used in [219, 220] for plotting the boundary of a solid solution of tungsten, molybdenum, and zirconium in niobium. The alloys of the secondary sections used for plotting the boundary of the quaternary niobium α solid solution each contained 5 wt.% W and up to 30% Mo and Zr (in the ratios 2.1:1, 2:1, and 1:3).

The preparation of the alloys and the method of bringing them into an equilibrium state were described in [219]. The microsections were prepared with coarse and fine emery paper and then polished on a cloth with a suspension of Cr$_2$O$_3$ in water. In order to remove the surface-hardened layer the microsections were etched in a mixture of concentrated acids HF and

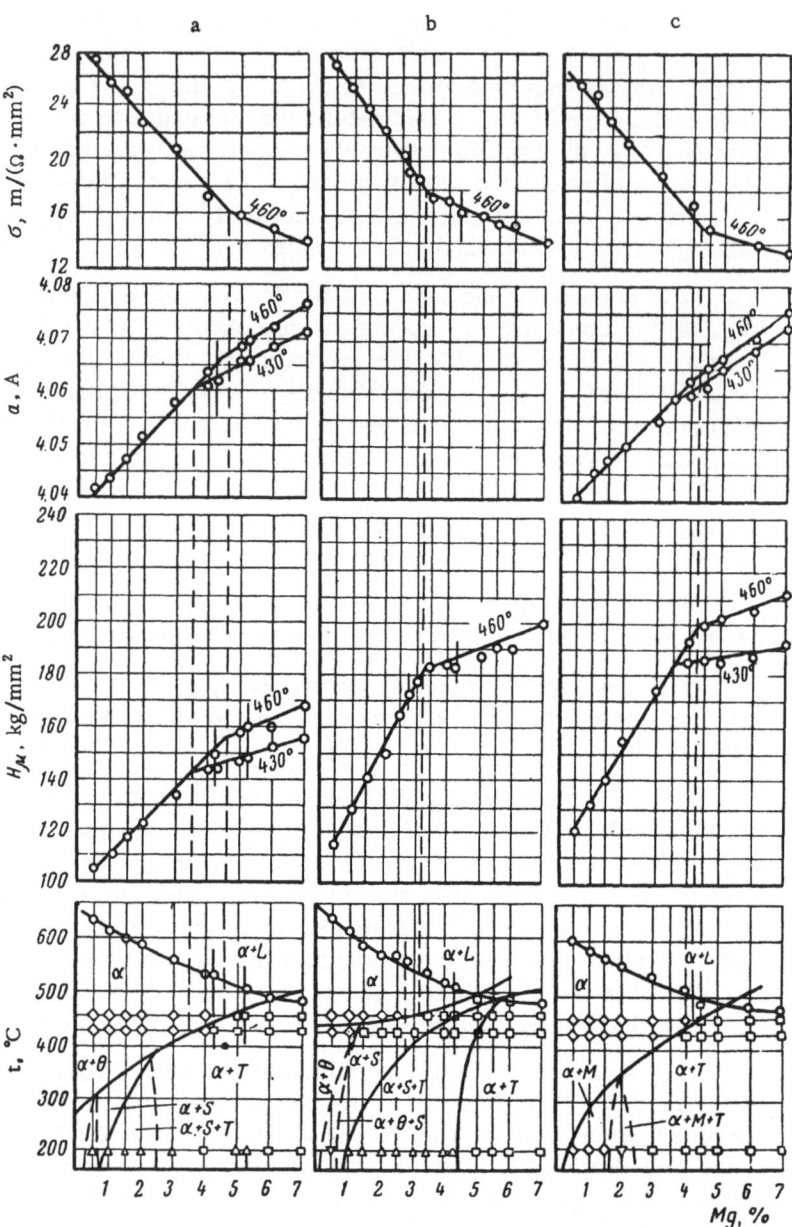

Fig. 76. Effect of magnesium on the microhardness, lattice con-
to their zinc and copper content, as follows (%): a) 4Zn + 0.5Cu;
f) 8Zn + 1.0 Cu.

HNO$_3$ (1:1) for 30 sec and again polished for 2 to 3 min. The microhardness was measured
with a PMT-3 under a load of 20 g; the loading time was 2 to 3 min. Microhardness was
measured with a PMT-3 under a load of 20 g; the loading time was 2 to 3 and the time under
load 7 to 8 sec.

The results obtained on plotting the boundary of the quaternary niobium α solution in the secondary sections of the Nb−W−Mo−Zr tetrahedron by the microhardness method
are shown in Fig. 77. The horizontal aspect of the microhardness isotherm beyond the bend
(Fig. 77a) indicates that the secondary section parallel to the Nb−Mo$_2$Zr section of the ternary

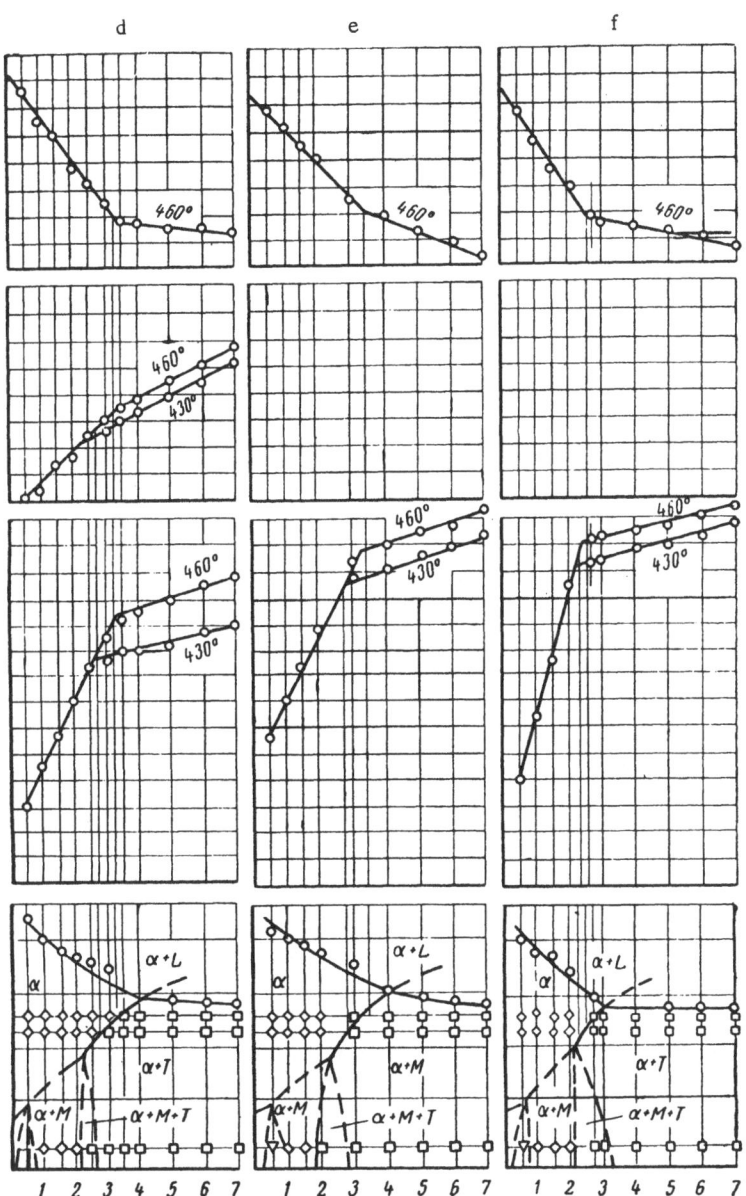

stant, and electrical conductivity of aluminum alloys in relation
b) 4Zn + 1.0Cu; c) 6Zn + 0.5Cu; d) 6Zn + 1.0Cu; e) 8Zn + 0.5Cu;

Nb − Mo − Zr system coincides with the direction of the conodes in the α + (Mo, W)$_2$Zr two-phase
volume. The more the secondary sections deviate from the direction of the conodes (Fig. 77b
and c), the more does the microhardness of the α crystals increase in the alloys of the two-
phase α + (Mo, W)$_2$Zr volume; however, the boundary of the α region is also established clearly.
The rise in the microhardness beyond the bend is explained by the fact that when the sections
deviate from the conodes the composition of the quaternary niobium α solid solution no longer
remains constant but is enriched with zirconium [219].

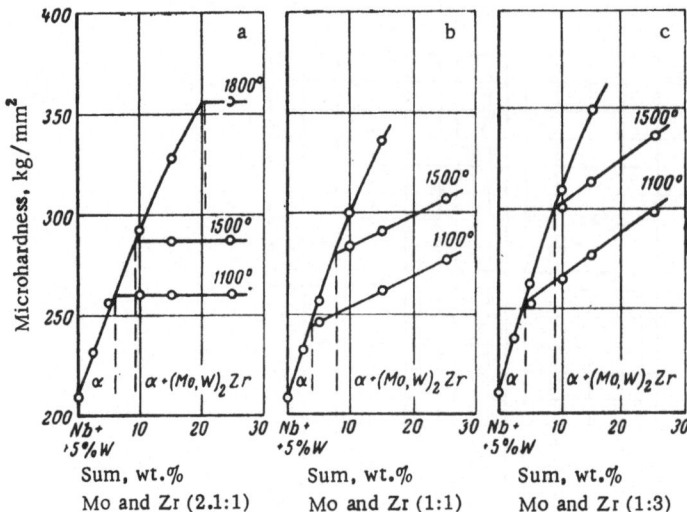

Fig. 77. Microhardness isotherms of a quaternary niobium α solid solution for alloys of the secondary sections
of the Nb−W−Mo−Zr tetrahedron containing 5 wt.%
tungsten and various amounts of molybdenum and zirconium in the ratios: a) 2.1:1; b) 1:1; c) 1:3.

The foregoing examples bear witness to the excellent prospects of using the microhardness method in studying the phase diagrams of four-component systems.

Plotting Conodes in the Two-Phase Volumes

of Quaternary Phase Diagrams

The possibility of plotting conodes in the two-phase volumes of the phase diagrams of
four-component systems was demonstrated by A. M. Zakharov in [217, 220]. We shall now
consider the results of these investigations.

As in the case of ternary systems, the principal assumptions amount to the following:
1) The microhardness of binary, ternary, and quaternary solid solutions is a linear function
of their concentration; 2) in the formation of ternary and quaternary solid solutions the additivity rule holds for the increments in microhardness.

It is well known that the representation of a quaternary phase diagram at constant temperature requires a three-dimensional space (the so-called isothermal tetrahedron). The complicated graphical constructions required to determine the positions of the conodes in the two-
phase volumes of this tetrahedron may be carried out far more simply on individual primary
sections (plane sections). In the general case the isotherms of the primary sections of the
tetrahedron cannot characterize the phase equilibria in the quaternary alloys, since the conodes
of the two-phase volumes of the isothermal tetrahedron may not occur in the plane of the primary section specified.

Hence the plane of a primary section in which it is possible to construct conodes in the
two-phase region must pass through the conodes in the analogous two-phase volume of the isothermal tetrahedron.

In a number of cases when studying real quaternary diagrams it may be quite easy to
choose such primary sections. Firstly these include the so-called quasiternary sections passing

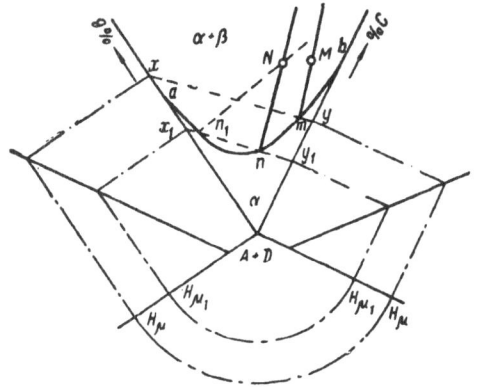

Fig. 78. Diagram illustrating the possibility of using the microhardness method for plotting the conodes in the two-phase $\alpha + \beta$ region of the primary section of the quaternary system A−B−C−D.

through the lines of the quasibinary sections in the faces of the tetrahedron and possessing the properties of true ternary phase diagrams. Clearly this requirement may also be satisfied by sections parallel to the faces of the tetrahedron, i.e., to the isothermal sections of the corresponding ternary systems. The quaternary alloys of such sections should not contain too large proportions of the quaternary component, since on moving the primary section away from the face of the isothermal tetrahedron the probability of the conodes of the two-phase volume lying within its plane will diminish.

In order to plot the conodes in the two-phase diagrams of the primary sections we may make use of the laws governing the changes in the microhardness of binary, ternary, and quaternary solid solutions. It is well known that the microhardness isosclers of the ternary α solid solution on the plane of the concentration triangle may be represented by straight lines joining the compositions of binary solid solutions with identical microhardness (see Chapter 6). Analogously, the microhardness isosclers of a quaternary α solid solution on the plane of the primary section will connect the compositions of ternary solid solutions with identical microhardness situated on the sides of the primary section (i.e., on the corresponding linear sections of the ternary systems). If the isoscler on the plane of the primary section intersects the solubility isotherm at a specified temperature, then the point of intersection will represent the compositions of the quaternary α solid solution which are completely saturated at the temperature specified and have the specified microhardness. Knowing the microhardness of the crystals of the quaternary α solid solution in an alloy situated on the plane of the primary section in the neighboring two-phase region, we may find the composition of this solid solution by reference to the point of intersection of the corresponding isoscler with the solubility isotherm, and this will enable us to determine the direction of the conode unambiguously.

The principal constructions required for finding the conodes in the two-phase $\alpha + \beta$ region on the plane of the primary section of the tetrahedron A−B−C−D at a specified temperature are indicated in Fig. 78. The quaternary alloys of this section contain a constant amount of component D; hence in the A−B−C−D tetrahedron the plane of the section passes parallel to the face of the ternary A−B−C system. To left and right of the triangular plane of the primary section appear the curves relating the microhardness of the ternary α solid solutions to the composition of the alloys lying in the corresponding sections of the ternary A−B−D and A−C−D systems. The solubility isotherm (i.e., the boundary of the region corresponding to the quaternary α solid solution based on component A) is represented by the curve an_1nmb, while the specified quaternary alloys in the two-phase region $\alpha + \beta$ are respectively represented by the points M and N.

For plotting the conodes in the two-phase $\alpha + \beta$ region one must proceed in the following way. First the microhardness is determined for crystals of the quaternary α solid solution, for example, in an alloy corresponding to the point M(H_μ), and then the isoscler corresponding to this microhardness value is plotted on the plane of the primary section. The point m of the intersection of the isoscler xy with the solubility isotherm an_1nmb indicates the composition of the quaternary α solid solution of limiting concentration in the alloy of composition M. By joining this point to the figurative point of the alloy M with the section mM we obtain the fixed position of the conode in the two-phase $\alpha + \beta$ region. It is clear that all alloys situated on the straight line mM will contain in their structure crystals of the quaternary α solid solution of the composition of the point M, the microhardness of which equals H_μ.

If the isoscler on the plane of the primary section intersects the solubility isotherm in two points, as for example the isoscler x_1y_1, then we have to decide which of the two straight lines passing through the figurative point of alloy N, n_1N or nN, is the conode. For this purpose we determine the microhardness of the quaternary α solid solution in alloys of both sections, n_1N and nN; the conode is the one for which the alloys contain crystals of the α solid solution with the same microhardness value.

As in the case of ternary systems (see Chapter 6), in solving the question as to which of the sections coincides with the conodal, attention must be directed toward conodes with unambiguously determined positions (for example, the conode mM of Fig. 78), since in general conodes should not intersect each other.* In addition, to this, the choice of conode is considerably simplified if we know the composition of the second equilibrium phase coexisting with the α crystals (in our example the β phase), and hence the general direction of the conodes in the two-phase region.

The accuracy of plotting the conodes in the two-phase $\alpha + \beta$ region is determined by the position of the alloys of types M and N relative to the boundaries of solubility. The accuracy should be high if the alloys in the two-phase $\alpha + \beta$ region are disposed at a considerable distance from the solubility limit and if the angle of intersection between the conode and the solubility isotherm is close to a right angle. If the conode intersects the solubility isotherm at an acute angle, the number of alloys used for studying the microhardness of the quaternary α solid solution in the two-phase region must be increased.

Thus if we know the relation between the microhardness of the quaternary α solid solution and the composition of the alloys and the position of the solubility limit on the plane of the primary section coinciding with the direction of the conodes in the two-phase $\alpha + \beta$ region, we may easily construct the conodes in this region.

In order to secure an experimental confirmation of the possibility of plotting conodes in the two-phase volumes of quaternary phase diagrams by the microhardness method, the primary section $Nb - W_2Zr - Mo_2Zr$ passing through the figurative points of niobium and the binary chemical compounds W_2Zr and Mo_2Zr in the $Nb - W - Mo - Zr$ tetrahedron was chosen. It was found in [219, 221] that the $Nb - W_2Zr$ and $Nb - Mo_2Zr$ sections of the ternary systems $Nb - W - Zr$ and $Nb - Mo - Zr$ practically coincided with the conodes in the two-phase regions $\alpha + W_2Zr$ and $\alpha + Mo_2Zr$; in addition to this, the chemical compounds W_2Zr and Mo_2Zr formed continuous series of solid solutions $(W, Mo)_2Zr$ in the ternary system $W - Mo - Zr$.

The alloys studied by A. M. Zakharov contained all together up to 30 wt.% of W, Mo, and Zr and were disposed on the following sections: $Nb - W_2Zr$ (W:Zr = 4:1), $Nb - Mo_2Zr$ (Mo:Zr = 2.1:1), I (W:Mo:Zr $\approx$ 2:1:1) and II (W:Mo:Zr = 1:1.6:1). The method of preparation and the subsequent treatment of the alloys for plotting the boundaries of the region of the quaternary niobium α solid solution on the plane of the $Nb:W_2Zr - Mo_2Zr$ section were described in [219]. The microsections of the alloys were prepared with coarse and fine emery paper and then polished on cloth with a suspension of Cr_2O_3 in water and etched in a mixture of concentrated acids HF and HNO_3 (1:3) to remove the surface hardened layer. The microhardness of the crystals of the quaternary niobium-base α solid solution was determined in a PMT-3 under a load of 20 g, the loading time being 2 to 3 sec and the time under load 7 to 8 sec.

---

*In a particular case the conodes in a two-phase region may "converge" to a single point corresponding to the composition of the second equilibrium phase.

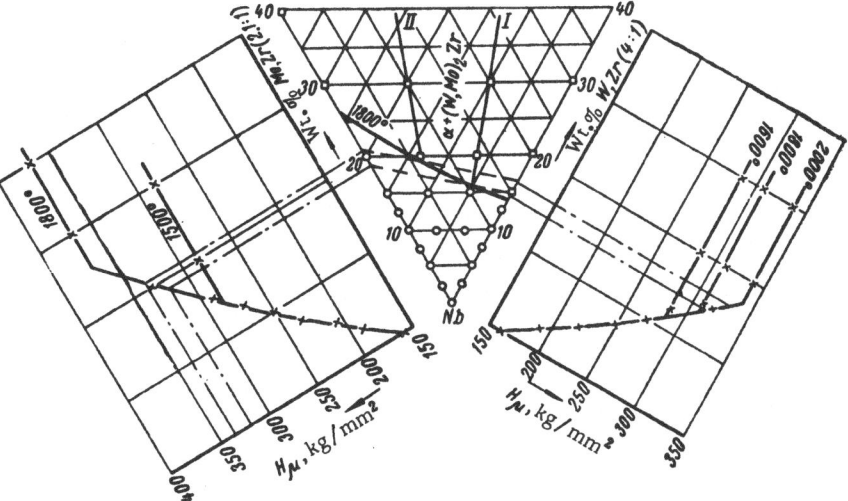

Fig. 79. Construction of conodes in the two-phase $\alpha + (W, Mo)_2Zr$ region of the primary $Nb-W_2Zr-Mo_2Zr$ section of the $Nb-W-Mo-Zr$ tetrahedron at 1800°C by the microhardness method.

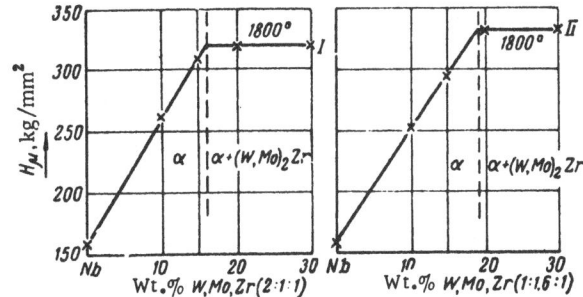

Fig. 80. Microhardness isotherms of the niobium-base quaternary α solid solution for alloys of the secondary sections I(W:Mo:Zr $\approx$ 2:1:1) and II(W:Mo:Zr $\approx$ 1:1.6:1) of the $Nb-W-Mo-Zr$ tetrahedron.

The position of the test alloys on the plane of the primary $Nb-W_2Zr-Mo_2Zr$ section and the graphical construction of the conodes in the two-phase $\alpha + (W, Mo)_2Zr$ region of this section at 1800°C are shown in Fig. 79. To the left and right of the isotherm of the primary section are the curves representing the microhardness of the α solid solution as a function of the composition of the alloys in the $Nb-W_2Zr$ and $Nb-Mo_2Zr$ sections of the ternary $Nb-W-Zr$ and $Nb-Mo-Zr$ systems. The bends on these curves correspond to the limiting concentration of the ternary niobium-base α solid solution at various temperatures, while the horizontal aspect of the microhardness curves beyond the bend indicates that the sections $Nb-W_2Zr$ and $Nb-Mo_2Zr$ in the two-phase regions $\alpha + W_2Zr$ and $\alpha + Mo_2Zr$ coincide with the directions of the conodes.

The same form applies to the microhardness isotherms of the quaternary niobium α solid solution at 1800°C constructed in Fig. 80 for alloys of sections I (W:Mo:Zr $\approx$ 2:1:1) and II (W:Mo:Zr $\approx$ 1:1.6:1). The microhardness of the crystals of the α solid solution in the quaternary alloys lying within section I in the two-phase region α + (W, Mo)$_2$Zr is 312 kg/mm^2, while in the analogous alloys of section II it is 335 kg/mm^2. The sharp breaks on the curves relating the microhardness of the quaternary niobium-base α solid solution to the composition of the alloys on sections I and II enable us to determine the boundary of the region corresponding to this solid solution at 1800°C; this agrees closely with the results of microscope analysis. The horizontal aspect of the microhardness curves beyond the bend also confirms the fact that the test sections I and II in the two-phase volume α + (W, Mo)$_2$Zr of the Nb−W−Mo−Zr tetrahedron practically coincide with the conodes.

It should be noted, in conclusion, that the microhardness method is also applicable for plotting the conodes in the two-phase volumes of quaternary phase diagrams in those cases in which the microhardness of different solid solutions (different with respect to the number of components) varies nonlinearly with their composition and disobeys the additivity law. The main difference in this case relative to the case considered earlier lies in the fact that the composition dependence of the microhardness of the ternary and quaternary solid solutions is represented by curved lines, so that for plotting the microhardness isosclers of the quaternary α solid solution in the plane of the primary section more alloys are required.

Finally, for an experimental construction of conodes in the two-phase volumes of quaternary phase diagrams we may of course use other methods of physicochemical analysis, for example, measurements of the crystal lattice constant, although this requires a great deal more time.

CHAPTER 8

USE OF THE MICROHARDNESS METHOD FOR STUDYING LIQUATION PHENOMENA IN ALLOYS

It is well known that in the crystallization of alloys under nonequilibrium conditions dendritic liquation develops, and that as a result of this the true picture of the equilibrium in the system is distorted so severely that the interaction between the components may appear completely different from that actually taking place, as described by the equilibrium phase diagram.

It was noted in Chapter 3 that when using the microhardness method for purposes of physicochemical analysis one of the main factors influencing the results of the investigation was chemical microinhomogeneity in the solid-solution crystals arising as a result of dendritic liquation so as to be able to allow for these when studying the corresponding phase equilibria.

The use of the microhardness method has played a major part in studying these laws.

Let us consider the process of dendritic liquation revealed in various experimental investigations, mainly carried out by the microhardness method.

Essence of the Phenomenon of Dendritic Liquation and Indicators of the Chemical Microinhomogeneity of Alloys

It is well known that, during the growth of a crystallite of a solid solution enriched with respect to one of the components, the layer of melt adjacent to the surface of phase separation is enriched with respect to the other component. The concentration gradient engenders a process of equalizing diffusion in the liquid phase.

Simultaneously with this process, a new layer of solid solution with a concentration differing by a specific amount from the concentration of the liquid phase is formed in the layer of liquid immediately adjacent to the crystallite. This process, associated with the redistribution of atoms of different kinds, may be arbitrarily called "separating diffusion." Separating diffusion is directed toward the creation of an equilibrium difference of concentration, which is determined by the difference between the liquidus and solidus in a horizontal direction.

Finally, the concentration gradient in the crystallites produces "equalizing diffusion" in the solid phase.

Clearly the character and degree of chemical microinhomogeneity arising in the course of nonequilibrium crystallization depends on the completeness with which these three elementary diffusion processes are executed. The essence of the dendritic liquation process lies

131

in the generation and development of this inhomogeneity during the period of crystallization.

The foregoing subdivision of the processes of atomic movements in the liquid and solid states is to a certain extent arbitrary.* However, it enables us to analyze the development of dendritic liquation, since the degree of completeness with which each of the three processes is executed differs in different ranges of cooling rates.

In view of the fact that dendritic liquation leads to the formation of a concentration gradient within the primary crystals of the solid solution, and also to the appearance, under certain conditions, of a second structural constituent (for example, a eutectic), it was suggested in [222] that two indicators or indexes of chemical microinhomogeneity of liquation origin should be distinguished.

The difference between the concentration of the component in the second structural constituent and its concentration in the crystallization center of the primary crystals was called the degree of general (or total) liquation microinhomogeneity. The difference between the concentration of the component at the periphery and in the center of the primary crystals was called the degree of intracrystallite liquation. We must accordingly not identify the concept of intracrystallite liquation with dendritic liquation, by which we mean the development of chemical microinhomogeneity. In the future, when analyzing investigations relating to dendritic liquation in alloys we shall use the concept corresponding to each specific case.

Starting from certain conditions (suppression of equalizing diffusion in the solid phase and unlimited diffusion in the liquid phase), it was shown in [194, 223-225] that, independently of the original concentration of the solid solution, the solidification of the lattice ended at the crystallization temperature of the more easily melting component, or at the minimum of the solidus (if any), or at the temperature of eutectic or peritectic crystallization. Under these conditions, the following expression was obtained for determining the amount of solid and liquid phases and the distribution of the second component at any moment in the crystallization process:

$$C = kC_0 (1 - g)^{k-1}, \tag{1}$$

where C is the amount of the second component in the solid phase at a particular instant of crystallization; g is the mass of the proportion of alloy which has crystallized at the particular instant; C_0 is the initial composition of the melt; k is the distribution coefficient, constituting the ratio of the concentration of the second component in the solid phase to its concentration in the melt.

Thus, in addition to the ratio of the diffusion velocities in the solid and liquid phases and the crystallization velocity, the development of dendritic liquation is also affected by the distribution coefficient.

For otherwise equal conditions, dendritic liquation arises the more easily, the greater the difference between the compositions of the solid and liquid phases in equilibrium at the particular temperature.

Study of Chemical Microinhomogeneity in Alloys by the Microhardness Method

The use of the microhardness method for determining the indexes of chemical microinhomogeneity in alloys is based on the existence of a strictly specific relationship between the microhardness and the concentration of the solid solution.

---

* This subdivision was proposed in [222].

TABLE 29. Zinc Distribution (wt.%) in the Solid-Solution Crystals of Aluminum−Zinc Alloys of Various Compositions (C_1 = Concentration at Periphery, C_2 in the Center of the Grain)

Average composition of alloy (wt. %)	Autoradiography results			Microhardness results		
	c_1	c_2	c_1-c_2	c_1	c_2	c_1-c_2
2.7	3.20	0.92	2.28	4.00	1.00	3.00
6.6	9.20	2.10	7.10	9.00	4.40	4.60
10.4	32.0	8.60	23.40	21.0	10.0	11.0

The microhardness method was first used for a qualitative estimate of chemical microinhomogeneity within a single crystal grain of a metal in [226]. The microhardness distribution was studied over individual crystal grains of copper samples in the cast state and α brass in the extruded state under a load of 2 g. Impressions were made in parallel series at equal intervals. As a result of the investigation it was established that the microhardness on the axes of the dendrites differed from that in the spaces between them. The microhardness measured along the axes of the dendrites remained constant.

An important conclusion drawn in [226] was that the microhardness method could be employed in order to study the character and degree of inhomogeneity in the structure of the grains in various metals and alloys.

The microhardness method was employed in [227] to study the grain inhomogeneity of aluminum and lead.

D. A. Petrov and L. A. Raikovskaya [228, 229] used the microhardness method to study the degree of liquation microinhomogeneity as a function of the composition of the alloys in the bismuth−antimony and aluminum−magnesium systems. In order to obtain a more coarsegrained structure the alloys were cooled with the furnace. As a result of the investigations it was established that the relation between the microhardness measured in the centers of the grains and at their periphery and the composition of the alloys in the systems under consideration was in fundamental agreement with theoretical predictions.

Any deviations (in the authors' opinion) were due to the fact that equalizing diffusion in the solid phase had not been entirely suppressed during the cooling of the alloys with the furnace.

Analogous investigations into aluminum−copper alloys were carried out in [223].

A quantitative determination of the degree of intracrystallite liquation in an aluminum alloy containing 10% Cu was carried out in [230]. A detailed quantitative study of the degree of intracrystallite liquation in alloys of the Duralumin type was made by Brenner and Kostron [231]. In order to determine the copper concentration at any point of the microsection by measuring the microhardness, the corresponding calibration graph was first plotted. Alloys cast in chill molds were studied as well as others obtained by continuous casting. The microhardness was measured with a load of 4.5 g.

These investigations established not only the microinhomogeneity within the bounds of a single grain but also the different degree of its development in different zones of bars cast by the two methods indicated. In addition to this, the change in the degree of microinhomogeneity in the course of annealing was studied and the time required to homogenize the alloys accordingly determined.

The study of intracrystallite liquation carried out by Jaffe and Bever [232] using a combination of microhardness measurements and quantitative autoradiography is particularly in-

TABLE 30. Scatter in Microhardness Values

System	Proportion of second component, %	Scatter in microhardness values (kg/mm^2) for alloys in the following states			System	Proportion of second component, %	Scatter in microhardness values (kg/mm^2) for alloys in the following states		
		cast	quenched	annealed			cast	quenched	annealed
Zn→Al	0.15	20	15	15	Zn—Pb	0.5	3	4	2
	0.25	25	20	20		1.0	5	5	4
	0.5	30	25	20		2.0	6	5	4
	1.5	30	25	25		3.0	6	5	4
	5.0	30	25	20		20.0	6	4	4
	5.5	35	—	—		38.0	4	4	2
Zn—Cu	0.2	20	15	5	Sn—Zn	0.5	6	4	4
	1.0	25	25	20		1.0	7	4	4
	2.0	25	25	20		2.0	7	6	4
	3.0	35	35	30		3.0	7	6	4
Cu—Ni	10	30	—	20	Sn—Sb	0.5	5	4	4
	30	50	—	40		1.0	5	4	4
	50	60	—	50		2.0	7	5	4
	70	40	—	—		3.0	7	5	4
	90	40	—	—		9.0	8	5	5
						10.0	8	5	5

teresting. Aluminum alloys containing 2.7, 6.6, and 10.4 wt.% Zn were examined. The microhardness was measured with a 50-g load using a Knoop pyramid. The results obtained on determining the degree of liquation inhomogeneity by reference to these two methods are presented in Table 29.

It follows from this table that the results obtained by the microhardness and quantitative-autoradiographic methods agree quite closely with each other. The discrepancies observed, particularly in the case of the alloy containing 10.4 wt.% zinc, were explained by the fact that the zinc concentrations were not determined in exactly the same parts of the sample. In addition to this, the differences may be due to the fact that grains intersected by the plane of the microsection in a manner differing from that associated with the autoradiographic method were studied by the microhardness technique.

The results obtained in [232] indicate that the use of microhardness measurements for studying the degree of microinhomogeneity in alloys is quite reliable.

I. N. Golikov [233] used the microhardness method to secure a quantitative estimate of the degree of intracrystallite liquation in steel. The microhardness was measured under a load of 100 g. Some 200 types of steel were studied.

The results indicate a high sensitivity of the microhardness method, sufficient to detect differences of 0.03 to 0.5 wt.% in carbon content.

A. A. Presnyakov and N. S. Sakharova [234] used the microhardness method for studying the degree of intracrystallite liquation in alloys of various compositions based on zinc and tin, and also in copper—nickel alloys.

The microhardness was determined with loads of 20 g for copper—nickel, 10 g for zinc, and 5 g for tin alloys. The measurements were made immediately after etching. Some 200 impressions were made on each microsection. Frequency curves were plotted from the results and subsequently the maximum, minimum, and most-frequently-appearing microhardness values were employed.

As a measure of liquation inhomogeneity the scatter in the microhardness values (i.e., $H_{\mu max} - H_{\mu min}$) was taken. The results appear in Table 30.

We see from Table 30 that the inhomogeneity increases with increasing degree of alloying up to a certain limit, after which it remains constant or increases very slightly.

When studying the intracrystallite liquation in two-phase alloys, the results of the microhardness measurements are affected by microinhomogeneity in the solid-solution crystals; this was noted in [234] when analyzing the concentration dependence of the minimum, maximum, and most-frequent microhardness in the systems under consideration. Comparatively brief annealing failed to eliminate the microinhomogeneity completely (Table 30).

The microhardness method was employed in [133-135] for checking the point at which germanium- and silicon-base alloys reached the equilibrium state in the course of a homogenizing anneal. The microhardness was measured in the centers of the crystallites and at their periphery. Equilibrium was considered to have been reached when the microhardness measured in different parts of the crystallites had become equal or nearly so, and also when the absolute microhardness remained constant on further annealing. This indicated that the consequences of dendritic liquation had been finally eliminated.

The details of this process were studied in [235] for the case of aluminum alloys.

The study of this process is of importance not only for analyzing phase diagrams but also in the technology of alloy production, in connection with the establishment of optimum conditions for the homogenization and prequench heating of alloys crystallizing in various ways. In studying these phenomena, the microhardness method plays a predominant part [235], and for this reason we shall pay a little more detailed attention to the results of the investigation in question.

In the opinion of Zolotorevskii et al., the obtaining of quantitative data regarding changes in structural characteristics* during the homogenization of an alloy of specified composition with various original structures facilitates the subsequent choice of diffusion-annealing conditions by direct reference to the structure of the cast alloy, without any prolonged, laborious experiments with the plotting of the properties as functions of time and temperatures.

The effect of prequench heating conditions on certain structural characteristics, particularly liquation microinhomogeneity, in binary alloys of aluminum with 3% Cu, 4.5% Cu, and 10.6% Mg was studied in [235]. The ordinates of the alloys in question intersect the limited-solubility lines in the phase diagrams of Al−Cu and Al−Mg and lie between the ordinate of pure aluminum and the point of limiting solubility at the eutectic temperature. The alloys were prepared from materials of the following purity: 99.99% Al, 99.95% Cu, and 99.92% Mg. Bars of each alloy were cast from a single earlier-prepared melt. The structure of the cast alloys was varied by using different cooling rates on crystallization of the alloys after casting into steels molds, heated to various temperatures, having an internal cylindrical cavity 200 mm long and 20 mm in diameter. The average cooling rates in the crystallization range, calculated from the thermal curves recorded on an electronic potentiometer, varied from 2 to 1000 deg · min^{-1} (the thermocouple junction was placed in the center of the casting). Immediately after the end of crystallization the castings were taken from the mold and cooled in air or water.

After cutting the castings into samples, these were heated for various periods in a salt-peter bath (Al−Cu alloys) or an air furnace (Al−10.6% Mg) at various temperatures and then cooled in water. The Al−Mg alloy was homogenized at 400, 420, and 440°C from 15 min to 20 h,

---

* By structural characteristics, Zolotorevskii et al. [235] mean the quantity and dimensions of the precipitates of excess structural constituents, the distribution of these constituents, the distribution of the alloying elements in the main body of the crystallites, the degree of intracrystallite liquation, the shape and size of the crystallites, and so forth.

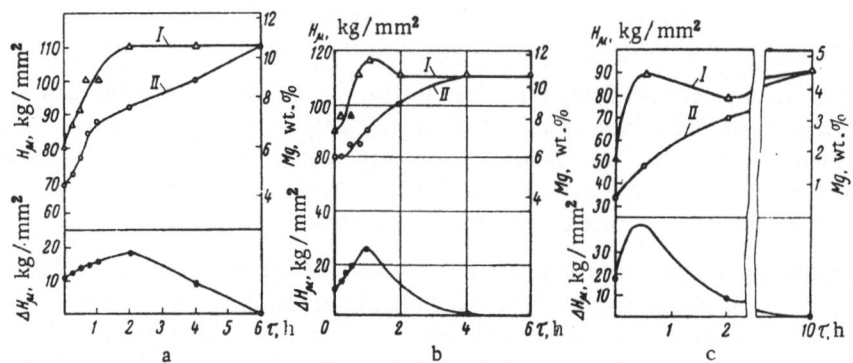

Fig. 81. Microhardness (H_μ) at the periphery (I) and center (II) of
the dendritic cells and degree of intracrystallite inhomogeneity
(ΔH_μ) as functions of prequench heating time: a, b) Al $-$ 10.6% Mg,
homogenization temperature 400°C [a) $v_{cool} = 19$; b) $v_{cool} = 150$
deg/min]; c) Al $-$ 4.5% Cu, homogenization temperature 500°C [v_{cool}
during crystallization 70 deg/min].

the alloy with 4.5% Cu at 500°C from 0.5 to 10 h and at 540°C from 10 min to 10 h, and the alloy
with 3% Cu at 530°C from 10 to 60 min.

The intercrystallite liquation, the amount of nonequilibrium eutectic, and the character
of the microstructure were studied for cast samples homogenized under different conditions.

The volume content of the eutectic was determined by linear metallographic analysis.
The degree of intracrystallite liquation was estimated in all the alloys by the microhardness
technique. For a qualitative estimate of the character of the copper distribution over the cross
section of the crystallites and the degree of homogenization in the Al $-$ Cu alloys, x-ray shadow
microscopy was employed.

In order to determine the degree of intracrystallite liquation by the microhardness method,
20 dendritic cells were examined in each sample. The microhardness was measured with a load
of 5 g for the Al $-$ Cu and 10 g for the Al $-$ Mg alloys, roughly in the geometrical center of the
dendritic cells of the primary solid solution, and also at their periphery, in the immediate vicinity of the inclusions of nonequilibrium eutectic, which gradually dissolved as the period of
homogenization increased. When the eutectic had completely vanished, the degree of intracrystallite inhomogeneity in the aluminum $-$ copper alloys was estimated from the difference
between the microhardness at the boundaries of the micrograins and in regions situated at a
distance roughly equal to half the diameter of a dendritic cell of the cast alloy away from the
micrograin boundaries, which were clearly visible in the microsection. In the Al $-$ Mg alloy the
center and periphery of the dendritic cells after dissolution of the β phase were located quite
easily by virtue of the fact that, after electropolishing and etching, the countours surrounding
the dendritic cell were clearly visible on the microsection.

The microhardness values obtained at the center and periphery of the dendritic cells were
converted into copper or magnesium concentration by using the corresponding calibration curves
relating the microhardness of the amounts of copper or magnesium in the aluminum-base solution, plotted on the basis of completely homogenized standards. The difference between the
microhardness or the concentration of the alloying element at the center and periphery of the
micrograins served as a measure of the degree of intracrystallite liquation. In addition to determining the degree of inhomogeneity, curves representing the distribution of the alloying
element over the cross section of the cells were plotted by the microhardness method for samples with fairly large dendrites.

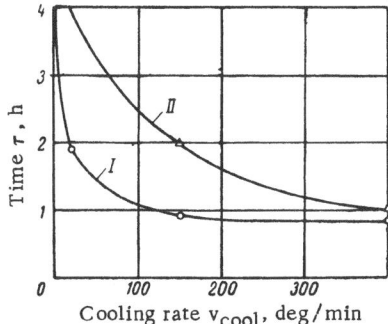

Fig. 82. Time (τ) for complete dissolution of the eutectic phase (I) and time required to reach the condition $\Delta H_\mu = 0$ (II) as functions of cooling rate during crystallization (v_{cool}) for Al−10.6% Mg (homogenization temperature 440°C).

By way of example, Fig. 81 shows some typical curves for the alloys in question, relating the change in composition at the center and periphery of the dendritic cells and the degree of intercrystallite liquation to the prequench heating period.

On increasing the heating time at any of the temperatures employed, the microhardness, and hence the concentration of the alloying element, in the center of the dendritic cells increases continuously up to the point of complete homogenization (II in Fig. 81). This rate is usually particularly sharp in the first minutes and hours of heating. The copper and magnesium concentration at the periphery of the dendritic cells also rises at first and then either remains constant (Fig. 81a) or falls slightly (I in Fig. 81b and c). The degree of intracrystallite liquation (ΔH_μ or ΔC) is in all cases represented by a curve with a maximum.

The time required to remove the intracrystallite liquation completely is different for samples crystallizing at different velocities. Figure 82 (curve II) relates to an Al−10.6% Mg alloy homogenized at 440°C, and illustrates the laws common for all the alloys and temperatures studied, according to which the time to reach the condition $\Delta H_\mu = 0$ diminishes as the crystallization of the casting accelerates. This time also naturally diminishes as the homogenization temperature rises.

The rise in the concentration of the alloying element in the middle of the dendritic cells on increasing the heating time is quite natural; it is due to the equalizing diffusion in the solid solution during the homogenizing anneal. The increase in homogenization may be explained by the change in the form of the curves giving the distribution of the alloying element over the cross section of the micrograins near their boundaries. Prior to the complete dissolution of the eutectic constituent, the true concentration of the second component at the boundaries of the solid-solution micrograins should remain constant and approximately equal to the concentration of the point of limiting solubility at the eutectic temperature. In measuring the microhardness, one establishes the average hardness of a layer having a thickness equal to the length of the diagonal impression (in the present case 10 to 12 μ). In Fig. 83, which shows some typical averaged magnesium-distribution curves for the Al−10.6% Mg alloy (Fig. 83a) and copper-distribution curves for the Al−4.5% Cu alloy (Fig. 83b), near the boundaries of the micrograins at various stages of homogenization, this layer is shaded. We see that at first an increase in heating time leads to a widening of the boundary layer enriched with the alloying element, causing a rise in the experimentally determined copper and magnesium concentration at the periphery of the dendritic cells (Fig. 81).

The widening of the enriched boundary layer in the first minutes of homogenization is due to the intensive development of diffusion within the solid solution (distribution curves I for the cast samples in Fig. 83), and hence (and this is very important) the distance from the eutectic, which contains a relatively large amount of the second component and constitutes its main "supplier" for the solid solution, is smaller. The increase in the width of the copper-enriched peripheral regions of the micrograins after brief homogenization is also clearly visible in the x-ray micrographs.

At a certain stage of homogenization, at which the amount of nonequilibrium eutectic has diminished, while the composition of the central part of the micrograins is approaching the aver-

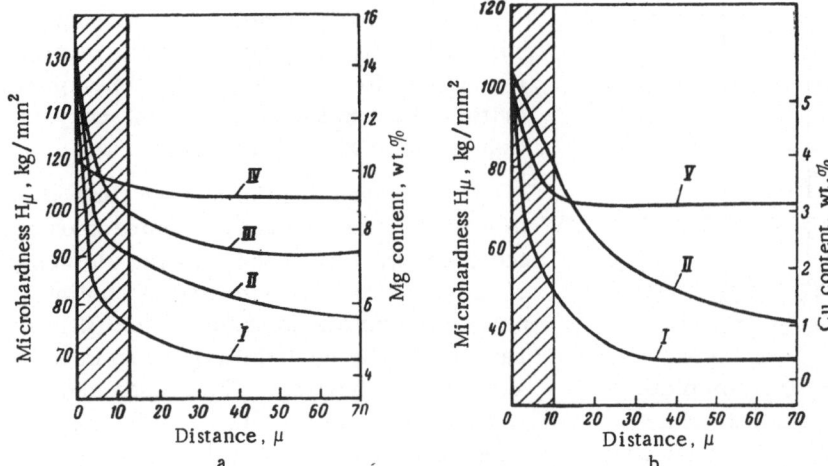

Fig. 83. Microhardness (H_μ) distribution curves over the cross section of the dendritic cells in Al−10.6% Mg (a) and Al−4.5% Cu (b) cooled during crystallization at 19 and 70 deg/min and homogenized at 400 and 500°C respectively: I) cast; II) 0.5 h; III) 1 h; IV) 4 h; V) 2 h.

age composition of the alloy, the enriched boundary layer contracts once again (III in Fig. 83b) and the microhardness at the periphery of the dendritic cells may fall (Fig. 81c), or at least stop growing. The growth in microhardness at the periphery of the dendritic cells may also cease owing to the complete dissolution of the eutectic (IV in Fig. 83 and I in Fig. 81).

For a comparatively brief heating period (first 1 or 2 h) the rise in the microhardness of the periphery associated with the widening of the enriched boundary layer is more intensive than in the central parts of the grain. This leads to an initial rise in the experimentally determined degree of intracrystallite liquation as homogenization proceeds (see Fig. 81). The true degree of concentration microinhomogeneity should fall continuously as heating time increases. The maximum on the curves relating the degree of intracrystallite inhomogeneity to the homogenization time (Fig. 81), associated with the insufficiently localized nature of the microhardness method, should also appear to a greater or lesser extent on using any other of the many modern methods of studying local variations in concentration.

In the initial period of homogenization of the alloys under consideration, the decomposition of the solid solutions proceeds in parallel with the diffusion equalization of concentrations within the crystallites in individual microvolumes. The inevitability of this process arises from the fact that, right up to the complete dissolution of the nonequilibrium eutectic, in the peripheral parts of the micrograins of the solid solution there are layers with concentrations lying in the two-phase region of the phase diagram at the homogenization temperature. For example, in the Al−10.6% Mg alloy, for a heating temperature of 400°C, there are layers of solid solution with a content of over 12% Mg in the two-phase ($\alpha + \beta$) region, at 420°C 13% Mg, at 440°C 14.5% Mg, and so on. On comparing Fig. 83a with the Al−Mg phase diagram it is easy to see that, the lower the homogenization temperature and the wider the boundary layer enriched with the alloying element, the greater volume of the peripheral regions of solid solution should be subject to decomposition. This decomposition was observed under the microscope, together with the attendant formation of secondary precipitates at the initial stages of homogenization (at 400 and 420°C) of a slowly cooled ($v_{cool} = 2.9$ and 19 deg/min) Al−10.6% Mg alloy. At the maximum homogenization temperature of 440°C no decomposition was observed in these samples.

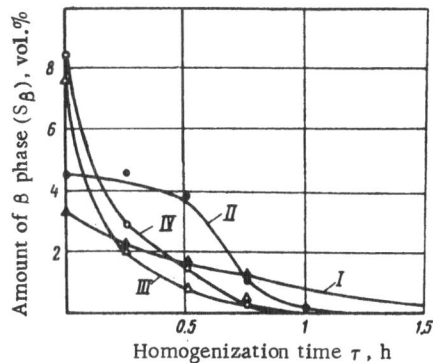

Fig. 84. Amount of eutectic β phase (S_β) as a function of the time of homogenization (τ) at 400°C for Al−10.6% Mg for v_{cool} values (deg/min) of: I) 2.9; II) 19; III) 150; IV) 400.

In the Al−Cu alloys no secondary precipitates of $CuAl_2$ could be found in any of the test samples under the optical microscope. This may be associated with the fineness of the decomposition products and also with the very small width of the layers in which it took place. The microheterogenization of the structure at the periphery of the dendritic cells limits the possibility of using the microhardness method for making a fairly accurate determination of concentration of the solid solution under these conditions. Clearly we are only free from appreciable error in neglecting the decomposition of the boundary layers of the α solid solution at the maximum homogenization temperatures, i.e., 440°C for Al−Mg and 540°C for Al−Cu. For lower homogenization temperatures, the error in determining the concentration at the periphery of the micrograins when using the microhardness method will depend on the width of the layer in which the decomposition has taken place and the degree of coagulation of the decomposition products. The appearance of these products, in turn, is due to the original cast structure and the homogenization conditions.

For a low homogenization temperature (400°C) of slowly cooled Al−Mg samples, decomposition of the supersaturated solid solution in the interaxial parts of the dendrites is also observed for a certain time (2 to 11 h) after the vanishing of the β phase. Only after 20 h of homogenization, when the magnesium concentration fell below 12% in every part of the solid solution, did the decomposition products vanish from these samples. The secondary β phase precipitates also appeared after quite prolonged (6 to 8 h) low-temperature homogenization of samples of this alloy crystallized very rapidly. The regions containing decomposition products here occupied an area of several micrograins. It is possible that this phenomenon is a result of the inheritance of zonal liquation arising during the crystallization of the casting.

By way of example, Fig. 84 shows the amount of the eutectic constituent as a function of the time of homogenization at 400°C for an Al−10.6% Mg alloy. We see that the rate of dissolution of the nonequilibrium eutectic is the greater, the more rapidly the casting was crystallized, and hence the smaller the size of the original dendritic cell in the actual eutectic precipitates. An analogous relationship was found in alloys containing 3 and 4.5% Cu. It is an interesting fact that the time which elapses before the vanishing of the eutectic in the Al−Mg alloy diminishes as the cooling rate rises (curve I in Fig. 84), even despite the fact that the quantity of eutectic in the original cast samples was the greater, the more rapidly these were solidified (see original points on the vertical axis of Fig. 84). The curves of Fig. 84 quantitatively illustrate the experimentally well-known fact that homogenization takes place more rapidly as the cast structure becomes finer (see, for example, [236, 237]).

Thus it was convincingly shown in [235] that the microhardness method could be used (in conjunction with careful metallographic analysis) for studying the course of the homogenization of alloys crystallizing under nonequilibrium conditions. The changes in microhardness thus observed at various stages of the process offer the possibility of not only rigorously solving the original problem associated with the progress of homogenization (namely, that of establishing the point of reaching the equilibrium state), but also drawing some important conclusions regarding the special characteristics of the diffusive displacement of the alloying-component atoms in the intermediate stages of the annealing of nonequilibrium alloys at various temperatures.

On the basis of the results obtained in [235] as well as [133-135], we may recommend the microhardness method as a fundamental procedure for monitoring the development of the equilibrium state, both when studying phase diagrams and when developing a technology for the homogenization of commercial alloys.

Dependence of the Indicators of Chemical Microinhomogeneity in Alloys on the Rate of Cooling during Crystallization

The chemical microinhomogeneity of an alloy arising on crystallization under nonequilibrium conditions depends considerable on the cooling rate.

Starting from very general considerations, A. A. Bochvar noted that chemical microinhomogeneity should be absent from alloys crystallized at very low and very high cooling rates, and that it should reach its maximum development at certain intermediate temperatures [238].

In his monograph D. A. Petrov [223] indicated that a great increase in cooling rate might lead to the complete suppression of diffusion in the liquid phase and to a vitreous solidification of the alloy.*

An analogous view was expressed by K. P. Bunin and Ya. N. Malinochka [239].

The idea of the fundamental possibility of the diffusion-free crystallization of alloys taking place at high cooling rates was developed in [240, 241]. It was shown on the basis of thermodynamic analysis that the supercooling of a melt below the temperature at which the values of the isobaric−isothermal potential of the solid and liquid phases were equal might lead to the formation of crystals of the same composition as the liquid phase from which they were formed, and hence to the elimination of the consequences of nonequilibrium crystallization.

The first quantitative measurements of the degree of intracrystallite liquation as a function of the rate of cooling were carried out by an x-ray diffraction method for Cu−Ni and Cu−Au alloys in [242].

The method of quantitative autoradiography was employed in [243, 244] for studying the microinhomogeneity in the distribution of a radioactive isotope of calcium in certain magnesium alloys as a function of the rate of cooling. It was shown experimentally that the degree of microinhomogeneity was a maximum for certain intermediate cooling rates.

The microhardness method was first used for studying the relation between the degree of intracrystallite liquation and the cooling rate by I. I. Novikov and G. A. Korol'kov [245].

The alloys studied included V95 (brand name) and Bi−25% Sb. The cooling rate during crystallization was determined from the cooling curve in each case.

One microhardness measurement under a load of 10 g was made in the center of the dendritic cell and two on the periphery at diametrically opposite points. In this way ten dendritic cells were studied in each microsection (at different points of the latter) and the average values of the microhardness at the center and periphery of the dendritic cells were calculated for a specific cooling rate.

---

* Naturally the question of the vitreous (amorphous) solidification of metallic alloys raises doubts, since no one has so far been able to demonstrate the vitreous solidification of metals. However, vitreous solidification has been observed in a number of semiconducting alloys.

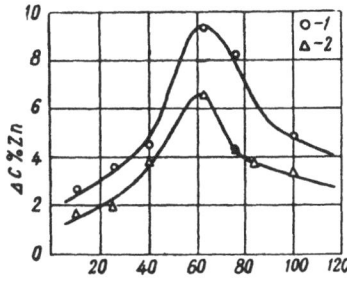

Fig. 85. Zinc concentration gradient (ΔC) in solid-solution crystals of Al-6% Zn as a function of cooling rate. 1) X-ray method; 2) microhardness method.

As a measure of intracrystallite liquation (ε), the percentage ratio of the difference between the microhardness at the periphery (H_μ^p) and center (H_μ^c) of the dendritic cell to the microhardness at the center was taken:

$$\varepsilon = \frac{H_\mu^p - H_\mu^c}{H_\mu^c} \cdot 100\%. \qquad (2)$$

It was accordingly found that the relation between ε and the cooling rate was expressed by means of a curve with a maximum. Such curves were of the same general character as those obtained in [254, 255] by the radiographic method.

Of particular interest is a paper by I. I. Novikov and F. S. Novik [246] in which the microhardness method was used in parallel with x-ray structural analysis in order to study the relation between intracrystallite liquation and cooling rate.

Alloys of aluminum with iron, silicon, and zinc were studied. Figure 85 shows as an example the zinc concentration gradient (ΔC) in the crystallites of an Al-6% Zn alloy as a function of cooling rate.

It follows from the figure that the results obtained by the two methods lead to the same conclusion as to the character of the relationship between intracrystallite liquation and cooling rate.

A detailed study of the relation between liquation microinhomogeneity and cooling rate in aluminum alloys using the microhardness method and quantitative microscope analysis was presented in papers by I. I. Novikov and V. S. Zolotorevskii* [247-250].

The experiments were carried out [247] with aluminum alloys containing 2 and 5% Cu and 6% Mg. The alloys were prepared from aluminum of 99.99% purity, copper 99.95%, and magnesium 99.92%.

In order to achieve different velocities of solidification, the samples were cooled (in graphite$-$chamotte or steel crucibles with different wall thickness) in the furnace, in air, or in cold water. The hot junctions of a Chromel$-$Alumel thermocouple without any protective casing were immersed in the melt, approximately in the center of the sample, at a distance of some 5 mm from the bottom of the crucible. For each sample the cooling curve was recorded on an automatic electronic potentiometer. The average cooling rate on crystallization was calculated as the ratio of the difference between the liquidus and solidus temperatures to the time of crystallization of the alloy (including the time of eutectic crystallization). The solidus temperature was established sharply as an arrest or break on the cooling curves; in all experiments it practically coincided with the eutectic temperature.

The plane of the microsections passed at the level of the hot thermocouple junction. The microsections of the Al$-$Cu alloys were prepared by electropolishing incorporating electrolytic etching. Thus the regions of solid solution enriched with copper were etched out in the form of yellowish regions, which gave a more accurate location of the "crystallization" center of the

---

* V. S. Zolotorevskii, Dissertation, Moscow Institute of Steels and Alloys (1963).

dendritic cell corresponding to the axis of the dendrite (in general this center does not coincide with the geometric center of the cell, depending on the particular point and the particular angle at which the plane of the microsection intersects the branch of the dendrite).

The Al−Mg microsections were prepared by mechanical polishing, using the method described in Chapter 3.

The microhardness was measured in the center and at the periphery of the dendritic cells in the immediate neighborhood of the eutectic inclusions with a load of 5 g. The measurements were made 1 to 2 h after casting the sample. On each microsection 15 to 20 dendritic cells were thus measured and the average microhardness at the center and periphery was calculated for a particular cooling rate. The difference between these values served as a measure of intracrystallite liquation.

By way of example, Fig. 86 shows the microhardness of the center and periphery of a dendritic cell in an Al−2% Cu alloy as a function of the cooling rate. We see that the composition of the center of the cell remains almost constant over a wide range of cooling rates. The slight rise in the microhardness of the center of the cell for very low cooling rates is due to the fact that here equalizing diffusion is able to take place to a certain extent in the solid solution and the center of the cell is enriched with copper.

Comparison of the graphs in Fig. 86a and b shows that the manner in which the degree of intracrystallite liquation depends on the cooling rate (Fig. 86b) almost entirely depends on the way in which the composition at the periphery of the dendritic cell varies with the cooling rate.

The effect of cooling rate on the degree of intracrystallite liquation was studied in [248] with the simultaneous employment of two methods: microhardness and x-ray shadow microscopy.

Aluminum alloys containing 2 and 5% Cu prepared from 99.99% pure aluminum and 99.95% pure copper were studied. The samples were 15 mm in diameter and 20 mm long; they were cast with cooling rates between 0.5 and 1000 deg/min in the crystallization interval. Different rates of solidification were achieved by cooling the castings (in graphite−chamotte or steel crucibles with various wall thicknesses) in the furnace, in air, or in cold water. For each sample a cooling curve was plotted. In order to prevent partial homogenization in the period of cooling below the solidus temperature, the samples were quenched in water immediately after the completion of crystallization.

For the x-ray shadow microscopy a disc 1 mm thick was cut from each sample; this was brought to a thickness of 0.05 to 0.1 mm by bilateral grinding and polishing. Special attention was devoted to uniformity of sample thickness and the absence of pores and scratches [251]. An x-ray microscope of the needle type with a resolving power of about 0.2μ was used [252]. Depending on the rate of cooling the alloys, different x-ray magnifications were used; these were determined by means of preliminary photographs of the samples at natural size. The photographs were taken on AGFA−Laue film, the exposure not exceeding 30 min, so as to obtain a photometric density corresponding to the rectilinear part of the photoemulsion curve. The exposure was monitored by photographing a standard sheet of aluminum foil together with the sample on a separate part of the x-ray photograph.

For quantitative determinations of the copper concentration in various microvolumes of the dendritic cells, calibration curves were plotted between the photometric density and the amount of copper in standard samples. The samples were pieces of aluminum and aluminum alloys containing 0.7, 2.0, 3.5, and 5% Cu, homogenized for 480 h at temperatures above 550°C.

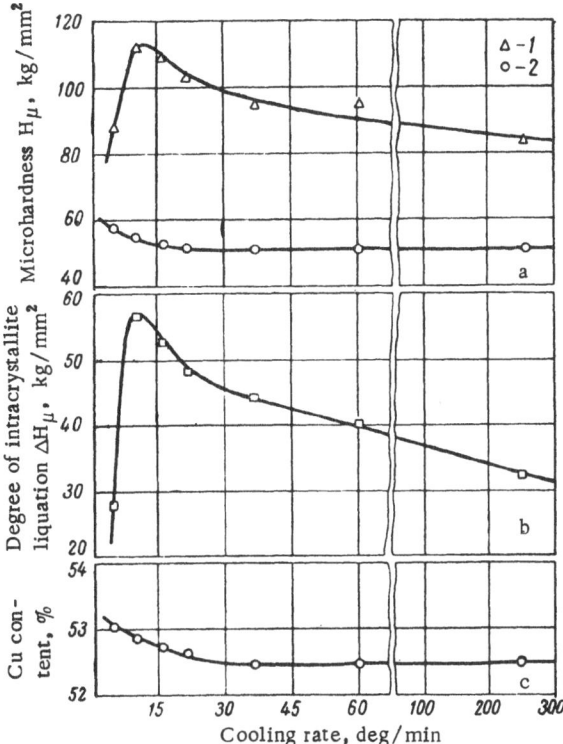

Fig. 86. Microhardness of the center (H_μ^c) and periphery (H_μ^p) of a dendritic cell (a) and degree of intracrystallite liquation ΔH_μ (b) as functions of cooling rate in an Al−2% Cu alloy. 1) Periphery; 2) center.

The thickness of the samples and standards was determined mechanically and also by the x-ray method on a Hilger diffractometer in monochromatic radiation.

The x-ray microscope photographs of the samples and standards were photometered on an MF-4 microphotometer. In determining the copper concentration from the photometric curves corrections were allowed for the differences in the thicknesses and exposures of the samples and standards. The locational precision in determining the copper concentration in the microvolumes was limited by the area of the microphotometer slit; it varied from 5 to 50 μ^2, depending on the magnification of the x-ray micrographs. The error in measuring the copper concentration was 0.2 to 0.5%.

The microsections for measuring the microhardness were electropolished with simultaneous etching for 20 sec at a current density of 0.9 A/cm^2, using a "Diza-Elektropol" (electrolyte composition: 200 ml chloric acid at a density of 1.20, 700 ml ethyl alcohol, and 100 ml glycerin). In order to determine the copper concentration at various points of the micrograin the microhardness was converted by reference to a microhardness/composition calibration curve. In plotting the microhardness distribution curves the impressions were made at 10 to. 20 μ from each other. The standards for plotting the calibration graph were annealed under the same conditions as the standards for x-ray shadow microscopy. The mode of electropolishing and the conditions for measuring the microhardness of the standards and samples were strictly alike. The microhardness was measured with a load of 5 g.

On the basis of the results obtained, copper distribution curves were plotted across the grain. The character of the distribution curves obtained by the microhardness and x-ray shadow-microscopy techniques was the same.

TABLE 31. Average Copper Concentration in the Center
of Aluminum – Copper Micrograms

Alloy	Concentration (wt.%) for cooling rate (deg/min)							
	0.5	7	13	29	52	82	180	410
Al+2% Cu	0.6	0.3	0.5	0.3	0.4	0.6	—	—
Al+5% Cu	3.5	1.6	—	—	1.3	—	1.7	1.4

Table 31 presents the average values of the minimal copper concentrations in the central parts of the dendritic cells, calculated from the photometric curves. In both alloys these values lay a long way from the original concentration of the liquid phase (2 and 5% Cu) and extremely close to the compositions of the solid solution determined by the points of the equilibrium solidus at the temperature of the onset of crystallization.

The copper concentration in the centers of different micrograins of the same sample varied substantially. In the center of individual dendritic cells the copper content was smaller than or greater than the concentration determined by the point of the equilibrium solidus at the temperature of the onset of crystallization. In the opinion of Novikov et al. [248], the excess of the copper concentration in the center of the micrograins over the concentration corresponding to the point of the equilibrium solidus at the temperature of the onset of crystallization was due to several causes.

Firstly, different branches of the dendrites may be formed at different temperatures, and hence the copper concentration in these determined by the points of the equilibrium solidus at the corresponding temperatures should be different. The lower the temperature corresponding to the formation of the dendrite branch, the greater is the copper concentration determined in the center of the micrograin.

Secondly, by analogy with the distribution of an element along a single crystal growing in a directional manner from some wall, the curve representing the distribution of the alloying element across the cross section of the crystallite may take different forms, depending on the relation between the crystallization velocity, the distribution coefficient, and the diffusion coefficient in the liquid phase. There may be a distribution such that the composition of many micrograins not constituting cross sections of the trunk of the dendrite will approach the composition of the original melt. In this way we may also explain the results of [253], according to which the composition at various points of the cross section of primary solid-solution crystals is close to the composition of the original liquid solution over almost the whole interval of crystallization.

Thirdly, the composition of individual microvolumes of the liquid solution may differ for various reasons, and this should lead to the formation in these parts of a solid phase with a composition lying to the right or left of the equilibrium solidus point.

Thus the experiments showed that over a wide range of cooling rates (from several to several hundred deg/min) the composition of the first portions of solid phase in the Al – Cu alloys was determined by the point of the equilibrium solidus curve at the temperature corresponding to the onset of crystallization. These results indicate that separating diffusion proceeds to completion in the liquid phase, while in the solid phase equalizing diffusion is almost entirely suppressed. Only in the alloy containing 5% Cu cooled at 0.5 deg/min is there an appreciable rise in the copper concentration in the center of the micrograins (Table 31), owing to the partial occurrence of equalizing diffusion in the solid phase over the crystallization interval. Data obtained by x-ray shadow microscopy agree with the determinations of the con-

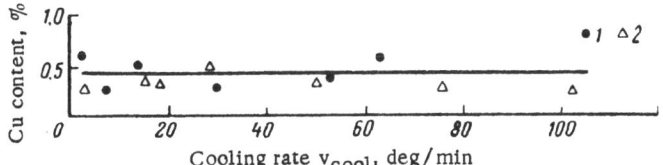

Fig. 87. Composition in the center of Al−2% Cu micrograins as a function of the cooling rate, according to results based on x-ray shadow microscopy (1) and microhardness (2).

centration of the solution in the center of the micrograins based on microhardness measurements and other methods.

Since the rate of crystallization (derivative of the amount of solid phase with respect to the temperature) in the alloys studied is very high near the liquidus temperature so that the greater part of the alloy solidifies at this point, the copper concentration is approximately the same over most of the micrograin section. Over a wide temperature range only a small amount of liquid phase crystallizes, forming a comparatively narrow copper-enriched peripheral zone of the dendritic cells. Over the whole range of cooling rates studied, the width of this zone diminishes with increasing rate of cooling [254]. At the actual boundary of the micrograins of the primary crystals, the copper content of the solid solution may reach the point of limiting solubility (5.65%) if the crystallization ends by the precipitation of a nonequilibrium excess of $CuAl_2$ of eutectic origin, or may fall below this concentration. In the second case there will be no $CuAl_2$ inclusions at the cell boundary.

The following are the average values of copper concentration in the Al−2% Cu alloy at the boundaries of the micrograins, as obtained by reference to the peaks on the photometric curves:

v_{cool}, deg/min	0.5	13	29	52	82	980
Cu, wt.%	2.9	3.5	2.3	3.9	4.1	4.6

The copper concentration at the boundaries of the solid-solution micrograins in all samples of the alloy containing 2% Cu was less than the limiting value; this was due to the low probability of intersecting the rarely-encountered fine isolated $CuAl_2$ inclusions when photometering, and also the low resolving power of the photometric procedure. In the alloy containing 5% Cu, eutectic inclusions form almost continuous veins along the boundaries of the micrograins, and it is thus very hard to make any experimental measurements of the concentration of the solid solution at the boundary, owing to the impossibility of separating it from $CuAl_2$ inclusions.

It was shown by various methods in the papers just mentioned that the relationship between the degree of intracrystallite liquation and the cooling rate was described by a curve with a maximum at an extremely low cooling rate (degrees to tens of degrees per minute). In the investigations in question the degree of microinhomogeneity was taken as the difference or ratio between the maximum and minimum concentrations of the alloying element at the boundary and center of the micrograins respectively. Since the composition of the center of the dendritic cells is almost constant over a wide range of cooling rates, the influence of these cooling rates on the degree of intracrystallite liquation is determined by the manner in which the composition of the solid solution varies at the boundary of the micrograin.

Owing to inadequate resolving power, none of the known methods of estimating intra-crystallite liquation enable us to measure the concentration of the solid solution directly at the actual boundary of the micrograin; they actually yield information regarding the composition of a peripheral layer of greater or lesser thickness. For example, when using the microhard-ness method, the average concentration in the boundary zone is measured. Since the slope of the copper-distribution curves near the boundaries of the micrograins differs for differing cooling rates, the experimentally measured concentration around the boundary of the micro-grain will first rise with increasing cooling rate and then fall. This explains the observed maxi-mum in the degree of intracrystallite liquation at a particular cooling rate. Actually the maxi-mum concentration of the alloying component at the boundary of the dendritic cell of the solid solution should remain constant for a wide range of cooling rates and be determined by the con-centration of the point of limiting solubility if the equilibrium crystallization ends by the pre-cipitation of an eutectic component. Hence the degree of intracrystallite liquation is indepen-dent of the cooling rate over a wide range of the latter, as demonstrated for alloys belonging to a system with a continuous series of solid solutions. The results obtained on determining the composition at the boundaries of the micrograins by x-ray shadow microscopy agree with this conclusion.

The copper concentration in the solid solution around the boundaries of the micrograins was approximately constant over the range of cooling rates usually yielding a maximum in the degree of intracrystallite liquation, since the x-ray micrographs of the alloys cooled at differ-ent rates were taken with different magnifications. Since these magnifications were chosen in such a way that the dimensions of the dendritic cells were roughly the same for all cooling rates, the width of the copper-rich boundary layers of the micrograins in different samples was also the same in the photographs, and even a little enlarged for high cooling rates. Hence the copper concentrations at the boundaries of the micrograins obtained after photometering the x-ray micrographs of alloys cooled at different rates were approximately equal and had a tendency to rise for high cooling rates. If, however, the liquation were studied at a constant magnification (as is frequently the case), the relationship between the copper concentration at the boundaries of the micrograins and the cooling rate would be described by a curve with a maximum, owing to the different widths of the copper-enriched peripheral zones of the micro-grains and the inadequate resolution of the method.

Thus, owing to the constancy of the composition at the center and boundaries of the solid-solution micrograins, the maximum degree of intracrystallite liquation in Al−Cu alloys re-mains constant over a wide range of practically realizable cooling rates. The experimentally determined degree of liquation, related to the cooling rate by way of a curve incorporating a maximum, is only of practical value as a characteristic of the width of the boundary layer en-riched with the alloying component.

The following should be noted in connection with the results just described. Firstly, the conclusion as to the constancy of the degree of intracrystallite liquation (as also the degree of total liquation microinhomogeneity) drawn on the basis of investigations into dendritic liqua-tion in alloys belonging to systems with a continuous series of solid solutions is of considerable significance, since it emphasizes the leading part played by the equilibrium phase diagram in interpreting such a "nonequilibrium" process as the crystallization of alloys at different cool-ing rates.

Secondly, the parallel conduct of experiments based on x-ray shadow microscopy and microhardness measurements clearly reveals the advantages of the microhardness technique.

We have not idly recounted all the details of the experiments based on the two methods carried out in [248]. By comparing these details, we reach the undoubted conclusion that the

TABLE 32. Microhardness of the Center and Periphery of Micrograins of an Al−5% Cu Alloy for Various Cooling Rates

v_{cool}, deg/min	H_μ of center of micrograin, kg/mm²		H_μ of periphery of micrograin, kg/mm²		H_μ av, kg/mm²	H_μ max, kg/mm²
	middle of macrograin H_1	boundary of macrograin H_2	middle of macrograin H_3	boundary of macrograin H_4		
2	32	39	44	50	11.5	18
8.5	36	41	74	93	45	57
20	34	48	78	100	48	66
50	38	50	59	84	27.5	46
150	34	37	48	52	14.5	18
1000	35	37	46	50	12	15

microhardness method is incomparably simpler and more efficient and enables us to solve exactly the same physicochemical problems.

This situation is illustrated extremely clearly by the graph obtained in [5] and presented in Fig. 87. It follows from this graph that the composition at the center of the grain determined by the microhardness method agrees with that based on x-ray shadow microscopy.

We must mention one further detail noted in [248]. In all papers published, the degree of intracrystallite liquation has been determined by reference to the difference or ratio between the concentrations of the alloying component at the center and periphery of the dendritic cells, without considering their arrangement with respect to the cross section of the macrograin. However, the concentration of the alloying element (for example, copper) in the trunk and branches of a dendrite formed at the onset of crystallization should be smaller than in the branches of the dendrite formed in the course of its growth at lower temperatures. Table 32 shows the microhardness values at the center and periphery of dendritic cells situated close to the boundary and center of the macrograins in an Al−5% Cu alloy cooled at various rates.

These results show that the amount of copper in the center (H_2) and at the periphery (H_4) of dendritic cells adjacent to the boundaries of the macrograins is considerably higher than in micrograins situated in the middle of the macrograin (H_1 and H_3). In view of this the experimentally determined mean-statistical degree of intracrystallite liquation $[(H_3 + H_4)/2] - [(H_1 + H_2)/2] = \Delta H_{\mu av}$ may be considerably smaller than the maximum $H_{\mu max} = H_4 - H_1$.

This should certainly be considered both when studying the actual process of dendritic liquation and also when using the method described in Chapter 5 in relation to the possibility of plotting the equilibrium solidus by studying the microhardness of the primary crystals of alloys crystallized in a nonequilibrium manner [143].

The results of investigations into the effect of cooling rate on the degree of intracrystallite liquation in Al−Zn, Al−Mn, Cu−Al, Cu−Si, Cu−Sn, Cu−Sb, and other alloys are presented in [5]. The results are in general similar to those just described.

The microhardness method was employed in [249] to study the width of the intergranular interlayers of liquid phase during the nonequilibrium crystallization of solid solutions at a specified cooling rate, which was varied in accordance with the conditions of crystallization. The method proposed in this investigation was based on the fact that over a wide range of cooling rates the composition of the layer of solid solution formed at each temperature was determined by the corresponding point on the curve of the equilibrium solidus and remained unchanged

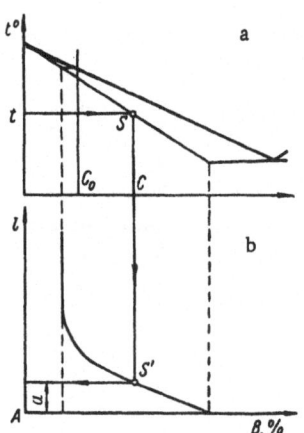

Fig. 88. Scheme illustrating the determination of the width of liquid intergranular interlayers from the phase diagram (a), and curve representing the distribution of the alloying element near the grain boundary (b): l = distance from grain boundary.

right down to room temperature. We have shown quite convincingly in the foregoing that this condition is realized over a wide range of cooling rates for a number of alloys. Thus, in the alloy C_0, at a temperature higher than t (Fig. 88) the concentration of component B is smaller than the concentration given by the point C in all parts of the crystallite. If the composition of each layer of solid solution is unable to alter after its formation as the alloy is cooled to room temperature, then it is easy by using the curve representing the distribution of the element over the cross section of the crystallite to determine the width of each liquid interlayer at any temperature in the crystallization interval as well. For this purpose we must plot an isotherm at the specified temperature on the phase diagram until it intersects the curve of the equilibrium solidus (point S) and note the resultant concentration on the curve giving the distribution of the alloying element over the cross section of a dendritic cell at the boundary of a macrograin (point S'). The distance from the boundary of the dendritic cell to the point with this concentration corresponds to the width of the part of the crystallite formed on cooling below the temperature t, since in this part the concentration of component B is higher than at the point C. Hence the intercept a determines the width of the liquid layer at the temperature t from which the surface layer of the grain of solid solution was formed (without allowing any correction for the difference in the specific volumes of the liquid and solid phases). If there is a component of eutectic or peritectic origin of width b along the boundaries of the primary solid-solution crystals, then the sum $(2a + b)$ gives the width of the liquid intergranular interlayer at the temperature t.

As a method for plotting the distribution of the alloying element over the cross section of the crystallites, the microhardness technique was employed.

The experiments were carried out on aluminum alloys containing 1.5 and 5% Cu. The aluminum was 99.96 and the copper 99.95% pure. Each alloy was cast into steel molds, either in the cold state or after heating to 500 or 700°C, the cooling rates being 150, 22, and 12 deg · min^{-1} respectively. The mean cooling rate in the crystallization interval was calculated from thermal curves recorded on an electronic potentiometer. Microsections were cut from the castings and prepared by electropolishing. The dendritic cells inside which it was desired to plot the distribution curves were selected near the boundary of a macrograin. The first impression was made in the immediate vicinity of the eutectic component at the boundary of the micrograin and the subsequent impressions at distances of 10 to 15 μ from each other. The microhardness values were converted into copper concentrations by reference to a calibration curve relating the microhardness to the composition of the solid solution. The conditions of preparing the microsections and measuring the microhardness when plotting the calibration graph and the distribution curves were exactly the same.

The curve given as an example in Fig. 89 reflects average data obtained by studying six or seven dendritic cells on the microsection, and characterizes the copper distribution near the boundary of a macrograin in the alloy under consideration for a specific cooling rate.

The resultant distribution curves and the phase diagram of the Al−Cu system [249] were used to plot the temperature dependence of the width of the liquid intergranular interlayer for

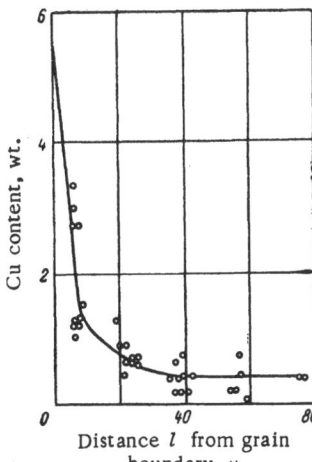

Fig. 89. Copper distribution curve near the grain boundary in an Al−1.5% Cu alloy cooled at 22 deg. min⁻¹.

the two alloys under consideration and three cooling rates. The isotherms were drawn at intervals of 10° on the phase diagram. By way of example, Fig. 90 shows this relationship for the Al − 1.5% Cu alloy. Analogous curves were also obtained for Al −5% Cu.

It appeared interesting to compare the resultant data with results obtained by measuring the width of the intergranular interlayers on microsections quenched from various temperatures in the crystallization range. For this purpose all the microsections used in plotting the distribution curves were heated in a salt bath to various temperatures in the crystallization range and then water-quenched. The total time for heating and holding the sample was no greater than 3 min.

The thickness of the interlayers clearly visible in the quenched samples after electropolishing and etching was determined as the arithmetic mean of 60 measurements with a micrometer eyepiece. Only the widest layers were measured, the crystallization of these presumably having ended in the formation of a eutectic component.

The comparison showed that the value determined by the microanalysis of the quenched samples was much lower than that found from the distribution curves.

It was suggested in [249] that the main reason for this was the crystallization effect in the course of quenching. This phenomenon was illustrated earlier by one of the authors in [283], using the microhardness method; it will be considered in detail at the end of the chapter.

Thus it was shown in [249] that the microhardness method could successfully be used for solving complex metallographical problems associated with the nonequilibrium crystallization of alloys at various cooling rates.

A detailed study of the relation between the degree of liquation microinhomogeneity and the cooling rate in alloys belonging to systems with an unlimited mutual solubility of the components in the solid state was made in [147].

The principal method of investigation was that of microhardness measurement.

Bismuth alloys containing 30, 40, and 50 wt.% Sb and germanium containing 4 and 8 wt.% Si were taken for study. Materials suitable for semiconductor technology of at least 99.999% purity were used to prepare the alloys.

The alloys were placed in sealed quartz ampoules evacuated to $1 \cdot 10^{-4}$ mm Hg with an internal diameter of 6 mm. A capillary for a thermocouple was sealed into the ampoule. After melting, the ampoules containing the alloys were cooled at various rates by transferring to other furnaces held at different (lower) temperatures. Cooling was also effected in various media. For each sample a cooling curve was plotted in the recording apparatus. The cooling rate in the crystallization interval (v_{cool}, deg/min) was calculated from this curve. For very high cooling rates it was difficult to plot the cooling curve, and the cooling rate within the crystallization interval was estimated approximately.

Microsections were made from the alloy samples thus prepared. The effect of surface hardening was eliminated by brief low-temperature vacuum annealing; this also served to eliminate internal stresses arising from the rapid cooling.

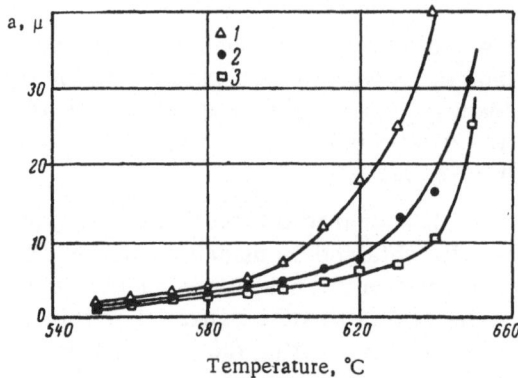

Fig. 90. Effect of temperature on the width of
the liquid-phase intergranular interlayer "a"
in Al−1.5% Cu cooled at the following rates
(deg/min): 1) 12; 2) 22; 3) 150.

The microhardness of the alloys was measured on a PMT-3 furnished with an automatic
loader. The load was 10 g for the Bi−Sb and 50 g for the Ge−Si alloys. Sometimes smaller
loads were used. The mean-square error in determining the microhardness of the Bi−Sb alloys was ±2 kg/mm^2 and in the case of Ge−Si ±25 kg/mm^2. Impressions were made concen-
trically over the whole grain, starting from the periphery and gradually approaching the center.
In each alloy 10 to 12 grains were usually examined, and in fine-grained alloys 25 to 30; the
average values were then taken.

As a measure of microinhomogeneity, the absolute difference ΔH_μ in microhardness values measured at the center (H_μ^c) and periphery (H_μ^p) of the grain was taken (the crystallization
center of the crystallite, not the geometrical center, is meant).

In those cases in which the deviation from equilibrium crystallization resulting from the
slowness of diffusion in the solid phase was so considerable as to lead to the precipitation of
the pure low-melting point component (in the present case bismuth or germanium) in the final
stages of crystallization, the difference between the microhardnesses measured in the center
of the grain (H_μ^c) and the microhardness of the precipitating pure component (H_μ^p) was taken as
a measure of microinhomogeneity. In these cases the structure appeared two-phased under the
microscope (although in principle this could not really be so) owing to the sharp change in etch-
ability experienced on passing to the regions essentially constituting the pure low-melting
point component (Bi or Ge).

In view of this, in those cases in which a structure of this kind was observed, the microhardness close to these regions ($H_\mu^{p'}$) was also measured, and the microinhomogeneity inside
the parts enriched with the more refractory component (Sb or Si, $\Delta H_\mu' = H_\mu^c - H_\mu^{p'}$), i.e., the
intracrystallite liquation, was calculated.

In each alloy the mean grain size was also determined for each specified cooling rate.

The results of all the measurements are presented in Tables 33 and 34.

For a correct treatment of the results obtained the general character of the relationship
between the microhardness and the composition in the two systems under consideration must
also be known. For this reason a series of binary Bi−Sb and Ge−Si alloys were prepared by
the method just described. Crystallization was effected by immersing the ampoules containing
the melts in cold water. After this the alloys were annealed at temperatures of 50 to 100°

TABLE 33. Results Obtained on Measuring the Degree of Micro-inhomogeneity ΔH_μ (kg/mm²) and Grain Size δ of Bi−Sb Alloys Crystallizing at Different Rates

v_{cool}, deg/min	$\lg v_{cool}$	H_μ^C	H_μ^P	$H_\mu^{P'}$	ΔH_μ	$\Delta H_\mu'$	δ, μ	Notes
				30 wt.% Sb				
4	0.60	90	15	70	75	20	500	
14	1.15	88	15	46	73	42	288	Pure bismuth in the structure
20	1.30	87	15	33	73	54	225	
30	1.48	88	18	24	70	64	170	
50	1.70	87	24	—	63	—	95	
100	2.15	89	60	—	29	—	70	
720	2.86	97	92	—	5	—	54	ΔH_μ could not be measured
1 800	3.25	101	—	—	—	—	28	
3 600	3.56	93	93	—	0	—	580	
10 000	4.00	93	93	—	0	—	750	
				40 wt.% Sb				
4	0.60	83	17	73	66	10	555	
14	1.15	81	16	61	65	20	300	Pure bismuth in the structure
21	1.32	80	17	42	63	38	260	
33	1.52	80	18	30	62	50	191	
60	1.78	80	35	—	45	—	94	
76	1.88	79	45	—	34	—	82	
160	2.15	80	59	—	21	—	75	ΔH_μ could not be measured
720	2.86	91	80	—	11	—	50	
1 800	3.25	96	—	—	—	—	18	
3 600	3.56	103	103	—	0	—	490	
10 000	4.00	104	104	—	0	—	640	
				50 wt.% Sb				
4	0.60	77	18	74	59	74	580	
14	1.15	75	19	65	56	65	380	Pure bismuth in the structure
20	1.30	74	22	56	52	56	300	
40	1.61	74	23	—	51	—	180	
60	1.78	74	36	—	38	—	108	
160	2.15	75	59	—	16	—	70	ΔH_μ could not be measured
720	2.86	86	84	—	2	—	50	
1 800	3.25	94	—	—	—	—	15	
3 600	3.56	102	102	—	0	—	374	
10 000	4.00	103	103	—	0	—	573	

below the equilibrium solidus for two months. Microsections were obtained from the annealed samples, and in order to remove surface hardening these were again annealed in vacuum for a week at temperatures of 250°C (Bi−Sb) and 900°C (Ge−Si).

The microhardness of the samples thus obtained was measured. The results of the measurements are presented in Fig. 91a and b and compared with the equilibrium phase diagrams of the Bi−Sb and Ge−Si systems. The resultant microhardness/composition relationships are described by curves containing maxima and are typical of systems having a continuous series of solid solutions. The maximum microhardness corresponds to 40 wt.% (55 at.%) Sb in the Bi−Sb system and 35 wt.% (50 at.%) Si in the Ge−Si system.

Figure 92a and b shows the degree of microinhomogeneity (ΔH_μ), the microhardness at the center of the crystallites (H_μ^C), and the mean crystallite size (δ) as function of the rate of cooling in the crystallization interval. For convenience of graphical representation the logarithm of the cooling rate is plotted along the horizontal axis.

The graphs presented in Fig. 92a and b are analogous in shape. This indicates the generality of the changes in the crystallization process as the cooling rate increases. Starting

TABLE 34. Results Obtained on Measuring the Microinhomogeneity ΔH_μ (kg/mm²) and Grain Size δ of Ge−Si Alloys Crystallized at Different Rates

v_{cool}, deg/min	$\lg u_{cool}$	H_μ^C	H_μ^p	$H_\mu^{p'}$	ΔH_μ	$\Delta H_\mu'$	δ, μ	Notes
				4 wt.% Si				
5	0.70	1036	600	920	436	116	410	Pure germanium in the structure
15	1.08	1008	590	814	418	194	250	
30	1.48	1020	596	710	434	310	156	
60	1.78	1036	650	590	386	346	130	
120	2.08	1036	743	—	293	—	118	
720	2.86	1020	970	—	50	—	60	
2 100	3.32	940	—	—	—	—	30	ΔH_μ could not be measured
3 600	3.56	790	790	—	0	—	300	
10 000	4.00	790	790	—	0	—	450	
				8 wt.% Si				
5	0.70	1182	590	1030	592	152	430	Pure germanium in the structure
15	1.18	1165	580	865	585	300	212	
35	1.65	1182	590	787	592	395	148	
60	1.78	1174	620	815	554	359	132	
120	2.08	1165	825	—	340	—	107	
720	2.86	1174	1090	—	84	—	64	
2 100	3.32	1080	—	—	—	—	35	ΔII_μ could not be measured
3 600	3.56	920	920	—	0	—	400	
10 000	4.00	920	920	—	0	—	417	

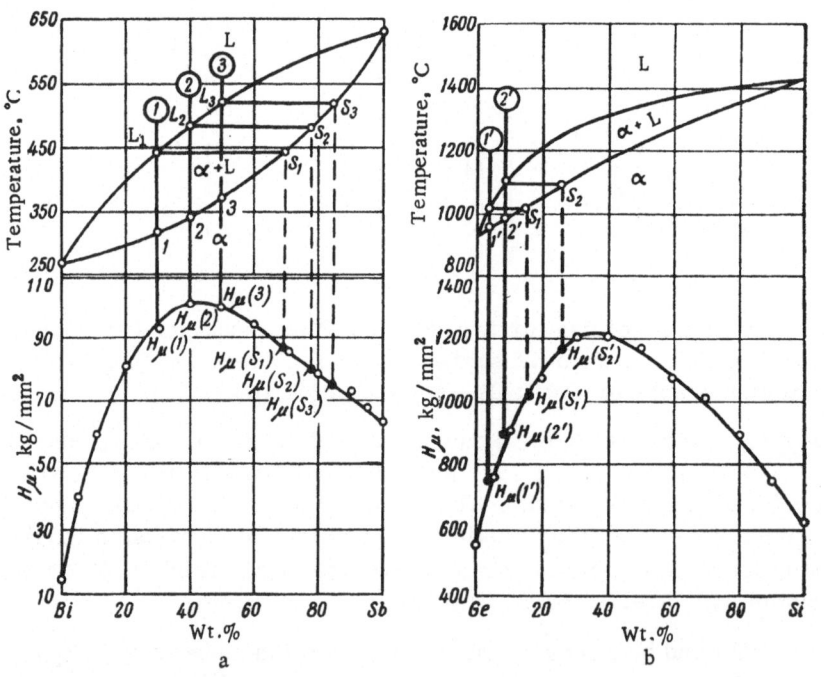

Fig. 91. Microhardness/composition relationship in (a) the Bi−Sb and (b) the Ge−Si system, shown in comparison with the corresponding phase diagrams.

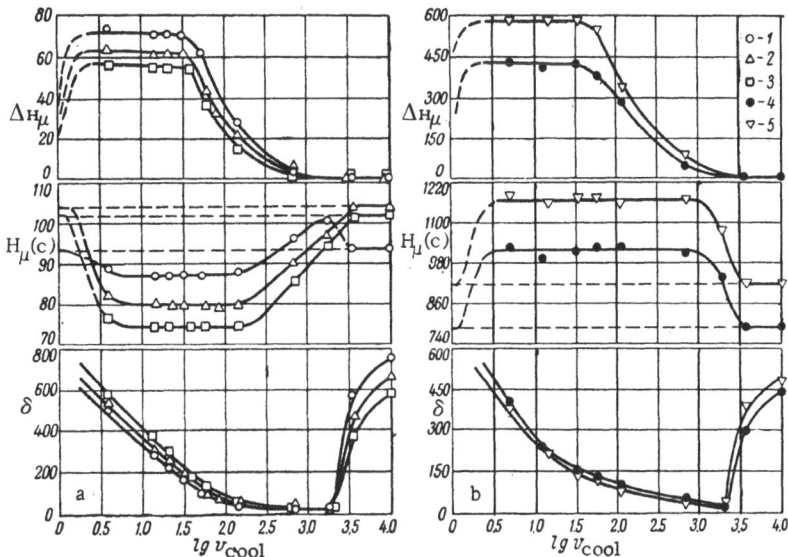

Fig. 92. Degree of microinhomogeneity (ΔH_μ), microhardness in the center of the crystallites [$H_\mu(c)$], and mean crystallite size (δ) as functions of the cooling rate in the crystallization interval for the following alloys: a) Bi−Sb; b)
Ge−Si; 1) 30% Sb; 2) 40% Sb; 3) 50% Sb; 4) 4% Si; 5) 8% Si.

from the lowest cooling rates employed (4 to 5 deg/min), the microinhomogeneity is at a maximum and remains roughly constant on increasing the cooling rate up to a certain limit, after
which it starts falling. All samples of the Bi−Sb alloys studied contain practically pure bismuth in the structure ($H_\mu \approx 15$-20 kg/mm^2), while the microhardness in the center of the crystallites enriched with antimony retains an almost constant value corresponding to the microhardness of alloys with compositions S_1, S_2, S_3 (Figs. 91 and 92a). In the Ge−Si alloys there is
a very similar picture (Fig. 91b and Fig. 92b).

After reaching a particular cooling rate characteristic of each alloy, the degree of microinhomogeneity starts falling and ultimately becomes negligible at very high cooling rates.

However, analysis of the curves relating the microinhomogeneity (ΔH_μ) to the cooling
rate does not enable us to draw any conclusions regarding the reasons for the fall in the degree
of microinhomogeneity after reaching a specific cooling rate.

Nor can the nature of the relationship between the microinhomogeneity and the cooling
rate be explained by comparing the results obtained with changes in grain size.

We see from Fig. 92a and b that in the range of cooling rates corresponding to a fairly
sharp fall in microinhomogeneity the grain size changes smoothly. For high cooling rates
the grain becomes so fine that it is practically impossible to establish any difference in composition between the center and the periphery. However, at still higher cooling rates the grain
size δ increases again, while the microinhomogeneity remains practically zero. Thus we find
that the microinhomogeneity is close to zero for comparatively high cooling rates, whereas the
grain size may be either very small or very large.

By comparing the dependence of the microinhomogeneity and the microhardness of the
center of the crystallites on the cooling rate in the crystallization interval (Fig. 92, Tables
31 and 33), we arrive at the important conclusion that the onset of the fall in liquation micro-

TABLE 35. Comparison of the Microhardness Measured in the Center of the Grains of Bi−Sb and Ge−Si Alloys Crystallized at Different Rates and the Theoretical Values of the Microhardness of These Grains H_μ (S_1, S_2, S_3) and H_μ (S_1', S_2') Obtained from the Graphs of Figs. 91a and b

v_{cool}, deg/min	Bi+30%Sb		Bi+40%Sb		Bi+50%Sb		Ge+4% Si		Ge+8% Si	
	H_μ^C	$H_\mu(S_1)$	H_μ^C	$H_\mu(S_2)$	H_μ^C	$H(S_3)$	H_μ^C	$H_\mu(S_1')$	H_μ^C	$H_\mu(S_2')$
12—15	87		81		75		1008		1165	
20—21	87		80		74		—		—	
30—35	88		80		—		1020		1182	
40	—		—		74		—		—	
50	87	85	—	77	—	72	—	1010	—	1160
60	—		80		74		1036		1174	
76	—		79		—		—		—	
120—160	89		80		75		1036		1165	
720	—		—		—		1020		1174	

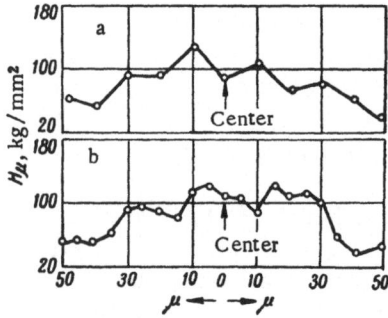

Fig. 93. Distribution of microhardness over the cross section of a crystallite in a Bi−30% Sb alloy crystallized at a cooling rate of 50 deg/min in the crystallization interval for loads of: a) 5 g; b) 2 g.

inhomogeneity is in no way associated with the onset of the suppression of separating diffusion in the liquid phase, since the microhardness in the center of the crystals (H_μ^C) and hence the composition remain constant at considerably higher cooling rates than those at which ΔH_μ starts falling. The microhardness values and hence the composition approximately correspond (see Table 35) to the microhardness $H_\mu(S_1)$, $H_\mu(S_2)$, $H_\mu(S_3)$, $H_\mu(S_1')$, and $H_\mu(S_2')$ and hence the composition S_1, S_2, S_3, S_1', and S_2' of the primary crystallites separating during the crystallization of the alloys 1, 2, 3, and 1', 2' (see Fig. 91a and b).

We see from Tables 33 and 34 that the fall in microinhomogeneity is associated with a change in the microhardness (and hence the composition) of the periphery of the crystallites (H_μ^P). An increase in cooling rate leads first to a fall in the proportion of the low-melting point component precipitating as a result of the suppression of diffusion in the solid phase in nonequilibrium crystallization (of which it is not hard to convince onself by studying the microstructure of the alloys), and then to its complete disappearance.

Thus in this range of cooling rates we have the following situation: Separating diffusion is able to proceed to completion in the liquid phase, as indicated by the microhardness results in the center of the grain, while equalizing diffusion is completely suppressed in the solid phase. At the same time, the degree of liquation microinhomogeneity determined from the microhardness measurements falls as a result of the change in the composition of the periphery.

It may be considered that this is due to the suppression of equalizing diffusion in the liquid phase during crystallization. An impurity displaced at the initial instant of crystallization cannot be reabsorbed; thus a layer enriched with respect to the second component is formed in front of the crystallization front and this then solidifies. The crystallization process is thus completely analogous to the process of pulling the solid phase from the melt at a high velocity.

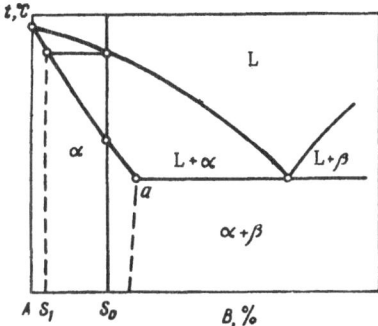

Fig. 94. Part of a eutectic-type phase diagram.

It has been shown in a number of papers (reviewed in [158]) that the formation of the so-called layer structures then becomes possible.

In view of this, the microhardness distribution was carefully measured over the cross section of Bi−Sb crystallites solidified at cooling rates corresponding to the fall in the degree of liquation microinhomogeneity revealed by microhardness data, using small loads (2 and 5 g). In a number of cases an undulating variation in microhardness was observed; this indicated a layer-like distribution of impurity across the crystallites. By way of example, Fig. 93 shows the microhardness distribution over the cross section of certain Bi−30% Sb crystallites solidified at a cooling rate of 50 deg/min.

On the basis of these results we may well consider that the supposition to the effect that the fall in the degree of liquation microinhomogeneity is associated with the retardation of equalizing diffusion in the liquid phase is justified.

It must be noted, however, that according to the theoretical considerations set out in I. N. Golikov's monograph [223], a limitation on equalizing diffusion in the liquid phase should not affect the ultimate result when there is a complete suppression of equalizing diffusion in the solid phase, i.e., in this case also, the pure low-melting point component should ultimately crystallize last. However, it would appear that, when diffusion is limited in the liquid phase so as to lead to the formation of layered structures, the amount of the pure low-melting point component may be so insignificant as to be almost impossible to detect by microhardness measurements and microstructural analysis.

Thus in this case the microhardness method reveals not the overall liquation microinhomogeneity in such alloys but a particular case of intracrystallite liquation related to the cooling rate in a more complicated manner.

There is thus a certain limitation imposed on the prospects of the microhardness method, as noted earlier in connection with the work of I. I. Novikov and his colleagues [245-250].

Kinetic Characteristics of Dendritic Liquation

in Metallic Alloys

On the basis of the foregoing investigations, as well as considerations as to the relative completeness of the diffusion processes taking place in the liquid and solid phases, we may enunciate some general laws relating the parameters of chemical microinhomogeneity to the cooling rate in the crystallization of various types of alloys, for example, those of the eutectic type and those with unlimited mutual solubility of the components in the solid state [222, 147].

Let us first consider the effect of cooling rate during crystallization on the concentration of the components in the center of a dendritic cell in an alloy S_0 (Fig. 94) in a system of the eutectic type. For very slow cooling all the diffusion processes proceed to completion, dendritic liquation is absent, and the composition of the initially separating solid-solution crystallite is characterized by the point S_0 (Fig. 95a) over the whole cross section after crystallization has been completed. Starting from a particular cooling rate, equalizing diffusion is unable to take place completely in the solid phase, and the composition of the center of the dendritic cell lies between S_0 and S_1 on the completion of crystallization. The faster the cooling, the less completely can the equalizing diffusion take place, and the closer to S_1 will the composition at the center of the dendritic cell be.

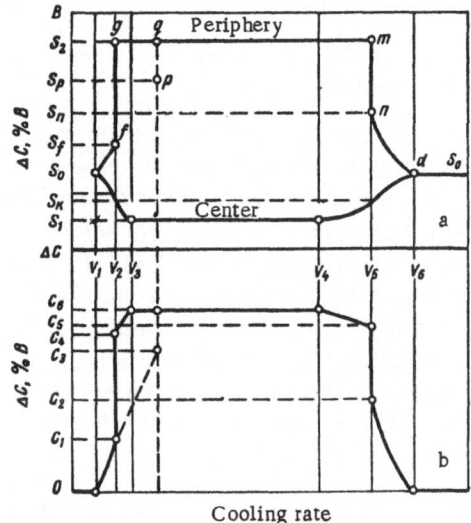

Fig. 95. Concentration at the center and periphery of the dendritic cell (a) and degree of overall liquation micro-inhomogeneity (b) as functions of cooling rate for alloys belonging to a system of the eutectic type situated (as regards composition) within the limits of the region of limited solubility (schematic) ($C_1 = S_0 - S_1$; $C_2 = S_n - S_k$; $C_3 = S_p - S_1$, etc.).

On reaching a cooling rate of v_3 the equalizing diffusion is almost completely suppressed in the central part of the growing crystallite. The composition at the center of the dendritic cell is then determined by reference to the equilibrium solidus (point S_1 in Fig. 94) and remains unaltered over a certain range of cooling rates (Fig. 95a).*

Starting from a cooling rate of v_4, separating diffusion is particularly suppressed. The composition of the center of the dendritic cell is then intermediate between the composition determined from the equilibrium solidus (point S_1) and the original composition of the melt (S_0). Finally, on reaching a cooling rate of v_6 separating diffusion is completely suppressed; diffusionless crystallization takes place, and the composition of the homogeneous solid solution is characterized by the point S_0 over the whole cross section of the crystallite.

Let us now trace the influence of cooling rate on the composition at the periphery of the dendritic cell, understanding periphery to mean the surface layer of the solid solution or inclusions of a second structural constituent if these have been formed by nonequilibrium crystallization from the melt. For cooling rates of under v_1 the composition of the peripheral part of the crystallite is the same as that of its center and is determined by the point S_0.

Starting from a cooling rate of v_1, the peripheral layers of the dendritic cells are enriched with component B as a result of the suppression of equalizing diffusion in the solid phase. At a certain cooling rate v_2 the eutectic component appears and the concentration of component B at the periphery of the dendritic cell suddenly rises, reaching the concentration of the eutectic (or the second phase if the eutectic is "degenerate"). The rate v_2 may be greater or smaller than v_3, which corresponds to the almost total suppression of equalizing diffusion in the central zone of the dendritic cells. If equalizing diffusion in the liquid phase is partly suppressed for cooling rates of less than v_3, then the layer of liquid solution adjacent to the separation boundary may become so enriched with the component B that a eutectic constituent will be formed (section fg in Fig. 95a). Actually another version is possible, i.e., the concentration may remain almost constant in the center of the dendritic cell (a rate v_3 being reached), while in the peripheral layer equalizing diffusion is still able to take place owing to the high concentration gradient, and the eutectic component will not be formed over a certain range of cooling rates greater than v_3. The second phase only starts crystallizing out from the liquid when equalizing diffusion is also almost entirely suppressed in the surface layer of the initially separating crystallite of the solid solution (section pq, Fig. 95a).

The eutectic component determining the composition of the periphery will be formed in the range of cooling rates between v_2 and v_5. The rate v_5 cannot be lower than v_4; this is indicated, for example, by the mathematical analysis of [233], from which it follows that for the

---

* A detailed quantitative analysis of the breakdown of equilibrium during the crystallization of solid solutions for cooling rates between v_3 and v_4 was given in [223-225].

complete suppression of equalizing diffusion in the solid phase and the unimpeded occurrence of separating diffusion, the limitation of equalizing diffusion in the liquid phase should have no effect on the degree of total liquation microinhomogeneity.

As noted earlier, on cooling the alloy at a rate greater than v_4, the center of the crystallite is formed with a composition between S_1 and S_0, owing to the suppression of separating diffusion. Thus a cooling rate v_5 may be reached such that the eutectic component is no longer formed and the concentration of the component at the periphery of the dendritic cell falls suddenly (section mn in Fig. 95a). The cooling rate v_5 lies between v_4 and v_6 or may in the limiting case coincide with v_6. For cooling rates of over v_6 (point d), for which diffusionless crystallization takes place, the composition of the periphery of the crystallite is the same as that of its center and is characterized by the point S_0.

Knowing how the cooling rate affects the composition of the center and periphery of the dendritic cells, it is easy to plot a graph relating the total liquation microinhomogeneity to the cooling rate for alloys belonging to a system of the eutectic type; this graph also relates to systems of the peritectic type (Fig. 95b).

There may also be some other particular cases; these are mentioned in the monograph by I. I. Novikov and V. S. Zolotorevskii [250] and we shall not consider them here.

The cooling rates v_1, v_2, v_3, v_4, v_5, and v_6 constitute critical characteristics of dendritic liquation in an alloy of specified composition for such systems.

In alloys belonging to systems with a continuous series of solid solutions, the degree of total liquation microinhomogeneity should vary with cooling rate almost in the same way as in systems of the eutectic type. The corresponding total liquation-microinhomogeneity curves will be distinguished, in the case of alloys belonging to systems with a continuous series of solid solutions, by an absence of jumps, since the composition of the peripheral parts of the dendritic cells should vary smoothly with cooling rate in a concentration range extending between that of the original alloy to the pure low-melting point component.

It should be noted however that it is difficult to plot the total liquation-microinhomogeneity/cooling rate curve by the microhardness method in alloys of this type in view of the earlier-mentioned difficulties in determining the true composition of the periphery.

Hence when establishing the kinetic characteristics of dendritic liquation in alloys belonging to systems with a continuous series of solid solutions by the microhardness method it is essential to study the relation between the composition in the center of the crystallites and the cooling rate. The curve expressing this relationship reproduces the corresponding curve of the total liquation microinhomogeneity and directly indicates the changes in the composition of the crystallite nucleus in the course of crystallization.

For an alloy of composition C_1 (Fig. 96) the relation between the composition of the center of the solid-solution crystallites and the cooling rate may be schematically represented as in Fig. 97.

For very slow crystallization, diffusion is able to take place completely in both the liquid and solid phases. Accordingly the composition of the center of the crystallites of alloy C_1 (Fig. 96) should be equal to S_2 (Figs. 96 and 97). Over a certain range of very low cooling rates this picture will evidently remain unaltered.

Starting from a certain cooling rate v_1, equalizing diffusion will take place completely in the solid phase. The composition of the center of the crystallites should vary between S_2 and S_1 (Figs. 96 and 97).

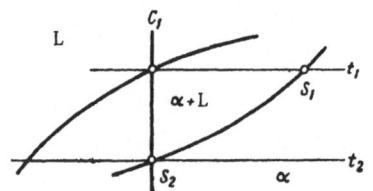

Fig. 96. Schematic representation of part of a phase diagram with a continuous series of solid solutions between the components.

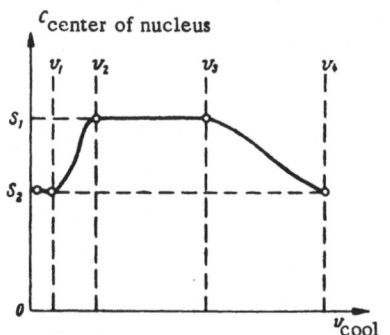

Fig. 97. Composition of a crystallite nucleus as a function of cooling rate (schematic representation of a typical dendritic-liquation kinetic curve).

On reaching a certain specific cooling rate the equalizing diffusion in the solid phase will be entirely suppressed, while separating diffusion will still occur fully in the liquid phase. The composition of the center of the crystallites should then be equal to S_1 (Figs. 96 and 97), and over a certain range of cooling rates this picture will remain unaltered.*

After reaching a certain third cooling rate v_3, the separating diffusion in the liquid phase will be partly suppressed, this process intensifying with increasing rate of cooling. Clearly for these cooling rates the composition of the center of the crystallites should move in the direction of S_2 from S_1.

After reaching a further cooling rate v_4 the separating diffusion in the liquid phase will be entirely suppressed, and a solid solution of the same composition will crystallize from the liquid.

Clearly, further increasing the cooling rate will not produce any more changes. The composition of the center of the crystallites should remain constant and equal to S_2 (Figs. 96 and 97).

Thus by analyzing the composition of the center of the crystallites as a function of the cooling rate (Fig. 97) we may distinguish four cooling rates constituting kinetic characteristics of dendritic liquation in alloys belonging to systems with a continuous series of solid solutions. By crystallizing the corresponding alloys in the range of cooling rates in advance of v_1 or after v_4, we may obtain these in equilibrium form as regards composition, without the consequences of nonequilibrium crystallization. Crystallization of the alloys at high cooling rates, exceeding the fourth critical value, is of particular interest.

Clearly in this range of cooling rates we reach a supercooling of the melt below the temperature corresponding to equal isobaric−isothermal potentials (Gibbs functions) of the solid and liquid phases, and, as a result of this, diffusionless crystallization may in principle take place [241].

It is not difficult to see that the relationships between the microhardness in the center of the crystallites and the cooling rate obtained experimentally for Bi−Sb and Ge−Si alloys (Fig. 92) correspond to the curve illustrated schematically in Fig. 97.

Certainly the initial parts of these relationships are not reproduced completely and they are therefore shown by broken lines; in practice it is difficult to achieve cooling rates low enough to render the initial parts of the curves completely.

In the range of cooling rates between about 5 and 200 deg/sec for Bi−Sb and between 5 and 800 deg/sec for Ge−Si alloys, the microhardness remains constant in the center of the crystallites and corresponds to the microhardness of primary crystallites of composition

_____

*For this range of cooling rates the deviation from equilibrium crystallization is analyzed in [223-225].

S_1, S_2, S_3 and S_1', S_2', which should be the first to crystallize according to the equilibrium phase diagrams (Fig. 91 and Table 35). Further raising the cooling rate leads to a rise in microhardness in the center of the Bi—Sb crystallites and to a fall in the case of Ge—Si, which in both cases indicates a fall in the concentration of the second component (Sb and Si) at these points. This is most probably due to the incipient suppression of separating diffusion in the liquid phase, when the composition of the crystallite nuclei no longer corresponds to the equilibrium solidus line. The maximum on the H_μ^c / v_{cool} curve in this region in the Bi—30% Sb alloy is associated with the fact that the composition of the center of the crystallites varies in the case of this alloy from S_1 to 1 (Fig. 91), and it is precisely in this range of concentrations that the relation between the microhardness and the composition of the solid solution passes through a maximum, varying from $H_\mu(S_1)$ to $H_\mu(1)$ (Fig. 91).

After reaching ~2100 to 2200 deg/min, further increasing the cooling rate produces no further change in the microhardness at the center of the grain. Clearly for these alloys this constitutes the fourth critical cooling rate for which the separating diffusion is almost entirely suppressed in the liquid phase.

In this range of cooling rates the liquid crystallizes into a solid phase of the same composition, as supported by the agreement between the microhardness of rapidly crystallized alloys after brief, low-temperature, stress-removing annealing and the microhardness of the equilibrium alloys $H_\mu(1\text{-}3)$ and $H_\mu(1\text{-}2)$ (see Fig. 91). Comparative data are presented in Table 36.

The development of crystallization by a diffusionless mechanism at the cooling rates indicated is also supported by the sharp increase in grain size (Fig. 92), which in fact ought to accompany diffusionless crystallization on theoretical grounds [241].

Thus the curves relating liquation microinhomogeneity and composition in the center of the crystallites to the cooling rate plotted by the microhardness method for Bi—Sb and Ge—Si alloys in the present case give a clear representation of the process underlying the development of microinhomogeneity and facilitate a qualitative discussion of the development of diffusion processes in the liquid and solid phases for various cooling rates.

Comparing the results for the two groups of alloys (Bi—Sb and Ge—Si on the one hand and Al-base alloys on the other), we may conclude that, in those systems in which the diffusion process occurs slowly (Bi—Sb and Ge—Si), high cooling rates should be used in order to obtain equilibrium alloys, while in systems with a high diffusion velocity (Al alloys) slow cooling is required.

Clearly the difficulties in obtaining equilibrium alloys of the aluminum type by diffusionless crystallization are also associated with their low inclination toward supercooling, which according to Yu. A. Krishtal [256] should be very considerable. Experiments carried out by D. S. Kamenetskii and colleagues indicated that diffusionless crystallization of solid solutions occurred on achieving very high degrees of supercooling.

The low tendency of aluminum alloys toward supercooling is due to the insignificant differences between the structures of these alloys in the solid and liquid states, whereas in alloys such as Bi—Sb and Ge—Si this difference is substantial [257, 258], and they are very inclined to undergo substantial supercooling.

TABLE 36. Comparative Data Relating to the Microhardness of Rapidly Cooled and Equilibrium Bi—Sb and Ge—Si Alloys

Composition of alloy, wt.%	Microhardness, kg/mm² for v_{cool}, deg/min		
	3600	10 000	equilibrium samples
Bi+30Sb	93	93	92
Bi+40Sb	103	104	101
Bi+50Sb	102	103	100
Ge+4Si	790	790	761
Ge+8Si	920	920	896

These circumstances must be borne in mind when choosing ways of achieving the equilibrium state in studying the corresponding systems.

The application of the microhardness method to the study of liquation phenomena in alloys has revealed the fundamental laws of dendritic liquation, as a result of which ways have been found for producing alloys with a specified degree of microinhomogeneity. This is very important in alloy production and in studying phase equilibria.

Effect of the Superheating of the Melt on the

Development of Liquation Microinhomogeneities

in Alloys

We have been considering investigations concerned with the use of the microhardness method for studying the effect of cooling rate during crystallization on the development of liquation phenomena in alloys. In this we have said nothing of the effect of the melt superheating temperature on the development of dendritic liquation, since in all the cases considered this temperature was constant.

However, a change in the superheating temperature may influence the crystallization of an alloy and hence the development of liquation phenomena. It has been known for a long time, for example, that on raising the superheating temperature the grain of castings coarsens; this is associated with the deactivation of suspended impurities, which in the case of low superheating values may serve as crystallization centers and promote the formation of a finer grain structure. V. I. Danilov [259] showed that increasing the superheating temperature promoted an increase in the degree of supercooling of the melt for fairly high cooling rates.

This effect was associated with the deactivation of suspended impurities and an increase in the difference between the structure of short-range order in the solid and liquid phases. Whichever cause was dominant, a change in the degree of supercooling of the melt should certainly influence the development of dendritic liquation in the alloys. A study of this question is of important practical significance, since a correct, scientific solution should help in establishing appropriate heat treatment for the melt and thus in a number of cases [121, 260, 261] facilitate changes in the structure and physicochemical and technological properties of castings.

Dendritic liquation after various degrees of superheating of the melt was considered in [262, 263].

All these investigations were based on the microhardness method, and we shall therefore now consider them more deeply.

A Bi−7% Sb alloy was studied in [262]. One batch of samples was heated to 700°C, held for 13 min, and cooled at 350 to 400 deg/min to the liquidus temperature, after which the samples were solidified at various rates. Another batch of samples was held for 15 min at 700°C, cooled to 330°C, held for for another 15 min at this point, and then solidified at various rates. It was found that in the first case the difference in microhardness values at the center and the periphery of the dendritic cells was higher than in the other case. This difference was due to a fall in the microhardness at the center of the dendritic cells. Even a 1-min holding of the melt at 330°C eliminated the effect of the previous superheating on the degree of intracrystallite liquation. Thus according to [262] an increase in the original superheating of the melt leads to a change in the composition of the central parts of the dendritic cells and to a reduction in the degree of intracrystallite liquation. These results are explained in [262] as being due to the fact that, when there is a slight superheating over the liquidus point, groupings close in composition to the solid-solution crystals exist even in the liquid solution. If, however, immediately

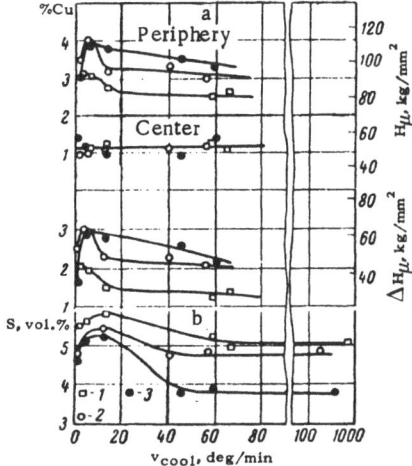

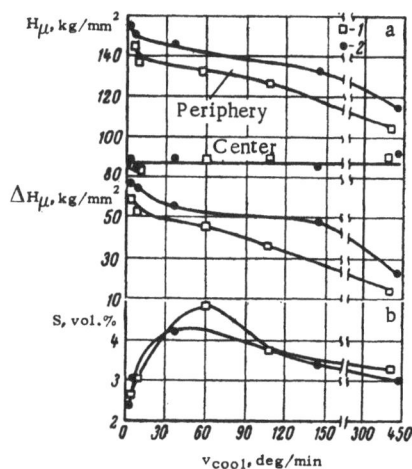

Fig. 98. Microhardness of the central and peripheral parts of the micrograins (a), and degree of intracrystallite liquation (b) as functions of the cooling rate after various superheatings of Al−5% Cu: 1) 680; 2) 750; 3) 900°C.

Fig. 99. Microhardness of the central and peripheral parts of the micrograins (a) and degree of intracrystallite liquation (b) as functions of the rate of cooling of Al−10% Mg after various superheatings: 1) 680; 2) 900°C.

before crystallization, the melt undergoes a substantial superheating over the liquidus point, then the probability of the formation of such groupings is greatly reduced, and for this reason there may be a reduction in the degree of intracrystallite liquation with increasing super-heating of the melt.

Data exactly opposite to the results of [262] were obtained in [263] for a number of aluminum alloys.

The effect of the temperature of preliminary superheating of the melt on the principal characteristics of dendritic liquation was studied in [263]; these characteristics were the composition of the central and peripheral parts of the micrograins of solid solution, the degree of intracrystallite liquation (difference in concentration at the periphery and in the center of the micrograins), and the amount of nonequilibrium excess of eutectic.

The experiments were carried out with binary aluminum alloys containing 2 and 5% Cu, 10% Mg, and 1.5% Mn. In order to protect the Al−Mg melt from oxidation 0.05% Be was added. For preparing the alloys, 99.99% pure aluminum, 99.92% magnesium, and electrolytic manganese were used. Each alloy was cast at six cooling rates after superheating to 680 and 900°C and holding for 30 min. The Al−Cu alloys were also cast after superheating to 750°C. The method of casting the samples and preparing microsections by electropolishing is described on p. 41.

Microscope analysis showed that in samples crystallized at the same cooling rate but after different intial superheatings of the melt the size of the micrograins was practically identical, and hence so was the mean crystallization velocity. Only for cooling rates of over 100 deg/min was there a marked tendency for the dendritic cells to become smaller as super-heating increased.

The intracrystallite liquation was studied by the microhardness method, while the volume content of eutectic was determined by the linear-analysis method from sections with a total

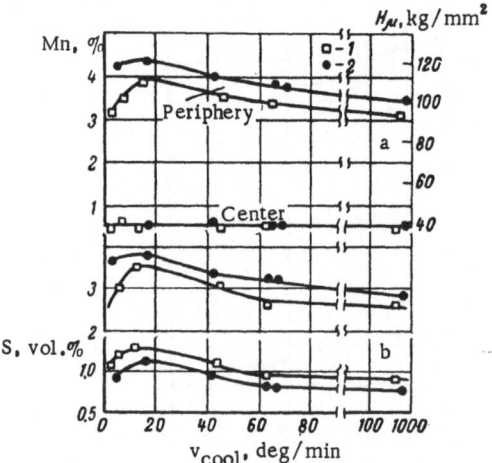

Fig. 100. Microhardness of the central and peripheral parts of the micrograins (a) and degree of intracrystallite liquation (b) as functions of the cooling rate after various superheatings of an Al−1.5% Mn melt: 1) 680; 2) 900°C.

length of 150 mm on each microsection. For the Al−Cu alloys the microhardnesses were converted to copper concentrations by means of a calibration graph relating the microhardness to the copper concentration in the solid solution plotted on the basis of homogenized standards. The mode of electropolishing and the conditions for measuring the microhardness were exactly the same for both standards and samples.

Figures 98 to 100 show the results obtained by studying intracrystalline liquation in aluminum alloys containing 5% Cu, 10% Mg, and 1.5% Mn. The results obtained for the Al−2% Cu were qualitatively analogous to those obtained for the Al−5% Cu alloy. The composition of the central parts of the micrograins of solid solution are almost independent of the melt superheating temperature in all the alloys studied (Figs. 98a, 99a, and 100). The composition also remains constant on varying the cooling rate.

It is most interesting that in the Al−Cu and Al−Mg alloys over a wide range of cooling rates the experimentally determined concentration of the alloying element at the periphery of the micrograins (near the eutectic precipitates) increases with increasing temperature of the original superheating of the melt (Figs. 98a and 99a). A change in the superheating temperature of the melt containing 1.5% Mn had practically no effect on the composition at the periphery of the solid-solution micrograins (Fig. 100). Since the composition of the center of the micrograins remains constant, the influence of the superheating of the melt on the degree of intracrystallite liquation is determined by its influence on the experimentally established concentration of the solid solution at the periphery of the micrograins.

Thus nonequilibrium crystallization of aluminum alloys takes place over the whole range of cooling rates studied under conditions of the almost complete suppression of diffusion in the solid phase and complete separating diffusion, a process ensuring the equilibrium concentration difference between the solid and liquid phases at the crystallization front.

Under these conditions the true composition of the boundary layer of the solid-solution micrograins near the second-phase precipitates should correspond to the concentration of the point of limiting solubility at the eutectic temperature.

Considering that for the practical range of cooling rates used in [263] and for any degree of superheating a second phase was always precipitated, the true degree of intracrystallite liquation should clearly be constant in each case.

However, the experimentally established degree of intracrystallite liquation depends on the superheating temperature and cooling rate; this is evidently associated with the varying character of the curve representing the distribution of the alloying component across the cross section of the crystallite, and with the inadequate resolving power of the microhardness method (the same features characterize all other current methods), which prevents the true composition of the solid solution from being determined at the phase boundary.

Figure 101 shows I. I. Novikov and V. S. Zolotorevskii's [263] distribution curves for copper across the solid-solution micrograins in samples of Al−5% Cu cooled at 5 deg/min

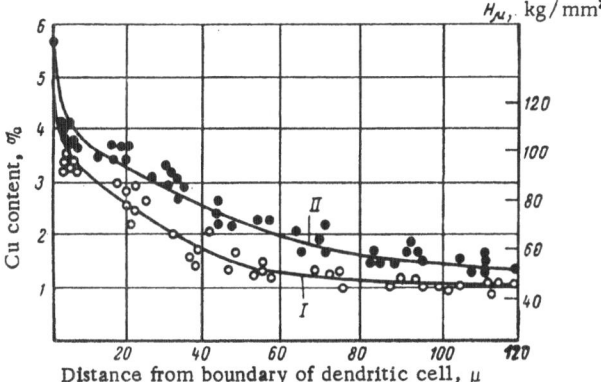

Fig. 101. Copper distribution curves over the cross section of micrograins in samples of Al−5% Cu after superheating to 680 (I) and 900°C (II).

after various superheatings of the melt. In plotting the distribution curves, impressions were made with a diamond pyramid at 10 to 20 μ from one another, while the center of the impression closest to the boundary with the second phase lay at a distance of 4 to 5 μ from this. The curves presented in Fig. 101 are averaged from the results of an examination of the dendritic cells in each sample. We see that raising the superheating temperature greatly increased the copper concentration, not only in the boundary layer but also over the greater part of the cross sections. In order to explain these results, concepts developed by Smith, Tiller, and Rutter [264] were employed in [263]; these were based on an analysis of nonequilibrium crystallization under conditions involving the complete suppression of diffusion in the solid phase, limited diffusion in the liquid phase, and the complete fulfillment of separating diffusion.

In an alloy with a concentration C_0 of the alloying element, the concentration C of the solid solution is expressed as a function of the distance x from the end of a single crystal grown in the directional manner from a plane wall by

$$\frac{C}{C_0} = 1 + 3\left(\frac{1-k}{1+k}\right)\exp\left(-2\,\frac{v_{cr}}{D}\,x\right) + 5\,\frac{(1-k)\,(2-k)}{(1+k)\,(2+k)}\exp\left(-6\,\frac{v_{cr}}{D}\,x\right)$$
$$+ (2n-1)\frac{(1-k)\,(2-k)\cdots(n-k)}{(1+k)\,(2+k)\cdots(n+k)}\exp\left[-n\,(n+1)\,\frac{v_{cr}}{D}\,x\right] + \cdots,$$

(3)

where k is the equilibrium distribution coefficient (k < 1), v_{cr} is the crystallization velocity, D is the diffusion coefficient in the liquid phase.

In the present case, on changing the superheating temperature of the melt, the manner in which the alloying component is distributed over the cross section of the solid-solution micrograins changes in such a way that any particular concentration C is reached at a greater distance (x_2) from the boundary in the case of substantial superheating than in the case of slight superheating (x_1).

It follows from the equation presented that, for constant C, C_0, k, and v_{cr}, the value of x will increase with increasing superheating, with a corresponding rise in the diffusion coefficient in the liquid D. Thus the change in the characteristics of dendritic liquation in relation to the original superheating of the melt may be due to its influence on the diffusion coefficient in the liquid phase at the time of crystallization.

The mechanism underlying this influence may be considered as follows. It is well known that, when the temperature of a melt is raised, one of the reasons for the rise in diffusion coefficient is the change in the structure of the liquid metal: a reduction in coordination number, an increase in interatomic distances, and a rise in vacancy concentration [265-267].

For the vacancy mechanism of diffusion, other conditions being equal, the diffusion coefficient is directly proportional to the vacancy concentration [268]. If the structure of the melt changed reversibly on heating and cooling, the diffusion coefficient in the melt in the liquid phase on crystallization would be independent of the preliminary superheating temperature. It is clear that, under practical conditions, after cooling the melt to the liquidus temperature, an excess vacancy concentration is to some extent established, and therefore the diffusion coefficient in the liquid phase on crystallization will be greater for a high initial superheating than for a low one. The higher the cooling rate of the melt from the superheating temperature, the more defects should remain in the structure of the liquid.

This proposition was confirmed in experiments with an Al−5% Cu alloy, which was cooled from 900 to 660°C at rates of 4 and 25 deg/min, and in the crystallization interval at an average rate of 15 deg/min. For this purpose some small crucibles containing the melt were raised to 900°C for 30 min and cooled in air until the end of crystallization, while others were cooled to 660°C in the furnace and then also in air. After this, the method described earlier was used to plot copper distribution curves over the cross section of the crystallites. As a result of this, it was found that the reduction of the rate of cooling between 900 and 660°C reduced the copper concentration over a large part of the solid-solution micrograins. This indicated a reduction in diffusion velocity in the liquid during crystallization.

If we superheat the Al−Cu alloy to 900°C, then cool it to 680°C and give a 30-min rest at this temperature before crystallization, then the copper-distribution curve over the cross section of the solid-solution micrograins will appear the same as in the case of an original superheating to 680°C. Evidently the isothermal rest at the lower temperature eliminates the effect of the earlier high superheating of the melt, since the structure of the liquid is able to rearrange itself completely and the vacancy concentration becomes the equilibrium one for the temperature in question.

Comparison of Figs. 98 to 100 shows that the influence of previous superheating on the compositions of the periphery of the solid-solution micrograins and the degree of liquation is different in the different alloys. This is primarily associated with the value of the distribution coefficient. For aluminum alloys containing 2 and 5% Cu, 10% Mg, and 1.5% Mn the ratio of the concentrations of the alloying element in the solid and liquid phases, calculated for a temperature in the middle of the equilibrium crystallization range, respectively equals 0.17, 0.35, and 0.7. Other conditions being equal, an increase in the distribution coefficient will lead to a contraction of the boundary layer of the solid solution enriched with the alloying element [264]. Hence the magnesium, and particularly manganese, concentration near the boundaries of the dendritic cells will rise in such a narrow layer on increasing the superheating of the melt that it will be hard to establish this rise experimentally.

The total increase in the content of alloying element in the primary crystals with increasing superheating of the melt becomes less significant on passing to alloys with a high distribution coefficient. Correspondingly the influence of the original superheating of the melt on the amount of nonequilibrium eutectic diminishes. Whereas in the alloys of the Al−Cu system this effect is very considerable (see Fig. 98c), in the Al−10% Mg alloy an increase in superheating only leads to a very small reduction in the amount of eutectic β phase, while in the Al−1.5% Mn alloy the superheating has no effect on the amount of eutectic constituent at all.

Thus the characteristics of dendritic liquation only depend appreciably on the original superheating of the melt in alloys belonging to systems with a large concentration difference between the liquidus and solidus lines.

Hence the conclusions of [262, 263] contradict each other. In the first of these the observed fall in the degree of microinhomogeneity on raising the superheating temperature is slight and lies within the limits of experimental error.

There is no doubt that we should accept the conclusion as to the constancy of the degree of intracrystallite liquation on raising the superheating temperature as being correct. We may only note that there is a change in the character of the curve representing the distribution of the alloying element over the cross section of the crystallites and that the microhardness is insufficiently localized. In this respect the facts mentioned in [262] may easily be explained.

In all probability, raising the superheating temperature leads to an increase in the degree of supercooling, as a result of which the micrograin becomes finer. The process invariably ends by the crystallization of the lower-melting point component (bismuth), which is also the softer. In addition to this, the character of the antimony distribution over the crystallite cross section also changes.

Owing to the insufficient resolving power of the microhardness method, impressions applied in the center embrace the contiguous softer parts, and this is ultimately interpreted as a fall in antimony content in the crystallization center.

I. I. Novikov and V. S. Zolotarevskii [250], explaining the results of M. V. Pikunov and A. I. Desipri [262], consider that on increasing the superheating of the Bi−7% Sb alloy the degree of supercooling on crystallization increases, and, in accordance with the equilibrium solidus curve in the Bi−Sb system, at lower temperatures the first portions of solid phase are formed with a lower antimony concentration and hence with a lower microhardness. Thus the increase in the degree of supercooling with increasing superheating of the melt may have the effect that the composition at the center of the dendritic cells approaches the original composition of the melt.

In the experiments with aluminum alloys [263], this factor had no marked effect on the composition at the center of the dendritic cells, since aluminum alloys are not inclined to severe supercooling on crystallization. However, the question as to the composition of the primary crystals on varying the degree of supercooling has not yet been elucidated. In particular, it is not yet clear whether the composition of the nucleus will be determined by the conode corresponding to the temperature at the intersection of the ordinate of an alloy of specified composition with the equilibrium liquidus, or by the conode corresponding to the temperature of supercooling with respect to the equilibrium liquidus. This is a fundamental question, to which as yet no unambiguous answer has been provided by experiment. Hence Novikov and Zolotarevskii's [250] explanation regarding the results of [262] are at the moment without foundation.

V. M. Glazov, L. A. Palelova, and S. A. Bezekovich* studied the effect of superheating temperature on the degree of liquation microinhomogeneity in Ge−Si alloys. It is well known that germanium and silicon form a continuous series of solid solutions. The experiments were carried out on a Ge−10% Si alloy. The original materials were germanium and silicon of $\sim 10^{13}$ and 10^{14} cm^{-3} purity respectively. In order to avoid as far as possible the entry of suspended impurities into the alloy, the original samples were obtained by pulling from the melt with continuous feeding of the latter, following D. A. Petrov's method.

In this way coarse-crystalline or single-crystal samples were obtained. The alloy samples were enclosed in sealed quartz ampoules evacuated to 10^{-3} mm Hg. After melting, the ampoules containing the alloy were heated to a specified temperature, held for a strictly-specified time, and cooled at three different rates calculated from the cooling curve. The superheating temperatures were $t_1 = 10$, $t_2 = 50$, $t_3 = 150$, and $t_4 = 250°C$ with respect to the equilibrium liquids. The cooling rates were ~ 18, 240, and 900 deg/min and were achieved by cooling the sample ampoules with the furnace, in air, and in water.

---

* The results of this investigation were presented to the Second Scientific-Technical Session of the Moscow Institute of Steels and Alloys on January 5, 1966.

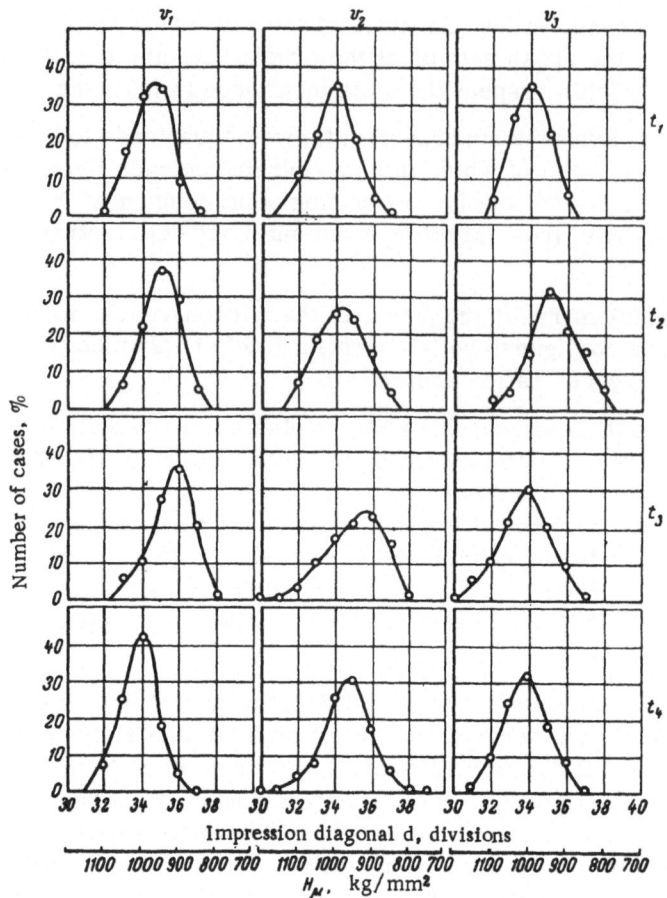

Fig. 102. Curves representing the microhardness values obtained in a Ge−10% Si alloy for various cooling
rates after various degrees of superheating.

Microsections were prepared from the samples thus obtained. In order to remove the
internal stresses arising on crystallization in a closed volume and also the surface hardening
due to grinding and polishing, the microsections were briefly vacuum-annealed at 500°C. The
degree of liquation inhomogeneity was studied by the microhardness method, using a load of
50 g and a loading time and time under load of 10 sec each.

Microstructural analysis of the samples obtained by the method described showed that
the center and periphery (the latter should essentially be pure germanium) differed in coloring
in by no means all samples. The peripheral parts could not always be sharply distinguished.
The following method was therefore used to study the microinhomogeneity in these peculiar
alloys. In each sample eight to twelve grains were studied, and in the coarse-grained samples
five to seven. The impressions of the diamond pyramid were made uniformly over the whole
area of the grain in such a way as to embrace both the "center" and the periphery. On each
sample 100 impressions were made or in some cases even more. The data thus obtained were
used in order to plot microhardness distribution curves and these are given in Fig. 102. From
these curves the minimum and maximum microhardness values were taken, and the absolute
difference between these served as a measure of the total liquation microinhomogeneity.

For control purposes, in some samples the center and periphery were studied separately
and microhardness distribution curves were plotted from the results. By way of example,

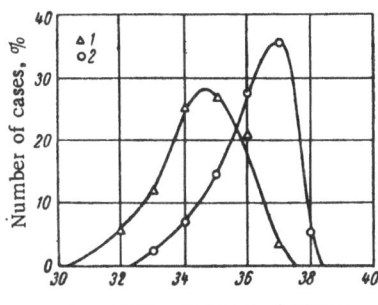

Fig. 103. Microhardness distribution curves obtained in a sample cooled at 240 deg/min after 150° superheating; curves shown separately for the center (1) and periphery (2).

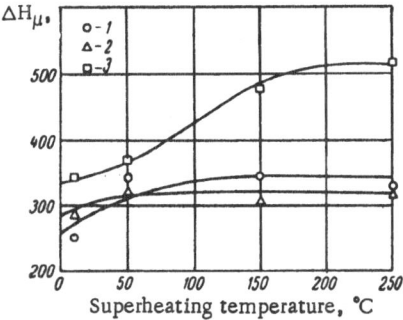

Fig. 104. Total degree of liquation microinhomogeneity as a function of the superheating temperature of the melt for samples of Ge−10% Si crystallized at three different cooling rates in the crystallization interval, deg/min: 1) 18; 2) 240; 3) 900.

Fig. 103 shows distribution curves for one of the samples obtained at a cooling rate of 240 deg/min after superheating by 150°C. This figure clearly shows that the curves diverge with respect to all microhardness values. The microhardness of the periphery is lower than the microhardness of the center; this is as it should be, if we remember that, according to theory, pure germanium or a very weak Ge-base solution should crystallize last.

The values of ΔH_μ obtained in Fig. 102 are shown in Fig. 104 as functions of the temperature of superheating relative to the equilibrium liquidus. We see from this figure that for cooling rates of 18 and 240 deg/min the degree of liquation microinhomogeneity is practically independent of the superheating temperature; such a temperature dependence only appears for a fairly high cooling rate, about 900 deg/min.

With increasing superheating temperature the experimentally determined degree of overall liquation microinhomogeneity increases. Let us analyze the nature of this rise. The maximum values of microhardness determined from Fig. 101 and taken as the microhardness of the crystallization center in our experiments may be regarded as constant and roughly equal to 1145 kg/mm², allowing for experimental error.

If we compare this value with the equilibrium curve relating microhardness to composition considered on p. 153, we find that the resultant microhardness value is close to the microhardness of an alloy having the composition of the equilibrium solidus for the case of our Ge−10% Si alloy.

Hence the composition of the central parts or the center is determined by the solidus and, within the limits under consideration, is independent of the superheating temperature and the rate of cooling. Owing to the fall in the microhardness of the periphery, the degree of overall liquation microinhomogeneity decreases; its dependence on the superheating temperature is detectable on using a fairly high cooling rate (~900 deg/min). The microhardness of the periphery reaches values of some 680 to 700 kg/mm², which is close to the microhardness of pure germanium. We may thus suppose that for low cooling rates after any superheating, and also for high cooling rates after slight superheating, the manner in which the alloy under consideration crystallizes is such that it leads to the formation of peripheral interlayers with a variable composition.

The resolution of the microhardness method is evidently inadequate to reveal parts corresponding to pure germanium in these interlayers, although such should certainly be present.

On crystallization at a relatively high cooling rate (900 deg/min) after considerable superheating, more substantial degrees of supercooling may be achieved. This may result in the formation of peripheral layers in which parts corresponding to pure germanium may be re-

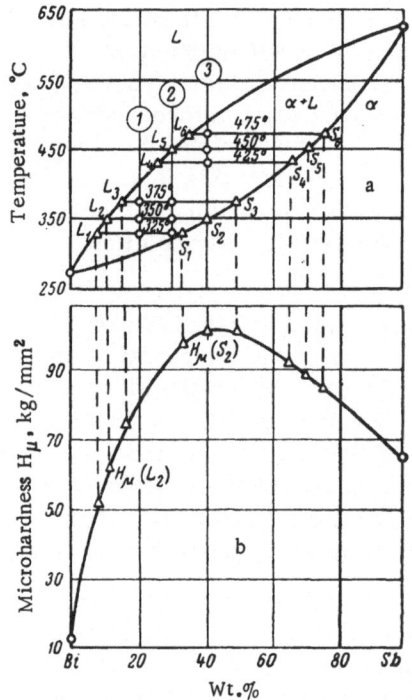

Fig. 105. Microhardness of Bi−Sb alloys as a function of composition, in comparison with the equilibrium phase diagram.

vealed by the microhardness method. In essence, the degree of overall liquation microinhomogeneity in the Ge−Si alloy under consideration should not depend on the superheating temperature for the cooling rates indicated; it should remain constant and equal to the difference 100% Ge−C_{so}, where C_{so} is the concentration of the solidus of the alloy in question.

The question as to the influence of superheating temperature on the degree of overall liquation inhomogeneity requires further, more detailed investigations using higher-resolution methods of determining compositions in microvolumes.

Crystallization Effect in the Quenching of Alloys from the Region of the Solid − Liquid State

It is well known that the quenching method is employed in order to fix the state in which an alloy exists at higher temperatures. This method is used in particular for fixing the structure of an alloy heated above the solidus temperature and thus existing in a semiliquid state. The question arises as to how far the final structure of an alloy quenched from temperatures corresponding to the two-phase region reflects the true picture of the equilibrium occurring prior to quenching.

Studying the structure of alloys quenched from temperatures corresponding to the two-phase region of the phase diagram, B. A. Movchan [269, 270] concluded that, for such temperatures, concentration layers 2 to 4 μ wide, with a higher concentration of the dissolved component, existed at the phase separation boundary. On the basis of this observation the author concluded that the phase diagrams only described the conditions of equilibrium in the alloys approximately. Movchan considered [280, 281] that the structure of the quenched alloy reflected the picture of the distribution of the components during a rest at a temperature between the liquidus and solidus points.

On the other hand, Ya. N. Malinochka [271] noted that, on quenching from temperatures of the semiliquid state, the primary crystals grew rapidly, as a result of which the thin layer of liquid adjacent to the crystal was enriched with the second component. The equilibrium distribution of the components in the semiliquid state could only be fixed by quenching if the possibility of the whole liquid undergoing diffusionless crystallization were ensured.

In order to resolve this question, V. M. Glazov [272] carried out an investigation by the microhardness method.

As object for study the Bi−Sb system was taken; the corresponding phase diagram is characterized by a continuous series of solid solutions.

The experiments were based on the following considerations. If diffusionless crystallization of the liquid is feasible on rapid cooling of the alloys from temperatures corresponding to the two phase (α + L) region of the phase diagram, then ultimately quenching should yield a structure with solid-solution crystals of two different compositions, determined by the ends of the conode characterizing the equilibrium between the solid and liquid phases during the isothermal holding period. For example, in the crystallization of an alloy 1 (Fig. 105a) initially held at 350°C, in the case of the diffusionless crystallization of the liquid of composition L_2, solid-solution crystals with compositions L_2 and S_2 (Fig. 105a) characterized by sharply differing microhardnesses $H_\mu(L_2)$ and $H_\mu(S_2)$ (see Fig. 105b) should occur in the structure of the alloy on quenching.

However, if there is no diffusionless crystallization on cooling the alloys after holding in the two-phase state, we have the ordinary process of nonequilibrium crystallization, with the suppression of diffusion in the solid phase, completed by the crystallization of the lower-melting point component (in the present case bismuth).

Bismuth alloys containing 20, 30, and 40% were studied. As original materials for preparing the alloys, extremely pure bismuth and antimony were used, the impurity content determined by spectral analysis not exceeding $1 \cdot 10^{-3}\%$.

The alloys were melted and after very slow cooling held at temperatures between the liquidus and solidus lines (Fig. 105a) for 10 h. In addition to this, the Bi−20% Sb alloy was held at 350°C for 10 days as a control. After the holding period crystallization was effected at various cooling rates.

The different cooling rates were achieved by cooling the alloys with the furnace, in air, in water, in liquid nitrogen, and also by casting into a copper mold cooled to −183°C. In the latter case, approximate calculation showed that the cooling rate reached $\sim(2-5) \cdot 10^5$ deg/min.

Each alloy was held at three different temperatures (Table 37). It follows from Fig. 105a that in some cases the amount of liquid phase was predominant in the alloys under consideration during the isothermal holding period, i.e., the alloy was situated in the region of the liquid−solid state [138, 139], and the possibility of forming narrow gaps between the primary crystals mentioned in [269] was excluded. In other cases the alloy was held at temperatures corresponding to the region of the solid−liquid state, and the possibility of such gaps being created therefore existed.

The microstructure and the distribution of microhardness over the crystallite cross sections were studied in the alloys so prepared. As etchant for revealing the microstructure, a 15% aqueous solution of HNO_3 was used. The microhardness was measured with a 10-g load in an apparatus furnished with a device for automatic loading.

Microscope analysis of Bi−Sb alloys crystallized from temperatures of the semiliquid state at various cooling rates after holding at various temperatures showed that in all cases the crystallization process was completed by the separation of almost pure bismuth. By way of example, Fig. 106 presents some micrographs of the structure of a Bi−20 wt.% Sb alloy crystallized at various rates of cooling after prolonged holding at 350°C. The micrographs clearly show dark regions constituting practically pure bismuth.

Table 37 shows the results of measuring the microhardness in the center and at the periphery of crystallites of the alloys under investigation.

It follows from the tabulated data that the microhardness of the dark parts (H_μ^p) corresponds to that of practically pure bismuth (which is equal to 10 or 12 kg/mm^2) or a dilute Bi-base solution.

The fact that the crystallization of the alloys, even on cooling very rapidly (v_3, v_4, v_5) from temperatures of the semiliquid state, in every case leads to the formation of practically pure bismuth, detectable both under the microscope and by means of microhardness measure-

TABLE 37. Microhardness of Bi−Sb Alloys Crystallized at Various
Cooling Rates after Isothermal Holding in the Semiliquid State

Sb content, wt.%	Temp., °C	v_1		v_2		v_3		v_4		v_5	
		H_μ^C	H_μ^P	H_μ^C	H_μ^P	H_μ^C	H_μ^P	H_μ^C	H_μ^P	H_μ^C	H_μ^P
20	375	102	12	103	18	102.0	17	101	49.0	101	19.5
	350	105	14	104	16	103.0	15	104	18.5	105	16.0
	325	97	13	99	13	95.0	15	97	17.0	—	—
30	375	100	15	102	15	100.0	16	101	19.5	102	18.0
	350	104	17	103	17	103.0	15	104	18.5	—	—
	325	99	14	98	18	95.5	17	98	22.0	—	—
40	475	78	22	79	19	81.0	18	80	15.0	81	21.0
	450	86	24	87	22	85.0	21	86	19.0	87	24.0
	425	92	23	92	24	89.0	21	91	18.0	92	27.0

Note. v_1, v_2, v_3, v_4, and v_5 are the cooling rates of the alloys with the furnace, in air,
in water, in liquid nitrogen, and in a cooled copper mold, respectively; H_μ^C is the microhardness of the center, H_μ^P is the microhardness of the periphery.

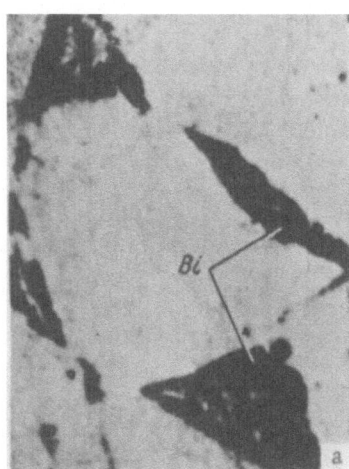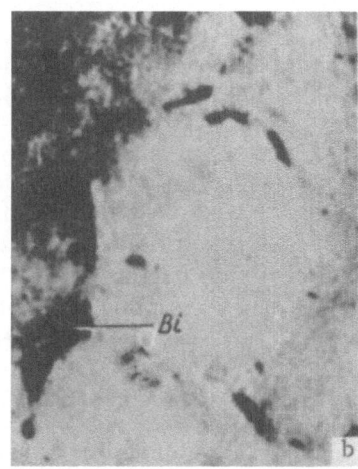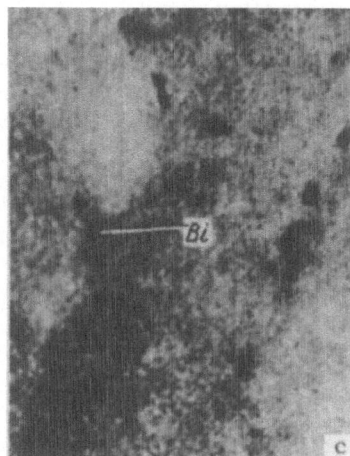

Fig. 106. Micrograph of the structure of a Bi−20% Sb alloy crystallizing after 240 h
at 350°C and cooling a) with the furnace, b) in water, and c) by casting into a cooled copper mold (×200).

ments, indicates that primary precipitating crystals are growing at the time of quenching, as a
result of which the melt is considerably enriched with the lower melting point component.
We may thus conclude that diffusionless crystallization cannot be achieved by using such cooling rates once crystallization by diffusion has started.

At the same time the microhardness of the centers of the crystallites corresponds to the
microhardness of an alloy having a composition determined by the corresponding point on the
solidus line at the holding temperature; this is confirmed by comparing with the measured
microhardness of alloys previously brought into equilibrium and having the composition S_1, S_2,
S_3, S_4, S_5, and S_6 corresponding to the composition of the solid phase in equilibrium with the
liquid at the temperature of the isothermal holding period (Fig. 105).

Comparison of the structure and microhardness data (Table 37) of the same alloy cooled at high cooling rates from the three temperatures indicated in Fig. 105 leads to the conclusion that, during the isothermal holding period in the semiliquid state, there is no special distribution of the components in the liquid, since crystallization from the region of the solid—liquid and liquid—solid states leads to essentially the same results. Furthermore, crystallization from the liquid—solid state, in which there are no narrow gaps between the primary crystallites containing (according to [269]) liquid substantially enriched with the second component (for example, from 375°C in the case of alloy 1 in Fig. 105), has the effect that the amount of bismuth in the structure of the quenched alloy is even larger than on crystallization of this same alloy at a lower temperature, for which the amount of the solid phase is greater and gaps of the kind mentioned may exist.

We may conclude from the foregoing that the equilibrium phase diagram completely characterizes the equilibrium in the alloy during the isothermal holding period. At the moment of quenching, on rapid cooling of the alloy after preliminary holding in the semiliquid state, dendritic liquation occurs, and owing to the suppression of diffusion in the solid phase this leads to the separation of the pure low-melting point component in the final stages of crystallization.

CHAPTER 9

MICROHETEROGENIZATION OF SOLID-SOLUTION CRYSTALS IN TWO-PHASE ALLOYS

The structure of the real crystals formed by primarily crystallizing solid solutions in two-phase alloys may be distinguished by a microinhomogeneity associated with the presence of submicroscopical second-phase particles, which are frequently not observed when studying the structure under the microscope.

This special microinhomogeneity, together with the microinhomogeneity of liquation origin considered in the previous chapter, may have a considerable effect on the results of investigations based on microhardness measurements as a method of physicochemical analysis.

In view of this we shall make a special study of investigations into this kind of microinhomogeneity, which is better called microheterogeneity.

Relation between the Microhardness of Solid-Solution
Crystals and the Composition of the Two-Phase Alloys

A. A. Bochvar and O. S. Zhadaeva [273] studied the microhardness of solid-solution crystals in cast Al−Cu and Al−Si alloys of various composition. The results are shown in Table 38.

It follows from these results that the microhardness of the primary crystals of the aluminum-base solid solution changes sharply with composition; it was accordingly concluded in [273] that even to the very first approximation the composition and properties of each of the phases in two-phase mixtures could not be regarded as constant.

After homogenization at 540°C for 10 to 18 h the difference in the microhardness values of the primary crystals diminishes, but nevertheless remains very considerable.

In addition to this, the absolute microhardness values of the aluminum solution in both systems under consideration were much higher for larger copper and silicon contents than might be expected even for the corresponding completely homogenized saturated solution.

On these grounds the authors of [273] proposed that those primary solid-solution crystals which appeared homogeneous under the microscope in fact had a complex internal structure owing to the branched skeleton of the primary dendrite, the branches of which enclosed very fine interlayers of a second phase, very hard to discern under ordinary microscope analysis.

With increasing content of the second component the number of such interlayers should increase and the microhardness of the solid-solution crystals should rise accordingly. It was

172

TABLE 38. Microhardness of Primary Solid-Solution Crystals in Al – Cu and Al – Si Alloys of Various Composition

Aluminum – copper		Aluminum – silicon	
Content Cu, %	Micro-hardness, kg/mm²	Content Si, %	Micro-hardness, kg/mm²
5.5	72	3.0	64
8.0	84	11.8	77
10.0	94	14.0	82
15.0	110	20.0	92
20.0	140	—	—
25.0	153	—	—
28.0	159	—	—

Note. For hypereutectic alloys in the Al – Si system the microhardness of the primary dendrites of the aluminum solid solution was measured in modified alloys.

noted in [273] that even Rosbaund and Schmid [274] and Straumanis [275] in the course of ordinary microscope study sometimes observed interlayers of a second phase within individual single crystals, in view of which the present microhardness data would suggest that the phenomenon is quite general.

In subsequent investigations Bochvar's suggestion [273] (also [8, 9]) was completely vindicated.

A systematic study of the relation between the microhardness of solid-solution crystals and the composition of a number of binary alloys was made in [113, 146].

The microhardness of crystals formed by a solid solution of copper in aluminum was studied as a function of the composition of the alloy for various quench temperatures in [146].

The alloys were prepared from 99.99% pure aluminum and electrolytic copper (99.9% pure) in corundum crucibles and cast into a cast iron mold. The cast samples, after 30% deformation, were homogenized at 520°C for 50 h and slowly cooled with the furnace to room temperature. The microhardness was measured with a load of 10 g. After measurement at room temperature the alloys were successively quenched from 350, 400, 450, and 500°C. At each temperature there was a rest of 50, 40, 30, and 20 h, respectively.

The results of the microhardness measurements are shown in Fig. 107 in comparison with the equilibrium phase diagram of the Al – Cu system.

We see from this figure that, after passing through the line of limiting solubility at the corresponding temperature, the microhardness remains constant over a range of concentrations ~2% Cu, the "plateau" (range of concentration with constant microhardness of the solid-solution crystals) moving regularly to the right with increasing temperature in accordance with the behavior of the limited-solubility curve.

After a certain concentration the microhardness of the solid-solution crystals increases substantially. Microscope analysis of the alloys in this range of concentrations reveals homogeneity of the solid-solution grains, even at large magnifications (×1000).

However, the rise in the microhardness of the solid-solution crystals in alloys to the right of the "plateau" and the rising branch of the microhardness/composition curve indicate the same lattice constant of the solid solution, corresponding to the value relating to a solution limitingly saturated at the temperature in question. The values of the limiting solubility of copper in aluminum at the corresponding temperatures, obtained from the breaks in the microhardness isotherms in Fig. 107, agree closely with data obtained by x-ray structural analysis. We may conclude that the rise in the microhardness of the solid-solution crystals of two-phase alloys on raising the proportion of alloying component is associated with the presence of submicroscopic particles of a second phase, heterogenizing the grains of the solid solution constituting the saturated solution at the temperature in question. It was furthermore noted that a reason for the development of microheterogeneity may be not only dendritic crystallization but also the decomposition of the saturated solid solution on cooling.

The number of second-phase particles heterogenizing the solid-solution crystals of two-phase alloys should increase with increasing content of the second component, independently

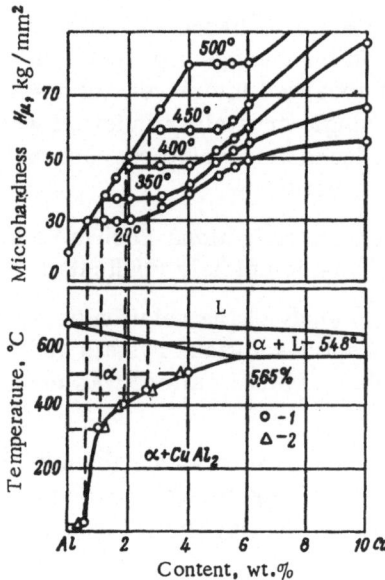

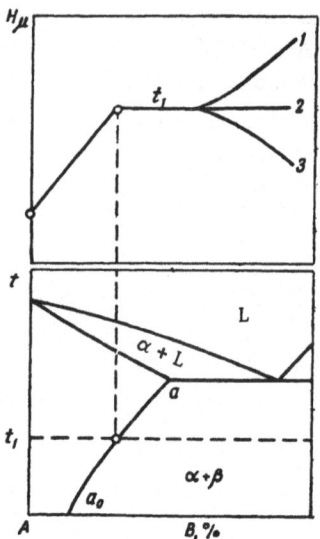

Fig. 107. Microhardness of the solid-solution crystals as a function of the composition of Al—Cu alloys quenched from various temperatures. 1) Micro-hardness data; 2) data obtained by x-ray diffraction.

Fig. 108. Typical relations between the microhardness of the solid-solution crys-tals and the composition of the two-phase alloys.

of their genesis. It is clear that, immediately after passing from the one-phase region into the two-phase region, the number of second-phase particles will be insignificant, and up to a cer-tain limiting concentration they have no effect on the microhardness. The "plateau" on the microhardness isotherms may be explained in this way.

After reaching a certain limiting concentration corresponding to the formation of a special kind of "framework," the microhardness starts rising sharply.

This kind of mechanism for the effect of structure on hardness was described and con-firmed experimentally for the case of lead/steel spheres and lead/wire models by A. A. Bochvar and R. G. Kozolupova [276].

The agreement between the results obtained for the solubility of copper in aluminum from the curves of Fig. 107 and the earlier data [119, 120] indicates that, even with microheterogene-ity in the solid-solution crystals of two-phase alloys, the position of the boundary must be de-termined from the transformation from the one-phase part of the phase diagram to the two-phase part.

There may be three kinds of curves relating the microhardness to the composition of two-phase alloys (Fig. 108) corresponding to cases in which 1) $H_\mu(\beta) > H_\mu(\alpha)$; 2) $H_\mu(\beta) \approx H_\mu(\alpha)$; 3) $H_\mu(\beta) < H_\mu(\alpha)$.

In the first case the microhardness of the solid-solution crystals of two-phase alloys should increase with increasing proportion of the second component, in the second it should re-main almost constant, and in the third it should fall. Subsequent experimental work confirmed the validity of the latter conclusion.

The concentration dependence of the microhardness of solid-solution crystals was studied in [113] for the Cd—Sn, Cd—Bi, Al—Si, and Al—Zn systems.

TABLE 39. Solubility of Tin and Bismuth in Cadmium in the Solid State

Temp., °C	Content, wt.%	
	Sn	Bi
170	4.15	—
130	2.10	3.55
100	1.00	1.75
20	0.35	0.45

These systems were chosen for study because in the first two the microhardness of the second phase was smaller than that of the solid-solution crystals and in the second two larger.

In addition to this, the choice of the first two systems was also determined by the fact that the existence of regions of Cd-base solid solutions in these systems was controversial. The Cd−Bi system serves in textbooks and monographs as one of the few chemical examples of a metallic system with practically no mutual solubility of the components in the solid state. In the Cd−Sn system the solubility of tin in cadmium has also been considered extremely low (under 0.25%). However, in the Cd−Sn and Cd−Bi systems an unusually high ductility appears in eutectic alloys, and this is explained by solution-precipitation processes at the phase boundary of the solid solutions. On this basis, I. I. Novikov [112] studied the hardness of alloys belonging to these systems in the solid and solid−liquid states and observed a considerable solubility of bismuth and tin in cadmium. This is also indicated by V. P. Shishokin and V. A. Ageeva [277], who observed a break on the hardness isotherm.

Cadmium−tin and cadmium−bismuth alloys containing 0.5, 1, 2, 3, 4, 5, 6, 7, and 10 wt.% of the second component were prepared for study, as well as aluminum alloys containing 2, 4, 6, 8, 10, 12, 15, 18, 20, 22, and 30 wt.% Zn and 0.05, 0.5, 0.8, 1.5, 2, 3, 4, 5, 6, 7, 8, and 10 wt.% Si. The Al−Si alloys were prepared in corundized and the rest in graphite crucibles under a layer of charcoal and were cast into a cast iron mold. The cadmium-base alloys were deformed by 30%, annealed at 130°C for 150 h, and cooled with the furnace. The surface-hardened layer on the microsections was removed by brief annealing and etching in a 5% alcohol solution of nitric acid, the etching products being carried away (see Chapter 3). In order to plot the microhardness isotherms the series of alloys in question were successively quenched from 100, 130, and 170°C (the latter only for the Cd−Sn system) after holding for 100, 75, and 50 h respectively.

The method of preparing the Al−Zn and Al−Si alloy samples was analogous to that indicated in the case of Al−Cu.

In order to plot the isotherms at 500°C for the Al−Zn system, series of alloys were water-quenched from the temperatures indicated* after holding for 30 and 60 h respectively at the temperatures specified.

After the foregoing treatment the microhardness was measured. The results of the measurements are given in Fig. 109a in the form of curves relating the microhardness of the solid-solution crystals to the composition of the alloy. The solubilities of tin and bismuth in cadmium obtained in these graphs are presented in Table 39.

The curves representing the limited solubility of tin and bismuth in cadmium obtained from these data are shown in Fig. 109.†

The character of the relationship between the microhardness of the solid-solution crystals and the composition of the alloys in the systems studied (Fig. 109) confirms the validity of the typical relationship indicated in Fig. 108. This may serve as support for the assertion that the solid-solution grains appearing homogeneous under the microscope actually have a more complicated internal structure, distinguished by microheterogeneity.

---

* The Al−Zn alloys were held at 200°C after heating to 400°C and cooling to 200°C slowly in the furnace.

† The solidus lines are taken from [112], which agrees closely with the results of [113].

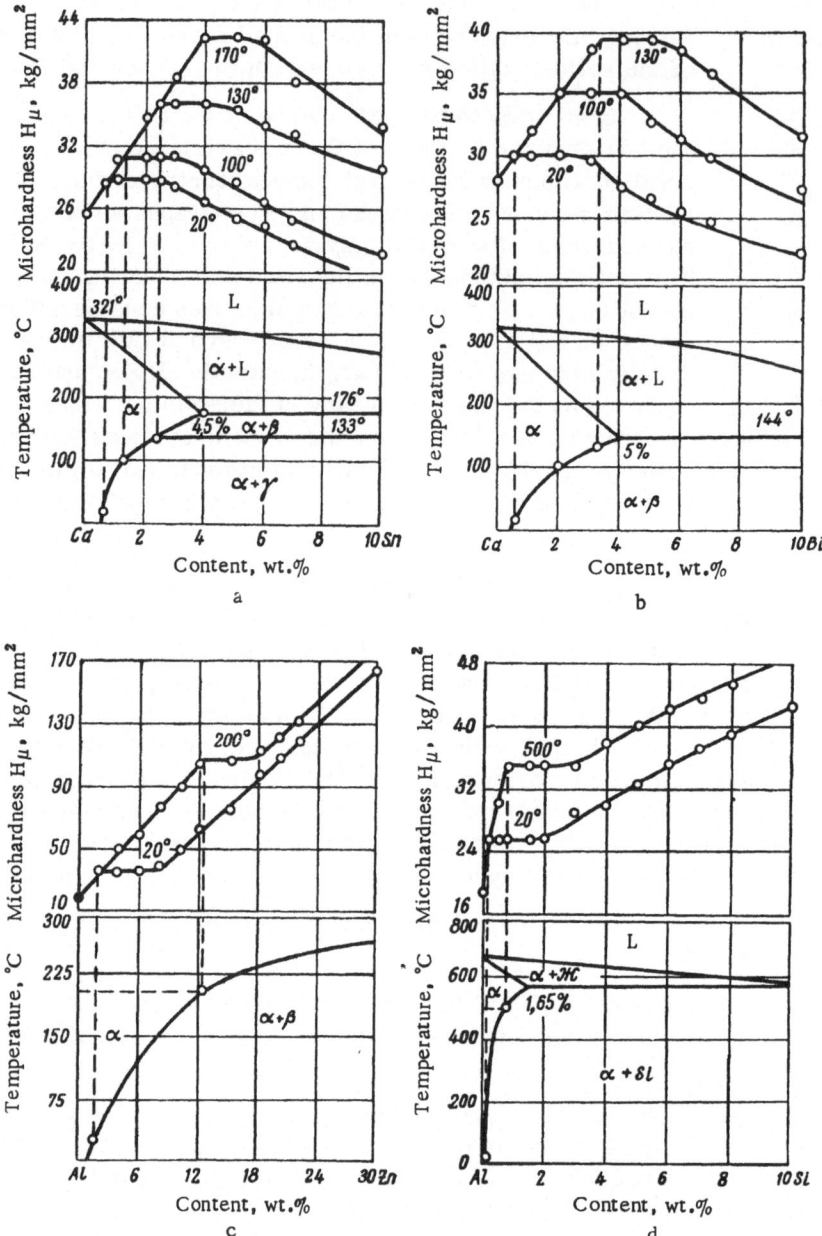

Fig. 109. Relation between the microhardness of solid-solution crystals and the composition of the alloys: a) Cd−Sn; b) Cd−Bi; c) Al−Zn; d) Al−Si.

The horizontal section ("plateau") on the corresponding curves and the smooth character of the transition from the plateau to the rise or fall in these curves show that the second-phase particles lying within the solid-solution crystals under consideration only start having a hardening or softening effect after they have been formed in sufficient quantities.

The results obtained on determining the solubility of silicon and zinc in aluminum at the corresponding temperatures, in the presence of microheterogeneity in the solid-solution crystals of the two-phase alloys, agree closely with earlier results, as in the case of Al−Cu alloys.

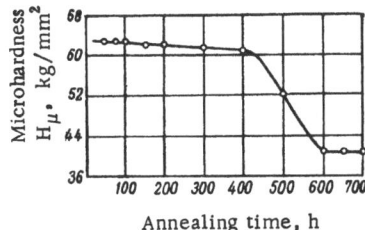

Fig. 110. Relation between the microhardness of the solid-solution crystals in an $Al-3\%$ Cu alloy and period of homogenization at 400°C.

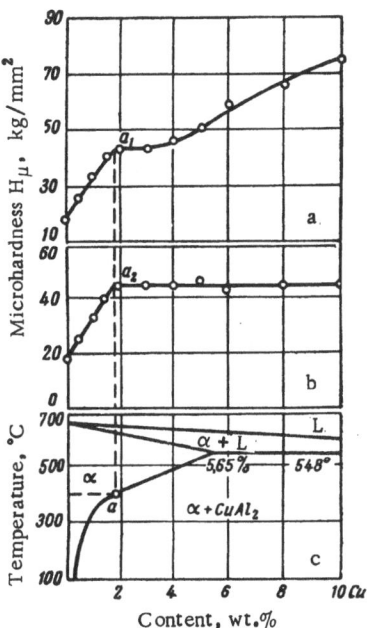

Fig. 111. Relation between the microhardness of the solid-solution crystals and the composition of the alloy at 400°C in the $Al-Cu$ system after homogenization for: a) 50 h; b) 650 h; c) equilibrium phase diagram.

Thus microheterogeneity does not prevent the correct determination of solubility by the microhardness method.* Experiment shows that in order to eliminate the effect of microheterogeneity on the microhardness of solid-solution crystals in two-phase alloys, very long homogenization indeed was required.

By way of example, Fig. 110 shows the relation between the microhardness of solid-solution crystals belonging to an $Al-8\%$ Cu alloy and homogenization time at 400°C. We see from this figure that homogenization at 400°C must last over 600 h in order to eliminate the microheterogeneity in the solid-solution grains entirely. In view of the falling diffusion velocity, still longer is needed at lower temperatures. In Fig. 111 the equilibrium phase diagram of $Al-Cu$ (Fig. 66) is compared with the relationship between the microhardness of the solid-solution crystals and the composition of the alloy obtained for the same series of samples but after different periods of homogenization at 400°C. We see from this figure that, in the $Al-Cu$ system, after homogenization for 50 (Fig. 111a) and 650 (Fig. 111b) h, the character of the microhardness/composition relationship at concentrations corresponding to the two-phase region of the phase diagram is completely different. However, the solubility determined is the same in both cases (points a_1 and a_2 in Fig. 111a and b).

It follows from Figs. 107 and 109 that the range of concentrations with constant microhardness of the solid-solution crystals is quite large after passing from the one- to the two-phase part of the phase diagram, and the corresponding breaks on the curves are quite sharp enough to determine the saturation boundary.

In view of this, the authors proposed "real" composition/microhardness diagrams, allowing for the phenomenon of the microheterogenization of the solid solution in two-phase alloys [110, 111]. Figure 112 presents these diagrams for several cases of interaction between the components.

Figure 112 shows the results of an investigation into the $Cd-Zn$ system carried out by V. N. Vigdorovich and B. I. Zharov, supporting the typical "real" microhard-

*V. M. Vozdvizhenskii made a number of investigations into the phenomenon of microheterogenization [278-281]. In addition to the microhardness method, Vozdvizhenskii also measured electrical resistance in order to reveal microheterogeneity [278-279]. In addition to this, the same author proposed determining the limiting solubility in the solid state in cases in which microinhomogeneity interfered with the determination by reference to the difference in microhardness values obtained under different loads [280, 281], this difference being greatest at the point of limiting solubility.

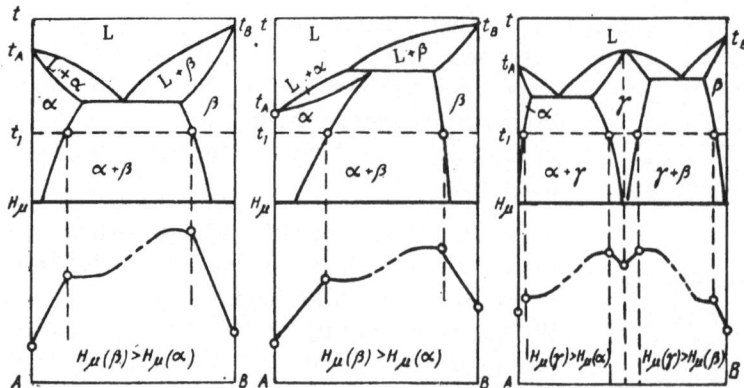

Fig. 112. "Real" microhardness/composition diagrams for
several cases of interaction between the components.

ness/composition diagram for a system of the eutectic type. On taking account of these diagrams, there is no longer any need to anneal for long periods in order to secure the theoretical relationships given in Fig. 33. It is sufficient to conduct the investigation close to the limiting-solubility line in the two-phase region, which is always possible, since the approximate position of this line appears even after preliminary microscope study of the corresponding samples. The use of these schematic representations is completely justified from the point of view of equilibrium, since the solid-solution crystals in alloys lying (as regards composition) in the two-phase region of the phase diagram become saturated after a comparatively brief period of annealing, while the required form of their microhardness/composition curves is explained by the special structural distribution of the second phase, the total amount of which is clearly to be determined from the lever rule.

Two Mechanisms of the Microheterogenization of Solid-Solution Crystals in Two-Phase Alloys

The results obtained by measuring the microhardness of solid-solution grains in two-phase Al−Zn alloys (Fig. 109a) show that the character of the microhardness/composition curves is the same as in the Al−Si system (Fig. 109d), although in the first case heterogenization of the solid-solution crystals as a result of crystallization was excluded. Hence the second-phase particles complicating the structure of the solid-solution grain of two-phase alloys belonging to the Al−Zn and similar systems are formed in the course of cooling as a result of the decomposition of the solid solution, i.e., ordinary aging.

In binary alloys analogous to the Al−Si system, the structure of the solid-solution grain is further complicated in two-phase alloys by a heterogeneity of second order formed in the course of dendritic crystallization.

Thus the heterogenization of the solid-solution grains in two-phase alloys may arise as a result of the precipitation of submicroscopic second-phase particles on cooling in the solid state, associated with a corresponding change in solubility (we provisionally call this the first process), and also as a result of their precipitation in the interaxial spaces of the solid-solution dendrites in the course of primary recrystallization (second process). Ultimately the solid-solution grain of the two-phase alloys consists of heterogenized particles of various origins. We may suppose that the particles arising in the course of crystallization will be more stable and will be able to preserve themselves longer in the course of protracted heat treatment than particles arising in the decomposition of the solid solution.

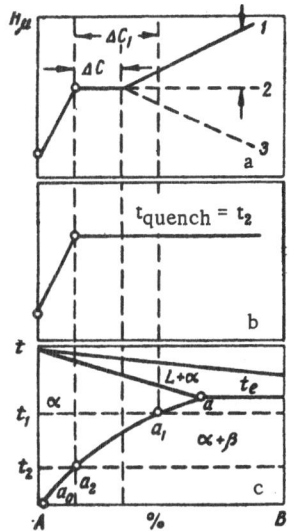

Fig. 113. Character of the relationship between the microhardness of solid-solution crystals and the composition of the alloy: a) when the two microheterogenization mechanisms are present; b) after removing the effect of microheterogeneity on the microhardness; c) phase diagram.

During prolonged annealing at high temperatures, the submicroscopic second-phase particles may coagulate, and this should clearly lead to the removal of their influence on the microhardness of the solid-solution crystals.

If we consider the combined effect of the two processes microheterogenizing the solid-solution crystals of two-phase alloys on the character of the microhardness/composition curve, we note, first of all, that in those systems in which the solubility is very low or depends little on temperature the first process in the heterogenization of the solid solution is excluded or only just perceptible.

When both microheterogenization processes occur, the microhardness/composition curves should have the form of curve 1 (or 3) as shown schematically in Fig. 113a. Such curves are found, clearly, when the homogenization period is sufficient for the solid solution to become saturated at the corresponding temperature, but too small for the coagulation of the submicroscopic second-phase particles heterogenizing the solid-solution grain in two-phase alloys.

In the same period, alloys situated (in composition) to the left on the point of limiting saturation at the temperature in question, in which the second phase appears as a consequence of dendritic liquation, should become single-phased.

In alloys lying to the right of the point (Fig. 113), the total microheterogenization is determined by the two processes considered; with increasing quench temperature the relative effect of the first process diminishes, since then the difference in the concentrations of the solid solution at the temperature of eutectic formation and the quench temperature becomes smaller.

If the alloys are homogenized at a temperature t_2 (Fig. 113) for a time sufficient for the coagulation of the submicroscopic second-phase particles, then the microhardness/composition curve will have the form indicated in Fig. 113b.

In order to fix the state at which the solid-solution grains of all the two-phase alloys only possess microheterogeneity arising as a result of the precipitation of second-phase particles from the lattice of the solid solution on cooling from t_1 to t_2 (Fig. 113), we must first provide long homogenization at t_2 (sufficient to eliminate all microheterogeneity), then heat the alloys to t_1 and hold sufficiently for the concentration of the solid solution to become equal to a_1, and then cool the alloys to t_2 and quench after a certain holding period at this temperature.

If in this case the difference ΔC_1 in the saturated solid solution at temperatures t_1 and t_2 is greater than the amount of second component ΔC required to create such a quantity of particles as will affect the microhardness, i.e., $\Delta C_1 = a_1 - a_2 > \Delta C$, then there should be a rise (or fall) on the corresponding curve relating the microhardness of the α crystals to the composition of the alloy after passing from the one- to the two-phase region immediately after the "plateau." Furthermore, with increasing content of the second component the microhardness should clearly only change in the range of compositions to the left of point a_1, after which the number of second-phase particles should remain constant, in view of which the microhardness of the α phase crystals should also remain constant and equal to the microhardness of crystals

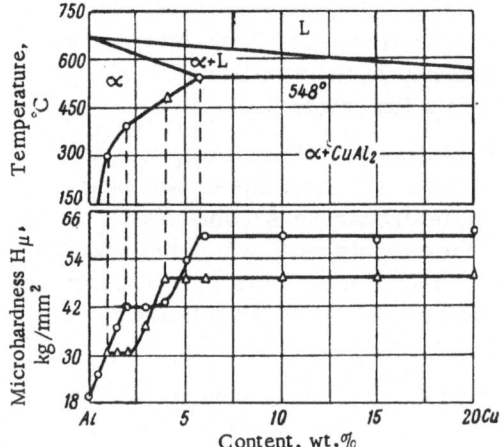

Fig. 114. Microhardness of solid-solution crystals as a function of the composition of Al−Cu alloys in the sole presence of the microheterogenization mechanism associated with the decomposition of the solid solution on cooling; upper curve: 500 to 400; lower curve: 550 to 300°C.

of this phase in an alloy containing a_1 (%) of the second component. If $\Delta C_1 \leq \Delta C$, then in this case the second-phase particles formed as a result of the decomposition of the solid solution will be few in view of the fall in the corresponding saturation concentration (from a_1 to a_2), and they will have no effect on the microhardness of the α phase crystals in the two-phase alloys.

In such cases, after corresponding repeated heat treatment, we should arrive at a relationship similar to that shown in Fig. 111b.

In order to secure an experimental verification of the foregoing, a series of Al−Cu alloys was studied [282]. This system is convenient because the solubility of copper in aluminum changes little between room temperature and roughly 300°C, while on further increasing the temperature it rises rapidly. We see from Fig. 111 that the extent of the plateau ΔC on the microhardness isotherms is about 2% Cu. Thus the condition $\Delta C_1 > \Delta C$ is easily satisfied in the Al−Cu system.

Aluminum alloys containing 0.5, 1, 1.5, 2, 3, 4, 5, 6, 8, 10, 15, 20, and 25% Cu were prepared.

The method of preparing the alloys and the initial materials were analogous to those described earlier (p. 64). Subsequently heat treatment based on the foregoing principles was carried out.

The resultant 100-g bars were each deformed by 20 to 30% and microsections were prepared from them; these were then annealed at 500°C for 10 h to remove the consequences of dendritic liquation. Clearly in this case the microhardness increment resulting from crystallization microheterogeneity was somewhat reduced, although, as indicated at the beginning of this chapter, this period was insufficient by a long way for the coagulation of the particles creating the microheterogeneity to take place.

After this the alloys were cooled with the furnace to 400°C, held for 50 h, and quenched in water. The microhardness of the samples thus obtained was measured. The results of the measurements are presented in Fig. 111a. We see from this figure that the relation between the microhardness of the solid-solution crystals and the composition of the alloys is a typical curve of the kind characterizing microheterogeneity in the solid-solution grains of two-phase alloys.

After plotting the relationship in question, alloy samples were subjected to continuous prolonged homogenization for 650 h at 400°C in order to eliminate the effect of microheterogeneity arising from the coagulation of the particles at the temperature under consideration. After this period the samples were quenched in water and their microhardness was measured after appropriate surface treatment. The results of the measurements, presented in Fig. 111b, indicated that as a result of the prolonged homogenization the microheterogeneity in the solid-solution grains had been eliminated. The relation thus obtained between the microhardness of the solid-solution crystals and the composition of the alloys agreed with theory. One series of samples, after the foregoing treatment, was placed in a furnace, heated to 540°C, held at this temperature for 50 h, and slowly cooled to 400°C, held for 5 h, and quenched in water, while another series underwent analogous treatment at 500 and 300°C, also followed by water quenching. After this the microhardness of the two series of samples thus treated was measured.

The results of the measurements (average of 15), shown in Fig. 114 together with the equilibrium phase diagram of the Al−Cu system, completely confirm the earlier-indicated changes in microhardness after the corresponding heat treatment; they also indicate that this kind of treatment (heating and slow cooling), applied to samples which have undergone preliminary long-term homogenization, leads to the appearance of only one mechanism in the microheterogenization of the solid-solution crystals, namely, that associated with the precipitation of submicroscopic second-phase particles from the lattice as a result of the reduction in solubility on reducing the temperature. In this way we may observe the phenomenon in pure form and establish the relative influence of the two mechanisms governing the development of microheterogeneity in the primary solid-solution crystals of two-phase alloys.

Clearly in these alloys the increment in the microhardness of the solid-solution grains relative to its equilibrium value will be composed of two terms:

$$\Delta H_\mu = \Delta H'_\mu + \Delta H''_\mu, \tag{1}$$

where $\Delta H'_\mu$ is the increment in microhardness associated with the microheterogeneity of crystallization origin or heterogeneity of the second order; $\Delta H''_\mu$ is the increment in microhardness associated with the microheterogeneity resulting from the decomposition of the solid solution on cooling due to the corresponding fall in solubility.

If we compare the curves relating the microhardness of the solid-solution crystals to the composition of the alloys given in Figs. 111a and 114, we easily see that in the alloy containing around 5.6% Cu, $\Delta H'_\mu = 0$, while in the alloy containing 15% Cu at 400°C, $\Delta H'_\mu = 20$ kg/mm²; in alloys containing over 15% Cu heterogeneity of the second order predominates at 400°C, while in alloys containing less copper the greatest effect is exerted by the microheterogeneity due to the decomposition of the solid solution.

On the basis of the foregoing data we may regard the existence of two mechanisms in the development of the microheterogeneity in the solid-solution crystals of two-phase alloys as being conclusively proved.

Possible Existence of a Third Mechanism for the

Microheterogenization of Solid-Solution Crystals

in Two-Phase Alloys

In the foregoing we have considered two mechanisms of microheterogenization in solid-solution crystals, respectively associated with the dendritic character of the crystallization of the primary crystals and with the decomposition of the solid solution in the presence of a temperature-dependent solubility in the solid state.

The microhardness of solid-solution crystals was studied in [283] for Ni−C alloys, and as a result it was suggested that there might also be a third mechanism of microheterogenization. Let us consider the results of this investigation.

The alloys were prepared in two stages: first electrolytic Ni was melted in a vacuum induction furnace, using a graphite crucible, in order to obtain an initial alloy. Then a mixture of nickel and the alloy was twice remelted in an arc furnace in an atmosphere of purified helium. The cast samples were annealed in vacuum (~1 · 10⁻³ mm Hg) at 1100°C for 10 h after pre-

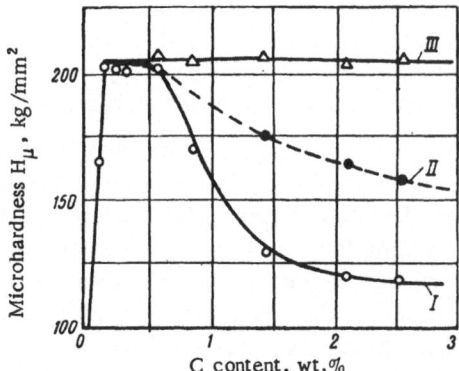

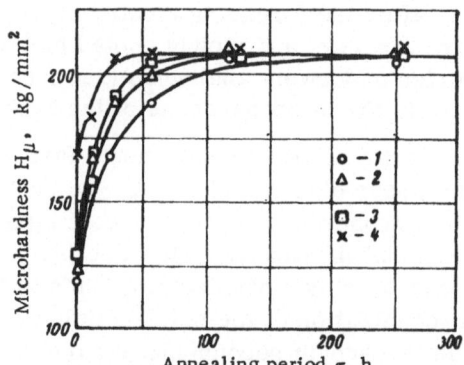

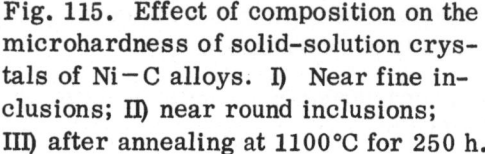

Fig. 115. Effect of composition on the microhardness of solid-solution crystals of Ni−C alloys. I) Near fine inclusions; II) near round inclusions; III) after annealing at 1100°C for 250 h.

Fig. 116. Effect of the period of annealing on the microhardness of a solid solution of carbon in nickel for C content (wt.%) of: 1) 2.55; 2) 2.13; 3) 1.46; 4) 0.82.

paring microsections, and then the microhardness of solid-solution crystals corresponding to alloys of different compositions was measured under a load of 20 g in a PMT-3 hardness tester with automatic application of the load.

Alloys containing 0.1, 0.16, 0.21, 0.33, 0.56, 0.82, 1.46, 2.13, and 2.56 wt.% C were studied.

Figure 115 shows the results of the measurements. We see from this graph that the microhardness/composition relationship for the solid solution is characterized by curves corresponding to the microheterogenization of the solid solution of two-phase alloys with a phase softer than the base component, i.e., this is analogous to the case considered on p. 175 for the case of the Cd−Sn and Cd−Bi systems.

The microheterogeneity in the solid-solution crystals of two-phase Ni−C alloys is confirmed by the results of an electron-microscope investigation. Comparative microstructural analysis showed that the solid-solution grains which appeared homogeneous under the ordinary microscope at magnifications up to 800 revealed second-phase particles at a magnification of 4000 in the electron microscope. Altogether some 30 photographs of the structure of solid-solution crystals in two-phase alloys, confirming microheterogeneity, were taken with the help of carbon replicas.

The resultant microhardness/composition relationship for the Ni−C system is distinguished by an unusually sharp fall in the microhardness in the two-phase region, particularly near the relatively fine graphite inclusions (Fig. 115, curve I).

In the neighborhood of the larger inclusions, the microhardness falls less sharply on increasing the amount of carbon in the alloy (Fig. 115, curve II); this is apparently associated with the coagulation of submicroscopic second-phase particles heterogenizing the solid-solution grains of the two-phase alloys.

This sharp fall in the microhardness of the solid-solution crystals on increasing the carbon content within the two-phase region of the Ni−C phase diagram indicates an extremely well-developed microheterogeneity. One notices that the extent of the "plateau" after passing from the one- to the two-phase region (2 to 2.5% in other systems) is here very insignificant. This may be associated with the extremely sharp rise in the number of submicroscopic second-

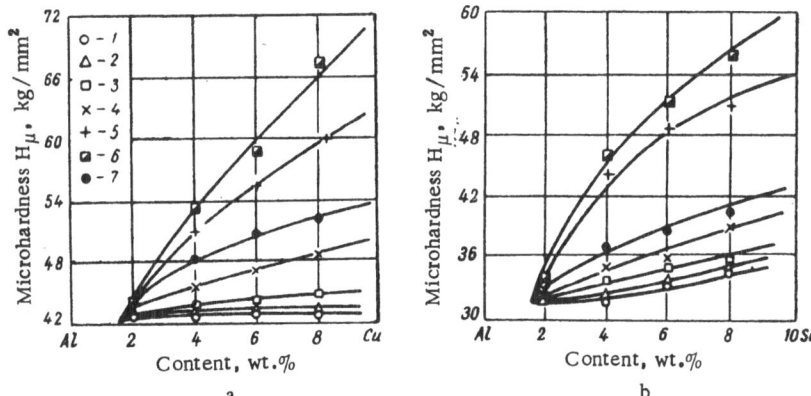

Fig. 117. Microhardness of solid-solution crystals as a function of the composition Al — Cu (a) and Al — Si (b) alloys for the following cooling rates on crystallization (deg/sec): 1) 0.02; 2) 0.5; 3) 2; 4) 20; 5) 89; 6) 180; 7) 400.

phase particles heterogenizing the solid-solution crystals of the two-phase alloys with increasing carbon content. It is furthermore important to remember that the heat treatment employed (annealing at 1100°C and water-quenching) practically excludes the possibility of microheterogenization by decomposition of the solid solution. Hence the origin of the microheterogeneity of the solid-solution crystals in two-phase alloys is almost entirely of a crystallization nature. In view of this, we may suppose that the observed rapid fall in microhardness and the small extent of the plateau are due to the possibility of microheterogenization, not only as a result of microheterogeneity in the actual melt, as indicated by the results given in [284].

The prolonged homogenization annealing of two-phase Ni — C alloys leads to the elimination of the influence exerted by microheterogeneity on microhardness. Figure 116 shows the microhardness of alloys containing 2.56, 2.13, 1.46, and 0.82 wt.% C as a function of annealing time at 1100°C. We see from this graph that the microhardness of the solid-solution crystals after annealing takes the same value as the microhardness of an alloy limitingly saturated at 1100°C (Fig. 115, curve III). The solid-solution crystals in alloys with a smaller C content reach the limiting microhardness much more rapidly than those in alloys with a larger content.

The foregoing investigation shows that in addition to the two earlier-mentioned mechanisms of microheterogenization there may also be a third mechanism associated with microheterogeneity in the corresponding melts.

Considering the analogy between the properties of liquid solutions of carbon in iron and nickel we may suppose that graphite also heterogenizes primary austenite. The submicroscopic graphite particles in primary solid solution of the metal — carbon type should exert an influence on graphitization processes.

Effect of Cooling Rate on the Microhardness

of Heterogenized Solid-Solution Crystals

of Two-Phase Alloys

Heterogeneity of the second order resulting from dendritic crystallization should depend (in the same way as microinhomogeneity of a liquation origin) on the cooling rate, which has a decisive influence on the formation of structure of one type or another. In view of this N. N. Glagoleva and V. M. Glazov studied the influence of cooling rate on the concentration dependence

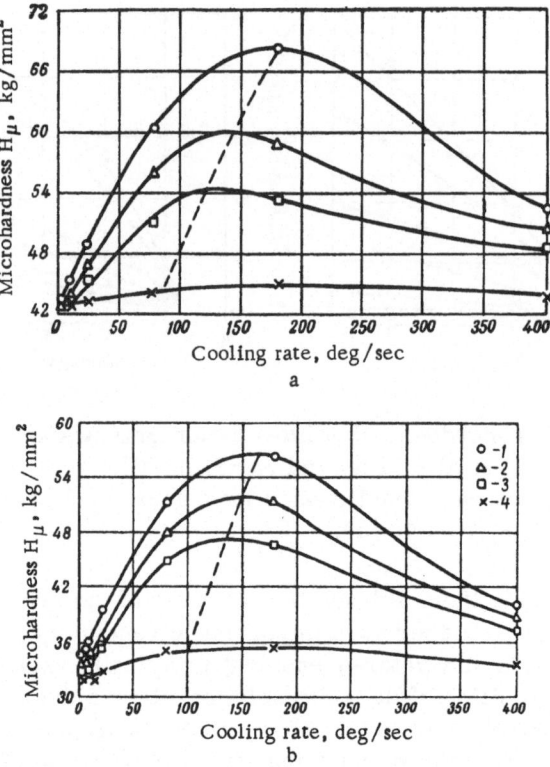

Fig. 118. Influence of cooling rate during crystallization on the microhardness of solid-solution crystals of two-phase Al−Cu (a) and Al−Si
(b) alloys, %: 1) 8; 2) 6; 3) 4; 4) 2.

of the microhardness of solid-solution crystals corresponding to two-phase alloys in the Al−Cu
and Al−Si systems.

Binary aluminum alloys containing 2, 4, 6, and 8 wt.% of copper and silicon were prepared. The alloys were crystallized at various cooling rates. For each sample a cooling curve
was recorded on the recording device and the cooling rate during crystallization was calculated from this.

For very high cooling rates approximate estimates were made.

Alloys crystallized at different rates were deformed by 20 to 30%, annealed at 400°C for
100 h (it was shown earlier that this period was insufficient to eliminate microheterogeneity),
and water-quenched from the temperature indicated. Then after surface preparation (Chapter 3)
the microhardness was measured.*

Figures 117a and b show the results of these measurements (average of 20 to 25) in the
form of a relationship between the microhardness of the solid-solution crystals and the composition of the Al−Cu and Al−Si alloys. We see from these figures that the character of the
relationships in question depends substantially on the cooling rate. This is determined chiefly
by the changes in the microhardness of alloys containing 4, 6, and 8% of the alloying component.
The microhardness of the alloys containing 2% of copper and silicon, however, remains almost

_____
* The influence of aging was excluded since the microhardness was measured immediately after
  quenching.

at the same level after the heat treatment in question; it is not difficult to see from the graphs of Figs. 107 and 109d that this level is determined by the limiting solubility of copper and silicon respectively at 400°C.

Figures 118a and b show the microhardness of the solid-solution crystals as a function of cooling rate for each of the alloys studied. We see from these figures that the relationships in question are represented by curves with maxima, the positions of which depend only slightly on cooling rate.

This kind of relationship between the microhardness of solid-solution crystals of two-phase alloys and cooling rate is evidently associated with corresponding changes in the character and degree of their microheterogenization.

If we start from the mechanism underlying the formation of microheterogeneity in the course of crystallization, we may conclude that the microheterogeneity of crystallization origin should primarily be associated with the degree of branching of the dendrites, that the microheterogeneity should be developed to a greater extent, and that hence this should have a stronger effect on the value of the microhardness.

In view of this the authors considered the effect of cooling rate in alloys of various composition on the degree of branching in the dendritic crystal growth forms, allowing for the two possible types of interaction between the components [285], on the basis of G. V. Kurdyumov's theory of diffusionless transformations [286-288].

It is well known that the main causes underlying the real process of crystal formation by dendritic crystallization are the nonuniform distribution of concentration gradient and degree of supercooling in various parts of the growing crystal [289-291].

It was concluded by analyzing the crystallization of solid solutions for different cooling rates that the relation between the degree of dendrite branching and the cooling rate might pass through a maximum.

The authors studied the relation between the microhardness of the solid-solution crystals and the cooling rate during the crystallization of aluminum—iron alloys of various compositions. The curves obtained were analogous to those shown in Fig. 118 as regards the shape and relative positions of the maxima, and it was concluded from this that the experimental results fell within the framework of the theoretical analysis [285].

Kinetics of the Transformation of Quasihomogeneous
Solid-Solution Crystals of Two-Phase Alloys into
the Truly Homogeneous State

It was shown earlier that the microheterogeneity in the solid-solution crystals of two-phase alloys was stable on brief (20 to 30 h) and even quite long (100 to 150 h) homogenization at high temperatures. It follows from Fig. 110 that in order to eliminate microheterogeneity completely in the solid-solution crystals an Al−8% Cu alloy had to be homogenized for over 600 h.

In order to elucidate the reasons for the stability of microheterogeneity and the nature of the processes taking place in microheterogenized solid-solution crystals during homogenization at high temperatures, the authors made a detailed study of the kinetics of the transformation of quasihomogenized solid-solution crystals of two-phase alloys into the truly homogenized state [292].

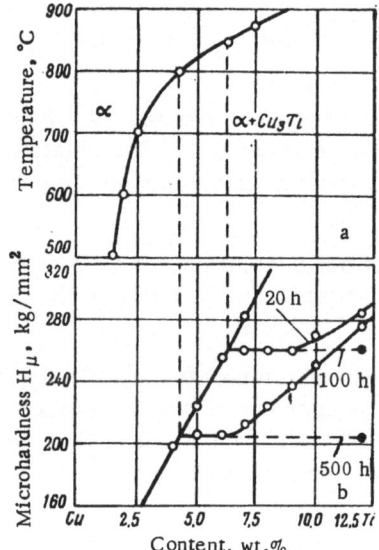

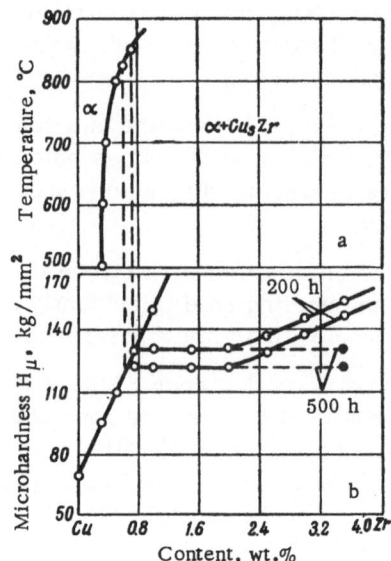

Fig. 119. Solubility of titanium in copper at various temperatures (a) and corresponding microhardness isotherms for various periods of homogenizing anneal (b).

Fig. 120. Solubility of zirconium in copper at various temperatures (a) and corresponding microhardness isotherms for various periods of homogenizing anneal (b).

As subject for study, binary alloys of copper with titanium and zirconium were taken; in these the submicroscopic particles of intermetallic compounds Cu_3Ti and Cu_3Zr in the solid-solution crystals had a considerable effect on the microhardness. Hence the principal method of studying the kinetics of the transformation of the quasihomogeneous solid-solution crystals into the truly homogeneous state was taken as that based on microhardness measurements.

The kinetics were studied at 850, 825, 800, 700, and 600°C for alloys of the Cu−Ti and 850, 825, and 800°C for alloys of the Cu−Zr systems. The temperature was regulated to an accuracy of ±3°.

In order to saturate the essentially α solid solution, cast samples were previously deformed by 50% and heat treated.

The mode of heat treatment was as follows.*

Heating to temperature, °C	850	825	800	700	600
Holding time, h	16	18	20	50	100

After this treatment the first microhardness measurements were made. The surface of the samples was prepared by the method given in Chapter 3 for studying the microhardness after quenching.

In almost all cases neither comparatively brief homogenization (16 to 20 h at 800 to 850°C) nor longer treatment (50 to 100 h at 700 to 600°C) caused coagulation of the colloidal solution of Cu_3Ti and Cu_3Zr in the solid-solution crystals, as indicated by the higher values of the microhardness of the α crystals as compared with the equilibrium values, which should correspond

---

*In all cases quenching was effected by cooling in water.

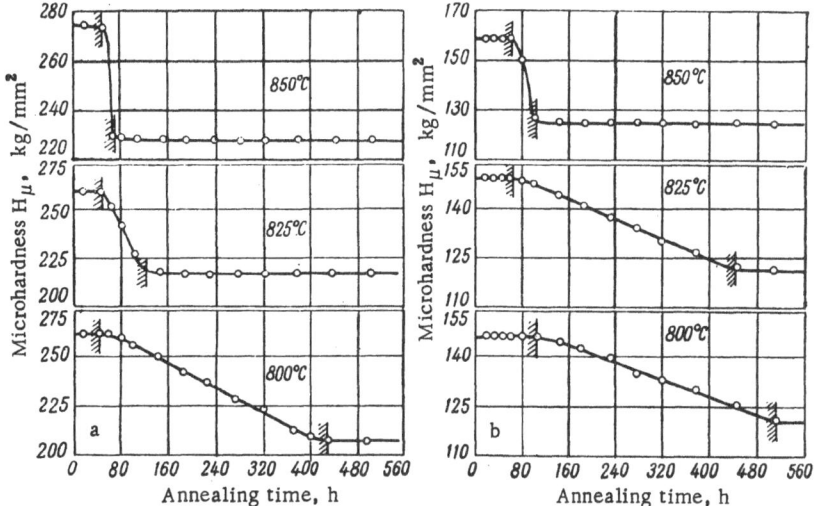

Fig. 121. Kinetic curves representing the transformation of solid-solution crystals from the quasihomogeneous to the truly homogeneous state: a) Cu−12% Ti alloy; b) Cu−3.5% Zr alloy.

to the absence of submicroscopic particles of the second phases in question. Longer holding periods were required for this. In order to illustrate the foregoing, Figs. 119 and 120 show the microhardness of the solid-solution crystals as a function of the composition of the alloy in the binary systems under consideration, corresponding to two different quenching temperatures. It follows from the data presented that the microhardness of the solid-solution crystals of a Cu−12% Ti alloy increased by 10 and 70 kg/mm² after holding for 20 h at the temperatures mentioned.

In Cu−3.5% Zr alloys after a 20-h holding period at 850 and 825°C the excess of the microhardness of the α crystals over the equilibrium value after 500-h holding was respectively 26 and 30 kg/mm².

After determining the initial points the alloys were annealed in stages, in the intervals between which quenching and microhardness measurements were carried out. The samples were held at the temperatures twice for 20 h each, then four times for 40 h, and finally five times for 60 h. Thus altogether holding periods of up to 600 h were studied. The results are shown in Fig. 121a,b. Despite such a long holding time, the effect of the particles heterogenizing the solid-solution grains on the microhardness of the latter at 700 and 600°C was never completely removed in the Cu−Ti alloy.

In all the remaining cases, the microhardness of the crystals constituting solid solutions of titanium and zirconium in copper, after prolonged homogenization, became equal to the microhardness of the solid solution belonging to the alloy completely saturated at the temperature in question.

Analysis of the resultant experimental data was difficult owing to the absence of a general criterion for the degree of microheterogeneity in the two systems considered, which made it possible to study samples in corresponding initial states. However, taking account of the chemical analogy between titanium and zirconium, we may suppose that the use of the same conditions in preparing and processing the Cu−Ti and Cu−Zr alloys would bring these into almost corresponding initial states.

The accuracy of determining the instant corresponding to the onset and completion of the transformation of the solid solution from the quasihomogeneous into the truly homogeneous state was of the order of 20 h. It should be added that the instant at which this process was completed could not be established absolutely accurately from microhardness measurements, since the microhardness reached a value corresponding to the microhardness of the solution saturated at the temperature in question before the last heterogenizing second-phase particles vanished.

On the basis of the foregoing considerations this investigation should be regarded as an approximate estimation of the influence of temperature and holding period at a given temperature on the stability of the microheterogeneity formed under the conditions in question. Nevertheless, an estimate of this kind is certainly useful, since a study of the behavior of the microheterogenized solid-solution crystals of two-phase alloys is, generally speaking, quite difficult.

Nature of Microheterogeneity

Careful microscope investigations into the structure of the solid-solution crystals in two-phase alloys of a number of systems enabled the authors to draw a number of conclusions regarding the nature of microheterogeneity [293].

In a microscope study of electropolished samples in two-phase Al−Cu and Al−Si alloys, no microheterogeneity could be seen at all in the solid-solution crystals in direct reflected light, even at high magnifications (1000 times). The grains of the solid solution appeared quite homogeneous.

However, a consideration of the same samples in the dark field of the microscope (Chapter 1) revealed an interesting effect, namely, that shining spots appeared uniformly distributed over the whole field of the black-looking grain, the number of these increasing somewhat near the grain boundary. At the same time no such effect was observed on studying samples subjected to prolonged homogenization (sufficient to eliminate microheterogeneity) in the dark field condition.

In dark-field examination the sample surface is illuminated from all sides by oblique rays falling at a very small angle, and this gives a better representation of surface tint and shape. Only a small proportion of the reflected and scattered rays falls into the objective.

In view of this the effect observed in our dark-field microscope may be regarded as a special kind of Tyndall effect.

On the basis of this idea as well as the facts listed earlier the phenomenon of microheterogenization may be explained as being a colloidal state of the solid solution.

We may suppose that the heterogenized grains of the primary solid-solution crystals constitute a colloidal solution. As in every colloidal system, in the present case there are two phases (there may be more); the heterogeneous system consists of a dispersion medium (solid solution) and a dispersed phase (second-phase particles distributed in the solid-solution crystals). The size of the particles of the dispersed phase distributed over the whole volume of the solid-solution grain is very small, and this leads to a considerable increase in the surface of separation between the dispersion medium and the dispersed phase. It is precisely this quantitative change in the extent of the internal surface which leads to the appearance of new qualities in the system, qualities which classify it as colloidal.

The two essentially differing mechanisms for the formation of microheterogeneity mentioned earlier may here be regarded as a dispersion (owing to dendritic crystallization) and condensation (owing to the precipitation of second-phase particles in the solid solution) mode of forming the colloidal solution in the solid-solution grains of two-phase alloys.

The reason for the formation of the colloidal solution by the dispersion process is the dendritic character of the growth of the primary solid-solution crystals. The effect of the dendritic crystallization is that portions of as yet uncrystallized melt are captured and surrounded on all sides by branches of the growing dendrite. Being still liquid and at the same time cut off from the main mass of melt, these portions of metallic liquid are crystallized independently. During crystallization the composition of these portions of liquid changes and becomes eutectic, after which second-phase particles are formed. According to the particular way in which the components interact, the melt may also ultimately attain peritectic composition, and this will also lead to the formation of second-phase particles. The subdivision of the melt by the branche of the dendrites on crystallization leads to the formation of particles of different sizes. Hence the colloidal solution formed on this principle may be regarded as polydispersional, i.e., containing particles of many sizes.

A study of microhardness over the grains of a solid solution constituting a colloidal solution formed by the condensation process leads to the conclusion that the second-phase particles are in this case precipitated and formed simultaneously over the whole volume of the solid-solution grain. On the basis of a somewhat simplified model of the process, we may consider that the second-phase particles are formed in places where the atoms of the second component have congregated (local fluctuations of composition) simultaneously over the whole volume, and grow at a constant linear velocity. Owing to the similar conditions of generation and growth, the size of the particles will in this case be approximately constant. Hence the colloidal solution formed by the condensation process may be regarded as monodispersoidal.

Clearly the greatest deviations from the assumption made will occur near the grain boundaries of the solid solution, where the material is characterized by a large number of vacancies and a high value of the isobaric-isothermal potential (Gibbs function), which promotes preferential formation of the second-phase particles.

It remains to consider the question as to the reasons for the stability and the possible mechanism responsible for the vanishing of microheterogeneity in the course of prolonged annealing.

It follows from the graphs presented in Fig. 121a,b (see also Fig. 110 that the relation between the microhardness and the holding time at various temperatures has a strictly regular form. These relationships obey the general scheme indicated in Fig. 122. It follows from this scheme that the stability of the colloidal solution at a specific temperature may be estimated by means of two quantities, of which one corresponds to the period of time (we call this the aggregate τ_{agg}) during which the microhardness remains practically constant, while the other is the period of the sharp fall in microhardness (we call this kinetic τ_{kin}) down to the equilibrium microhardness values.

Evidently in the period of aggregate stability we have relaxation of the internal stresses in layers surrounding the second-phase particles, constituting a result of the difference in the specific volumes of the coexisting phases. In this same period, the least stable second-phase particles heterogenizing the solid-solution grain dissolve.

In the period of kinetic stability, as a result of diffusion processes, the second-phase particles dissolve and the surface of separation between the phases within the solid-solution crystal vanishes. Considering that the transformation of the quasihomogeneous crystals into the truly homogeneous state most probably takes place by way of diffusion, we may make a comparative estimate of this process for the two systems indicated by using methods well known from chemical kinetics [284] and diffusion theory [295].

In the case under consideration the rate of diffusion may be estimated by the reciprocal of τ_{kin}.

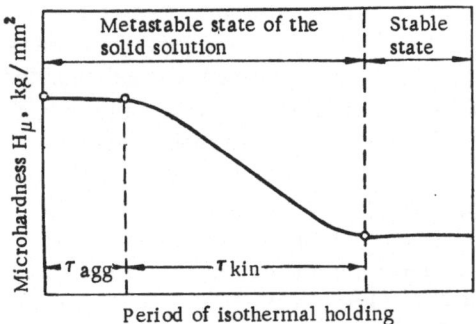

Fig. 122. General form of the relation between the duration of the isothermal homogenizing anneal and the micro-hardness of the solid solution of two-phase alloys when these transform from the quasihomogeneous to the truly homogeneous state.

Figure 123 shows the logarithm of the reciprocal of the period of kinetic stability as a function of the reciprocal of the absolute temperature. It follows from the graph that the relationship in question is excellently described by the equation

$$\ln \frac{1}{\tau_{kin}} = \frac{\Delta E}{RT} + \text{const},$$

where ΔE is the activation energy of the transformation of the quasihomogeneous solid-solution crystals into the truly homogeneous state; T is the temperature; R is the universal gas constant.

By determining the slope of the $\tau_{kin} - f(T)$ relationship, expressed in coordinates of $\log(1/\tau_{kin}) - (10^3/T)$, we find the activation energy for the transformation of the solid-solution crystals of two-phase alloys of the Cu−Ti and Cu−Zr binary systems as 140,500 and 261,300 cal/mole respectively.

It would be interesting to compare these quantities with the activation energy for the diffusion of titanium and zirconium in copper, and also with the heats of dissolution of the same elements in copper. However, there are no data regarding the diffusion of titanium and zirconium in copper in the literature. Nevertheless, we may note that the values obtained exceed the usual activation energies of the diffusion of one metal in another by a factor of two or three.

As regards the heats of dissolution of titanium and zirconium in copper, these may be estimated by mathematical analysis of the temperature dependence of the solubility of these elements in copper, using the Schroder−Van't Hoff equation [296]:

$$\ln x = \frac{Q}{RT} + \text{const},$$

where x is the atomic concentration of the dissolved component; Q is the molecular heat of dissolution.

For this purpose the solubility data obtained in [128, 297] for titanium and zirconium in copper were analyzed. Figure 124 shows the logarithm of the atomic concentration of the dissolved element as a function of the reciprocal of the absolute temperature.

The heats of dissolution of titanium and zirconium in copper thus determined were respectively −4200 and −7430 cal/g-atom.

By comparing the activation energy for the transformation of Cu−Ti and Cu−Zr solid-solution crystals from the colloidal to the truly homogeneous state with the heats of dissolution of these elements in copper in the solid state (Table 40), it is easy to see a specific correlation between the quantities in question for each of the systems under consideration.

To a certain approximation (say, by limiting consideration to binary alloys formed between one particular element and other elements all belonging to the same subgroup in the Mendeleev Table), we may accept the following relation between the quantities in question:

$$\frac{\Delta E_1}{\Delta E_2} \approx \frac{Q_1}{Q_2} = \text{const}$$

TABLE 40. Comparison between the Thermodynamic Characteristics
of the Cu−Ti and Cu−Zr Systems
(Quenching Temperature Range 800 to 850°C)

System	ΔE, cal/g-atom	Annealing temp., range, °C	Q, cal/g-atom	$\dfrac{\Delta E}{Q}$	$\dfrac{\Delta E_{Ti}}{\Delta E_{Zr}}$	$\dfrac{Q_{Ti}}{Q_{Zr}}$
Cu—Ti	−147 500	500—875	−4200	35.1 ⎫		
Cu—Zr	−261 300	400—980	−7430	35.2 ⎭	0.56	0.57

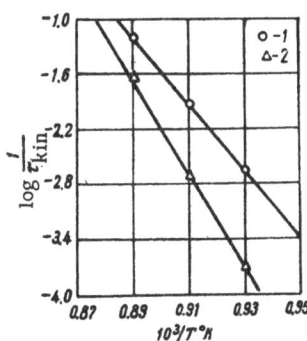

Fig. 123. Logarithm of the reciprocal of the period of kinetic stability of colloidal solutions of Cu$_3$Ti and Cu$_3$Zr in crystals of the α solid solution as a function of the reciprocal of the absolute temperature: 1) Cu−Ti; 2) Cu−Zr.

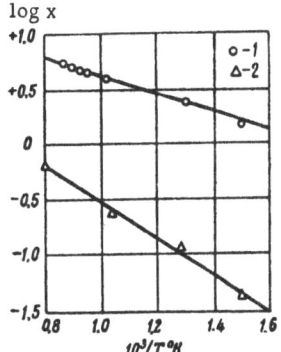

Fig. 124. Representation of limiting solubility curves of titanium and zirconium in copper in the solid state (see Figs. 119a and 120a) in semilogarithmic coordinates for determining the heats of dissolution of titanium and zirconium in copper: 1) Cu−Ti; 2) Cu−Zr.

or

$$\frac{\Delta E_1}{Q_1} \approx \frac{\Delta E_2}{Q_2} = \text{const,}$$

i.e., the ratio of the activation energy for the transformation of the quasihomogeneous solid-solution crystals into the truly homogeneous state to the heat of dissolution is a constant quantity.

Let us consider the mechanism underlying the process in hand.

It is quite obvious that the dispersed second-phase particles heterogenizing the solid-solution crystals of two-phase alloys are dissimilar in size. Any dispersity (particle size) of the inclusions is characterized by a certain distribution curve.

The difference in particle size is associated with the different mechanisms of their formation and also with the statistical character of their size deviations for each of the two groups of particles formed in the two ways indicated on p. 189. In view of this, the stability of some particles will be greater than that of others.

In addition to this, the stability of the second-phase particles formed in the course of dendritic crystallization will be greater than that of particles formed by the decomposition of the solid solution, since clearly in the latter case diffusion processes will be made easier in the regions adjacent to the second-phase particle.

The size difference in the finely dispersed second-phase particles has a decisive effect and constitutes the reason for the growth of the larger particles at the expense of the dissolution of the smaller ones. It is well known that the solubility of the interacting components depends on the size of the dispersed second-phase inclusions [298-301]. The greater the dispersion of the crystallites of the heterogenizing phase, the greater is the supersaturation of the solid solution surrounding it.

In this connection we may suppose that, owing to the presence of particles of different sizes, chemical inhomogeneity is created in the solid-solution crystal, and this leads to the formation of concentration fluctuations near the surface of the dispersed phase. As a result of this, at high temperatures there is a spontaneous, uncompensated diffusive transfer of atoms, directed toward the equalizing of the chemical composition in the solid-solution crystal. The regions closest to the smaller particles are impoverished with respect to the second component and become unsaturated (allowing for the effect of particle size on solubility), while the regions close to the larger particle supersaturate. The impoverishment of the regions surrounding the fine particles leads to their dissolution, as a result of which the surrounding layer of solid solution acquires a limiting concentration characteristic of the particular particle size and temperature specified. On the other hand, the supersaturation of the layer surrounding the larger particles leads to the growth of the latter, as a result of which this layer is impoverished with respect to the second component and also takes on a concentration characteristic of the particular particle size and temperature. This concentration is smaller than that of the layers surrounding the finer particles, so that the solid-solution crystal again becomes chemically inhomogeneous. Hence two processes take place in the solid-solution crystals heterogenized by submicroscopic second-phase particles: the equalization of chemical inhomogeneity as a result of diffusion, and the creation of such an inhomogeneity as a result of the different solubilities of the second component in the layers surrounding particles of different sizes.

Both these processes lead to the dissolution of the finer particles and the growth of the larger ones. Thus the transformation of quasihomogeneous solid-solution crystals into the truly homogeneous state is completed by the diffusion redistribution of the second-component atoms by way of dissolution and precipitation.

The foregoing process is thermodynamically spontaneous, since it involves a reduction in the store of surface energy of the system, the latter passing into a state with a lower Gibbs free energy.

A study of the kinetics of the process described and the hidden link between its activation energy and the heat of dissolution confirm the fact that the process takes place by the dissolution of atoms of the second component in some parts of the crystal and their deposition on the large second-phase inclusions. Since the process described had a diffusion character, it may clearly take place with an appreciable intensity when the system reaches a specific temperature, the intense process setting in after a certain latent period, in the present case constituting the quantity τ_{agg} (Fig. 122).

The higher the temperature, the smaller is the latent period and the more vigorous is the transformation of the quasihomogeneous solid-solution crystals into the truly homogeneous state (Figs. 119 and 120). In addition to this, comparatively small temperature changes have a very considerable influence on the velocity of the process, which also confirms its diffusion character.

In conclusion, it should be said that the questions considered in this chapter relating to the stability of microheterogeneity and the mechanism underlying its development, and the light thus thrown on the nature of this phenomenon, are of interest, not only in connection with the use of the microhardness method for purposes of physicochemical analysis, but also in establishing crystallization and heat-treatment conditions for metallic alloys (particularly those of the heat-resistant type).

By creating a stable microheterogeneity in the base of an alloy, we may increase its heat resistance as a result of the blocking action of the submicroscopic second-phase particles, tending to pin the dislocations.

Investigations into the degree of microheterogenization of the solid-solution crystals of two-phase alloys at various cooling rates in the crystallization period may cast light on further unsolved problems in connection with the hot shortness of alloys, since microheterogeneity is connected with the degree of branching of the dendrites, which has a specific influence on the position of the line corresponding to the onset of linear shrinkage.

CHAPTER 10

MICROHARDNESS OF INDIVIDUAL PHASES AND STRUCTURAL CONSTITUENTS OF METALLIC ALLOYS

At the present time the microhardness method tends to be used in the following ways for studying individual structural components of metallic alloys:

1. The identification of individual phases in the structural constituents of alloys. The microhardness method supplements earlier-known metallographic methods of recognizing the structural elements of alloys with various etchants and special ways of illuminating the object.

2. The study of the properties of individual structural constituents of alloys which for some reason or other cannot be obtained (at least without excessive demands on time and labor) in adequate quantity. Of particular value are investigations into the nature and properties of the complex compounds encountered in multicomponent alloys.

3. The study of the part played by (and the nature of the participation of) individual structural constituents of an alloy on heating, cooling, annealing, deformation (working), and other processes due to the intervention of external factors, leading to changes in the state and structure of alloys.

The Microhardness Method for Studying the Structural

Components of Nonferrous Alloys

A number of papers have been written in relation to the microhardness of individual phases and structural constituents of nonferrous alloys.

M. E. Drits and N. V. Dokukina [4] studied the macro- and microhardness of individual structural constituents of alloys belonging to the Pb−Sb, Pb−Sn, Cu−Sb, and Sn−Sb systems. These systems constitute the base of standard bearing alloys. The microhardness was measured on a PMT-2 with a load of 10 g and a holding period of 15 sec.

The investigations showed that in alloys of the Pb−Sb and Pb−Sn systems of various compositions the microhardness of the eutectic remained almost constant. In the Cu−Sn system the microhardness of solid solutions of tin in copper and copper in tin were measured as well as the microhardness of compounds.

The microhardness of all the phases indicated respectively equalled 100, 12 to 20, 560, and 340 to 370 kg/mm^2. A change of composition in this system had practically no effect on the microhardness of the chemical compounds.

Investigations into the microhardness of the compound Cu_2Sb in alloys of the $Cu-Sb$ system of various compositions showed that with increasing copper content the microhardness of this compound diminished.

The microhardness of phases in the $Sn-Sb$ system was also measured; in two-phase alloys it was practically independent of composition.

In order to study the effect of alloying additives on the character of hardening in tin- and lead-base bearing alloys, alloys of the $Sn-Sb$ and $Pb-Sn-Sb$ systems with various copper contents (0.5 to 6%) were taken.

The results showed that with increasing copper content in these alloys the microhardness of the compound SnSb and also the ductility of the base increased slightly; the authors explained this as being due to the dissolution of copper in these phases.

It was established from the foregoing investigations that the hardening of bearing alloys was mainly associated with the hardening of the primary precipitates of solid solution and the eutectic.

M. E. Drits [302] studied the effect of the grain size, temperature, and speed of conducting the tests on the microhardness of individual structural constituents of standard bearing alloys. Alloys taken for testing included B83, B16, BN, and BK. The results showed that the grain size had little effect on the microhardness of the individual structural constituents of Babbit metals. A study of microhardness as a function of time showed that in tin Babbits the β phase lost about 70% in 1 h, the ductile base 35%, and the overall hardness 30% of the initial value; the ductile base of alloy B16 proved to be the most stable.

The hard component of alloy BK lost 23% in this time and the ductile base 37% of its original hardness.

The hardness of the β phase of high-tin and low-tin Babbits fell by 30 to 40% at 75°C. Comparison of micro- and macrohardness measurements showed that the kinetics of the change in macrohardness were more determined by the corresponding changes in the microhardness of the ductile base than by changes in the microhardness of the β phase.

A. M. Korol'kov and É. S. Kadaner studied the microhardness of individual structural constituents in binary alloys based on aluminum and various other metals [4, 303].

The results established a qualitative link between the hardening of the alloy and of the individual structural constituents.

The overall rise in the hardness of the alloys with increasing alloying arose from the increase in the microhardness of the primary precipitates constituting primary solid solutions, and also the increase in the eutectic constituent.

The microhardness of the eutectic was independent of the composition of the alloy and was determined by the hardness of that constituent which constituted the main bulk of the eutectic.

The microhardness of the eutectic in such alloys as $Al-Mn$ and $Al-Ni$ is almost independent of the high hardness of the chemical compounds entering into the eutectic mixture, despite the fact that the quantity of such compounds in the eutectic extends to 18 vol.% and the hardness of these compounds is 8 to 10 times greater than that of the other eutectic constituent. It is clear that the microhardness of the eutectic may only be determined for a high degree of dispersion of the phases entering into the composition of the eutectic mixture. In this case the microhardness of the eutectic will depend on the quantitative relationship between the eutectic phases, the values of their microhardnesses, and the degree of dispersion.

The presence of mutually isolated crystals of very hard chemical compounds in the structure of alloys gave no appreciable hardening of the alloys in the range of concentrations within which the contribution from these crystals reached 25 to 35 vol.%. In order to explain this situation one clearly has to employ the principle of the so-called "framework" mechanism of hardening.

The annealing of cast alloys leads to a fall in the hardness of the alloy and the microhardness of the structural constituents; the microhardness of the eutectic falls most, possibly because of coalescence of the composite particles of the eutectic.

E. M. Savitskii and his colleagues showed [304, 305] by means of the microhardness method that intermetallic compounds in such systems as $Mg-Zn$, $Mg-Al$, and $Cu-Zn$ were particularly sensitive to the effects of temperature. Usually compounds of this type are extremely hard and brittle; however, on being heated to temperature approaching the melting point they behave as very ductile materials.

It was shown in [304] that on heating complex alloys containing inclusions of intermetallic compounds the relative hardness and strength of individual structural constituents might become redistributed owing to their different reactions to rising temperature. It may prove that the hardest structural constituent of the alloy becomes the softest at a higher temperature. Thus in the $Cu-Zn$ system the β phase softens much more on heating and becomes softer than the neighboring solid solution. Whereas, at room temperature, the α solid solution has a microhardness of around 80 and the β phase around 100 kg/mm^2, at a high temperature (300°C) the microhardness of the α phase is about 75 and that of the β phase about 55 kg/mm^2.

The reason for this effect does not apparently lie simply in the increasing amplitude of the thermal vibrations of the atoms. It was suggested in [304] that there were some qualitative changes in the structure of the crystal lattices of the intermetallic phases on heating, thus facilitating the occurrence of deformation. The use of the microhardness method for a more detailed study of this phenomenon may greatly ease the solution of the problem.

Z. A. Sviderskaya [306] studied the softening of certain compounds encountered in Mg-base alloys by the microhardness method. Later an analogous method was used by G. V. Zakharova for determining the heat resistance of the excess phases of certain cast aluminum alloys [307].

This question was studied most systematically in [308], which was devoted to an investigation into the microhardness of intermetallic phases in magnesium alloys at high temperatures.

For a comparative estimate of the heat resistance of the various phases a PMT-2 was employed. The microhardness was determined with a load of 20 g and a holding period under load of 0.5 to 60 min. The microhardness values were determined as the mean of three to five impressions.

Compounds studied included $Mg_{17}Al_{12}$, $MgZn$, Mg_2Ca, Mg_9Ce, Mg_5Th, Mg_xNd_y, Al_2Ca as well as Mn crystals in $Mg-Mn$ alloys, which enter into the composition of certain magnesium alloys and play an important part in their hardening. The microhardness of the phases was measured at temperatures of particular interest in connection with the use of magnesium alloys: 20, 150, 200, 250, and 300°C.

The results indicate that at room temperature the phases under consideration lie in the following order of diminishing microhardness (values given in kg/mm^2): Mn (980) $-Al_2Ca$ (402) $- MgZn$ (271) $- Mg_5Th$ (256) $- Mg_xNd_y$, $Mg_{17}Al_{12}$, Mg_9Ce, Mg_2Ca (186-156).

Heating to even 150°C leads to a change in the order of the phases. The compound Mg_5Th moved from fourth to third place, while MgZn moved from third to last. At 200°C an intensive softening of the compound $Mg_{17}Al_{12}$ sets in. Further heating has no effect on the order of the phases as regards microhardness.

Thus the phases may be divided into two groups as regards their inclination to softening on heating:

1. Heat-resistant phases such as the compounds Al_2Ca, Mg_5Th, Mg_9Ce and Mn crystals. The Al_2Ca and Mn phases ensure the heat resistance of alloys belonging to the Mg−Mn−Al−Ca system intended for working at 200 to 250°C; the Mg_9Ce and Mn phases enable Mg−Mn−Cu alloys to be used up to 200°C. The Mg_5Th and Mn phases give alloys of the Mg−Th−Mn system creep resistance at 300 to 340°C.

2. Non-heat-resistant phases: the compounds $Mg_{17}Al_{12}$, Mg_2Ca, and MgZn, which on heating lose 70 to 90% of their original hardness. Alloys of the Mg−Al−Zn system are not used above 150°C owing to the severe softening of the $Mg_{17}Al_{12}$ and MgZn phases on heating.

It has already been noted that alloys of the Mg−Nd−Mn system possess a high creep strength, but only up to temperatures of 200 to 250°C. The Mg_xNd_y phase present in these alloys may also only be regarded as heat-resistant up to these temperatures.

Hence the data regarding the heat resistance of the chemical compounds present in magnesium alloys obtained by the microhardness method agree with the behavior of these alloys at high temperatures.

A. A. Bochvar and Z. A. Sviderskaya used the microhardness method to obtain comparable results on the softening of eutectic alloys [320]; they studied the systems Pb−Sn, Sn−Bi, Bi−Cd, Cd−Sn, Sn−Zn, Zn−Cd, Cd−Pb, and Pb−Bi. The measurements were made on a PMT-2 with a load of 20 g, the samples being held under load for periods of 30 sec and 1 h.

As a result of this work it was established that in the cast state all the alloys studied lost hardness considerably with the passage of time, i.e., they had a tendency to creep under a steady load at room temperature.

After pressing, the tendency of the alloys toward softening increased considerably as compared with the cast state.

A slight increase in pressing temperature leads to a reduction in the scale of softening, although the reduction in microhardness is nevertheless considerably higher than for the cast state.

Cast metals have a tendency to soften under the action of a steady load; this tendency is not simply related to the melting point or to the simplicity or complexity of the crystal lattice, and the manner in which the microhardness diminishes is almost the same for all metals studied.

After pressing cast metals, no sharp increase in their creep is observed.

As a result of prolonged aging after pressing (four months), the resistance of the alloys to creep increases slightly, though the time dependence of the softening remains unaltered, diminishing slightly on raising the test temperature to 50°C (after reaching 75°C the rate of softening increases again).

Preliminary annealing at comparatively high temperatures (50 h at a temperature 20°C below the melting point) sharply reduces the tendency of pressed alloys toward creep.

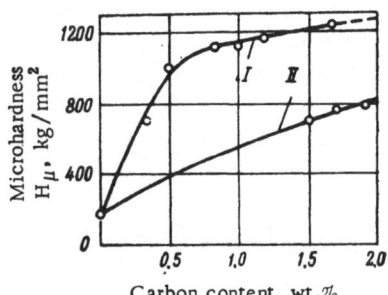

Fig. 125. Effect of carbon on the microhardness of martensite. 1) Hardness of martensite which has undergone self tempering; II) hardness of martensite which has not undergone this process.

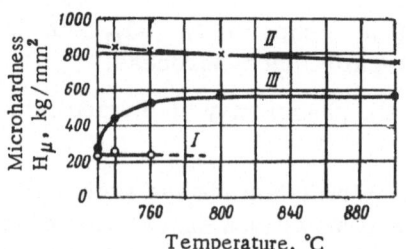

Fig. 126. Effect of the quench temperature of hypoeutectoid steel on the microhardness (H_μ, kg/mm²) of ferrite (I) and martensite (II) and on the macrohardness H_R of the steel (III).

These results may be explained as a manifestation of the solution-precipitation mechanism of ductility [309]. The large and well-developed interphase surface plays a special part in the intensive occurrence of creep processes in eutectic alloys.

The Microhardness Method for Studying the

Structure and Properties of Steels and Cast Irons

It is well known that the heating of alloys is usually accompanied by diffusion processes directed at bringing the alloy into an equilibrium state. There may thus be either the ordinary equalization of the composition (homogenization) or else structural changes depending on the concentration and initial distribution of the alloy components and also the annealing time and temperature. After processes taking place in alloys the microhardness of the structural constituents sometimes alters sharply. One may also follow processes of precipitation from supersaturated solid solutions on dispersion hardening or on the formation of new phases.

An interesting investigation into the effect of the carbon content and the quench temperature of steel on the hardness of martensite was carried out by A. N. Rozanov [310]. Rozanov's attention was attracted by the discrepancy between the two following laws: 1) On increasing the carbon content of martensite to 0.6% its hardness rose sharply, while further increasing the carbon content between 0.6 and 4.0% had comparatively little effect on the martensite surface; 2) the degree of tetragonality of the crystal lattice of the martensite varied in direct proportion to its carbon content.

The following experiments were set up. The first series of steel samples with differing carbon content were heated in evacuated quartz ampoules to 1100°C and cooled in water. Microscope study showed that, after quenching, martensite needles with various dark tints (brown, yellow, and so on) were principally obtained. The greatest microhardness of these needles varied with carbon content in accordance with curve I of Fig. 125. The second series of samples were heated to 1200°C, held for 2 h, and cooled in water. After quenching the samples, white martensite needles were chiefly obtained. The microhardness varied with carbon content as in curve II.

It was suggested that the reason for the different microhardness of the dark and light martensite needles was associated with the dispersion hardening of the martensite taking place during the self-tempering of the steel. The light martensite needles were formed at a lower

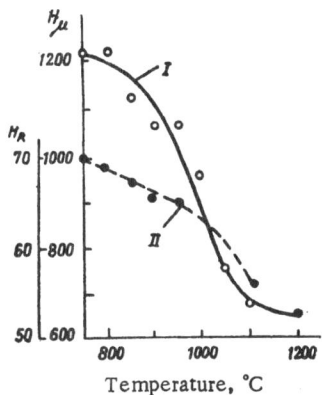

Fig. 127. Effect of the quench temperature of hypereutectoid steel on the microhardness (H_μ) of martensite (I) and the macrohardness (H_R) of steel (II).

temperature and for this reason the self-tempering process did not occur in these. The distance along the vertical between the points of curves I and II (Fig. 125) determines the effect of the self-tempering of the martensite for various amounts of carbon in the steel, the path of curve II agreeing with the laws governing the variation in the degree of tetragonality of the martensite with varying carbon content.

In another series of experiments A. N. Rozanov showed [310] that the well-known fact of the increase in the hardness of hypoeutectoid steel with increasing quench temperature, associated with the reduction in the proportion of the soft constituent of ferrite, was accompanied by a fall in the microhardness of the martensite (Fig. 126), while the fall in the hardness of hypereutectoid steel on raising the quench temperature was associated not only with an increase in the amount of residual austenite but also to a certain extent with a fall in the microhardness of the martensite (Fig. 127). In steel quenched from a very high temperature only light martensite needles with the lowest microhardness were formed.

A. A. Zhukov [311, 312] studied the structural anomalies and liquation of silicon in cast iron by means of microhardness measurements. The microhardness was measured with a PMT-3, 10 to 20 tests each being made with loads of 20 and 50 g.

The results obtained by measuring the microhardness of the pearlite in cast iron fluctuated over a wide range within a single microfield; this was explained as being due to the substantial influence of the pearlite particle size. For this reason microhardness analysis was only applied to the ferrite which occurred as ordinary dendrites in the structure of the cast irons studied. It was found that the microhardness of anomalous ferrite was much higher than that of the normal kind.

By normal ferrite we mean the inclusions precipitated near the graphite inclusions as a result of the eutectoid decomposition of the austenite; by anomalous ferrite we mean the inclusions precipitated along the axes of the austenitic dendrites as a result of the liquation of silicon, which stabilizes the ferrite. In view of the fact that in cast iron with the normal ferrite structure the ferrite inclusions occurred mainly near the graphite inclusions, it was suggested that this might influence the fall in the ferrite microhardness values. In order to eliminate this factor, a separate cast iron sample with the normal structure was subjected to a graphitizing anneal at 700 to 720°C for 8 h in order to decompose the cementite in the pearlite. However, the microhardness of the ferrite in the annealed cast iron was practically the same as before annealing. The excess of 30 kg/mm^2 (or more) of the microhardness of the anomalous ferrite over the ordinary type was explained as being due to the fact that the anomalous ferrite was enriched by ~0.5 to 1% of silicon relative to its mean content in the cast iron.

The Microhardness of Refractory Compounds

By refractory compounds we usually mean a large group of intermetallic phases of metallic systems or compounds of metals of the transition groups of the Periodic System with light metalloids (metal-like compounds). Many papers have been written on the properties of refractory compounds, since these compounds are widely used in modern technology as the main components of hard and also wear-resistant heat-resistant alloys.

The high hardness of intermetallic and metal-like compounds is the most characteristic property of these substances. In view of their great brittleness the only acceptable method of studying their hardness is the microhardness method.

The first investigation into the microhardness of the refractory hard constituents of met-alloceramic (cermet) alloys was undertaken by A. N. Zelikman and S. S. Loseva [313], who used the Hanemann apparatus with a load of 40 g and determined the microhardness of tungsten, titanium, and tantalum carbides (1760, 2000, and 1200 kg/mm^2) and also WC−TiC solid solutions (the microhardness/composition curve for this system had a maximum at 20 to 30% WC).

The building of a Soviet microhardness tester served as a beginning for a systematic investigation into the microhardness of refractory compounds and their solid solutions.

A. E. Koval'skii and L. A. Petrova [4, 314] studied the microhardness in continuous series of solid solutions of isomorphic carbides TiC−TaC, TiC−NbC, TiC−ZrC, TaC−VC, TaC−NbC, TaC−ZrC, and NbC−VC and limited solid solutions of the carbides WC and Mo$_2$C in TiC, ZrC, NbC, and VC.

In addition to other properties the microhardness of the TiB$_2$−CrB$_2$, TiB$_2$−W$_2$B$_5$, and ZrB$_2$−CrB$_2$ systems was studied in detail in [315].

Titanium and zirconium carbides were also studied in detail and their microhardness was related to the amount of carbon in the carbides [4, 316]. This relationship proved to be linear and indicated a rise of microhardness with increasing carbon content. The extrapolation of this relationship to zero concentration yielded the microhardness of pure titanium and zirconium. It was thus concluded that the high hardness of the carbides was associated with the intrusion of carbon into the lattice of the metal.

In a paper by G. V. Samsonov and V. B. Rukina [317] experimental data were obtained regarding the concentration dependence of microhardness in the tantalum−carbon system within the ranges of homogeneity of the carbides Ta$_2$C and TaC.

In addition to many other physical and chemical constants of a wide variety of refractory compounds, considerable experimental material relating to the microhardness of these was gathered together in [318, 319, 107]. The microhardness was studied with a PMT-3 using loads between 20 and 200 g.

The experiments showed that the microhardness depended on the load employed, first rising and then falling and becoming almost constant for a 60 to 200 g load. The measurements were therefore based on this range. The following gives series of microhardness numbers for a load of 120 g, taken from [107] (values in kg/mm^2):

		Carbides		
TiC, VC	ZrC	Cr$_3$C$_2$	WC, Mo$_2$C	TaC
~2600	2400	2160	1800	1629

	Nitrides
TiN	ZrN
2100	930

		Borides		
TiB$_2$	Mo$_2$B$_5$	NbB$_2$	W$_2$B$_5$	CrB$_2$	TaB$_2$
3400	2950	2900	2500	2150	2000
ZrB$_2$	CaB$_6$	BaB$_6$	LaB$_6$	CeB$_6$	
1500	3150	2900	2500	2350	

			Silicides			
WSi$_2$	TaSi$_2$	TiSi$_2$	CoSi	MoSi$_2$	ZrSi$_2$, NbSi$_2$	CrSi$_2$, NiSi$_2$
1430	1200	1100	1030	950	900	800

Thus the microhardness of the metal-like compounds in question rises in general in the order silicide, nitride, carbide, boride.

At the present time a great deal of experimental material has also been accumulated in relation to the microhardness of alloys in the binary carbide [314, 315, 318, 320], boride [318, 321, 322], and nitride [318, 316] systems of refractory metals and also in mixed carbide-nitride, carbide-boride, and other systems.

It is found that usually the microhardness/composition curve of systems constituting continuous solid solutions has a maximum; the height is governed by the difference in the microhardness of the original components, and the maximum lies on the side of the harder component along the concentration axis.

CHAPTER 11

MICROHARDNESS OF SEMICONDUCTING MATERIALS

Microhardness of Semiconducting Materials

of Various Groups

In order to secure a full understanding of the processes taking place in semiconducting materials it is important to take all the properties of the materials, electrical, physicochemical, mechanical, and so forth into account. The combined witness of all these properties should characterize the arrangement of the particles involved, their nature, and the bond between them.

One of the properties associated by very general laws with the character of the chemical bond in matter is the hardness, and the most reliable method of determining this in view of the great brittleness of many semiconductors in the microhardness method.

Up to the present there has only been very limited information regarding the microhardness of semiconducting materials. After V. M. Gol'dshmidt [323], who determined the Mohs hardness of a number of compounds with the zinc blende structure, Liu and Peretti [324, 325] determined their Brinell hardness.

The first systematic investigation into the microhardness of semiconducting materials with the zinc blende structure was carried out in [326]. The measurements were made on the PMT-3 with loads between 10 and 50 g. The brittleness of a large proportion of the materials prevented the use of heavier loads. The effect of the purity of the materials was determined in control experiments; impurities had little effect on the results. The microhardness method was used to monitor the degree of homogenization of the compounds synthesized for the investigation. The following are the hardness numbers determined for a load of 20 g from the results of [326] (in kg/mm^2):

InP	InAs	InSb	
430	330	225	
GaP	GaAs	GaSb	
—	721	469	
AlP	AlAs	AlSb	
—	—	400	
In_2Te_3	Ga_2Te_3	Ga_2Se_3	
180	237	316	
ZnTe	ZnSe	CdTe	CdSe
82	137	56	90
CuTe	CuBr	AgTe	
19.2	21.2	7.3	

These data were supplemented and confirmed in subsequent investigations relating to solid solutions ZnSe−GaAs [327], AlSb−GaSb [328], InAs−InP [329].

In Chapter 3 we considered the most important investigations into the microhardness of germanium, silicon, and certain $A^{III}B^V$ compounds in connection with an analysis of the possible influence of anisotropy of the properties and so forth.

A detailed study of the microhardness of semiconducting materials by the Knoop method [330] agrees excellently with the data obtained in [326-329]. The measurements were made with a load of 25 g. The following represent the Knoop microhardness numbers:

C	SiC	Si	Ge
8820	2880	1150(1)*	780(1)
(load 110 g)		1330(2)	845(2)
InBi	ZnS	CdS	CdSe
12(3)	178(4)	55(5)	44(5)
(load 10 g)		80(6)	66(6)

Twelve compounds of the $A^I B^{III} C_2^{VI}$ type possessing semiconducting properties were synthesized in [331] and their properties were studied. In the formula $A^I B^{III} C_2^{VI}$ the letter A corresponds to Cu and Ag, B to Al, Ga, and In, and C to Se and Te. In addition to other properties of these compounds (melting point t_m, lattice constants a in Å and c/a, some electrical properties such as the width of the forbidden band ΔE in eV, the temperature dependence $\sigma = f(T)$ of the electrical conductivity, the temperature coefficient α of the thermo-emf, and so on), their microhardness was measured. A PMT-3 with loads of 10 and 20 g was employed. There were no serious differences between the two series of measurements. The results are given in Table 41.

A study of the properties of the compounds in relation to their chemical composition revealed certain specific laws. All twelve compounds may be divided into triads (see Table 41) in which only the B atoms (Al, Ga, In) change. Within each triad the properties vary regularly but not a great deal.

On replacing Al by Ga and by In the microhardness diminishes and at the same time so do the melting point and forbidden band width. On replacing Se by Te the microhardness correspondingly diminishes (the forbidden band width does likewise); this is evidently associated with a reduction in the ionic component of the bond in compounds with tellurium. The replacement of Cu by Ag is also accompanied by a fall in the microhardness of the compounds, although this replacement affects the electrical properties more.

Taking all together and considering the relation between the microhardness and other properties of the compounds and their composition, we observe a general regularity according to which, as the metallic nature increases, the mechanical strength (H_μ), heat resistance (t_m), and activation energy of the intrinsic electrons (ΔE) diminish. The compound most metallic in this case has the lowest microhardness, melting point, and forbidden band width. This compound possesses the most weakened rigidity of the homopolar bonds of all compounds considered.

N. A. Goryunova and É. S. Osmanov[†] obtained some systematic data regarding the microhardness of semiconducting compounds of the $A^{II} B^{IV} C_2^V$ type; these were correlated with their other properties. Table 42 shows some of the results obtained.

These results indicate that the properties of compounds of the $A^{II} B^{IV} C_2^V$ type are very similar to those of elemental semiconductors and compounds of the $A^{III}B^V$ type. On quenching,

---

* Microhardness measurements were made: 1) on the (110) plane; 2) on the (112) plane; 3) on the (001) plane; 4) on the (011) plane; 5) on the (10$\bar{1}$0) plane parallel to the [0001] direction; 6) on the (10$\bar{1}$0) plane parallel to (12$\bar{1}$0).

[†] É. S. Osmanov, Candidate's Dissertation, Institute of the Chemistry of Silicates, Leningrad (1966).

TABLE 41. ·Some Properties of Compounds of the $A^I B^{III} C_2^{VI}$ Type

Compound	t_m, °C	H_μ, kg/mm²	ΔE, eV	α, µV/deg	a, Å	c/a
CuAlSe₂	1000	210	1.10	—	—	—
CuGaSe₂	861	197	0.96	+4.6	6.00	1.97
CuInSe₂	980	185	0.86	⁻+613	5.76	2.03
AgAlSe₂	950	160	0.70	—	—	—
AgGaSe₂	883	143	0.66	—	5.96	1.91
AgInSe₂	780	127	0.62	—226	5.94	2.18
CuAlTe₂	890	182	0.88	+14	5.93	2.01
CuGaTe₂	872	180	0.82	+235	0.09	1.81
CuInTe₂	791	152	0.67	+237	0.17	1.98
AgAlTe₂	729	149	0.56	+321	6.24	1.92
AgGaTe₂	714	135	0.52	+346	6.50	1.61
AgInTe₂	692	118	0.52	+298	6.46	1.87

TABLE 42. Some Properties of Compounds of the $A^{II} B^{IV} C_2^V$ Type

Compound	t_m, °C	H_μ, kg/mm²	Density, g/cm³		ΔE, eV	Effective mass of electrons		a, Å	c/a
			x-ray	pyknom.		exp.	theor.		
ZnSiP₂	1370	1100	3.390	3.35	2.3	—	0.096	5.399	1.933
CdSiP₂	—	—	3.998	3.97	2.25	—	0.092	5.678	1.837
ZnSiAs₂	1038	920	4.720	4.69	1.7	—	0.071	5.606	1.943
ZnSnAs₂	775	455	5.533	5.53	0.65	—	0.029	5.852	2.000
CdGeAs₂									
Crystalline	665	470	5.612	5.60	0,53	0.027	0.02	5.943	1.888
Vitreous	410 (soft-ening)	320	—	5.35	—	—	—	—	—

the compounds $CdGeAs_2$ and $CdGeP_2$ were obtained in a vitreous state, not characteristic of substances with a diamond-like structure. However, the compounds in question have a considerable tetrahedral compression. The vitreous phases differed sharply from the crystalline as regards microhardness and density.

Some interesting investigations into vitreous semiconductors based on the microhardness method were carried out by R. L. Myuller [332, p. 18]. It was suggested in connection with these experiments that vitreous and crystalline materials differed comparatively little in respect of the average energy of the chemical bond between the atoms. The energy of the transformation from the vitreous to the crystalline state is usually about 0.01 of the energy of atomization. This suggests the maintenance of short-range order in such a transformation; at the same time long-range order possesses a relatively insignificant energy contribution.

R. L. Myuller [332, p. 18] showed that confirmation of this point of view might be secured if the microhardness H_μ (kg/mm²) characterizing such substances with covalent directional chemical bonds between neighboring atoms depended additively on the volumetric content of elementary structural chemical units $[X_i]$ (mole/cm³), i.e., we should have

$$H_\mu = \sum_i h_i [X_i],$$

where $h_i = \dfrac{\partial H_\mu}{\partial [X_i]}$ is the so-called partial molar-volume microhardness of the structural unit X_i, the dimensions of h_i being kg · cm³/mm² · mole.

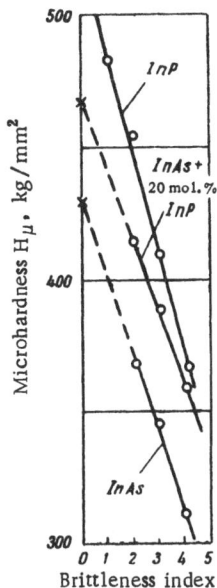

Fig. 128. Relation between the results of microhardness calculations and the nature of the damage associated with the impression under a load of 20 g for the cases of InAs, an alloy of InAs with 20 mol.% (16.1 wt.%) InP, and InP itself.

Such an additive dependence was confirmed (to a first approximation) by R. L. Myuller and his colleagues [332, p. 21; 333] when studying the microhardness of semiconducting systems As−S, As−Se, Ge−Se, Cu $Se_{1.5}As/(x = 0-0.60)$ and Ge Se As(x = 1.70-4.48; y = 0.10-1.99) in the vitreous state, and was also supported by electrical-conductivity data obtained for these systems.

Recently the possibility of a fall in microhardness taking place on forming solid solutions has been established for a number of systems. A. E. Koval'skii and L. A. Petrova [4] observed an almost linear law relating microhardness to composition in the TiC−NbC system of continuous solid solutions, and discovered a fall in microhardness when WC dissolved in NbC and Mo_2C in ZrC and NbC.

In the InAs−InP system of continuous solid solutions studied in [329, 334], the microhardness/composition relationship varied monotonically from the value for the harder compound InP to the value for the compound InAs. Ya. A. Ugai and colleagues [335] also initially observed the anomalous character of the microhardness/composition relationship in InSb−CdSb and InSb−ZnSb solutions.

In addition to this, a linear character of the variation in hardness and microhardness was also established in the continuous series of solid solutions formed between Ge and Si, the authors being unable to explain this on the basis of the theory of the hardening of solid solutions of ordinary metallic systems [336].

In view of the foregoing we might well suspect deviations from the laws of N. S. Kurnakov in quasibinary sections if it were not for the absence of corresponding anomalies on the concentration dependences of other properties in the systems under consideration.

In view of the importance of solving the problem as to the validity of the laws formulated by N. S. Kurnakov, a more detailed study of the concentration dependence of the microhardness of InP−InAs solid solutions was carried out as a continuation of [329]. As a result of this, a reliable method was found for discovering the true picture of the changes in microhardness with composition. This method may be recommended for studying the microhardness of other brittle materials.

The microhardness was measured in a PMT-3 with loads of 10 to 100 g (loading time 10 sec, time under load 5 sec). At the same time the microbrittleness was estimated. For this purpose the earlier proposed five-point scale for estimating the brittleness of refractory compounds [319] was used, suitably modified for the compounds under investigation (InAs and InP).

The tests were carried out under identical conditions and the microhardness and microbrittleness were estimated 10 to 15 sec after making the impression, since it was noted that the formation and growth of cracks, and in some cases the appearance of branches, took place some time after removing the load (8 to 10 sec).

By varying the loading conditions, impressions corresponding to all the brittleness indices were obtained. For example, the following was a typical law governing the change in the number of impressions corresponding to difference brittleness indices. For a composition of

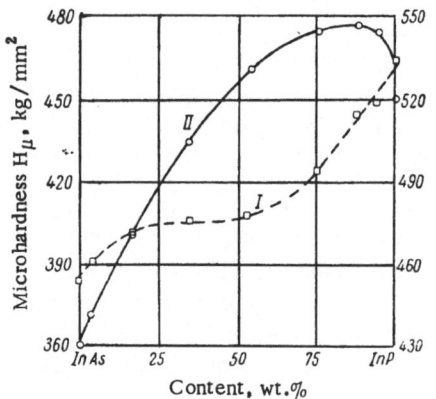

Fig. 129. Microhardness as a function of the composition of solid solutions of the InAs−InP system for a load of 20 g. I) Calculated as the arithmetic mean over the results of all the tests (left-hand scale); II) determined by extrapolation to zero brittleness number (right-hand scale).

40 mol.% (33.9 wt.%) of InP and a load of 20 g, the greatest number of impressions belonged to indices 2 or 3 and there were no impressions with index 5; for a load of 40 g the number of impressions with index 3 was the greatest and there were none with 0 and 1; for a load of 70 g the number of impressions with index 4 was the greatest and there were none with 0, 1, and 2. The frequency curves usually had the form of ordinary distribution curves, even for a small number of impressions (25 tests).

One of the convenient aspects of the selected brittleness scale appeared after the impression had been grouped into brittleness numbers (see p. 56) and the average value of microhardness corresponding to a particular group of impressions had been calculated. It was found that the scale had the important property that the resultant microhardness value varied linearly with the brittleness number. The more disrupted the impressions, the lower was the microhardness (Fig. 128).

If in calculating the microhardness we use only those impressions which have no cracks, as was done in [147] when studying microhardness in the Ge−Si system (Fig. 91b), we obtain the ordinary curve of solid solutions. However, this is not always acceptable when studying brittle materials, since, firstly, the number of tests required to give a reliable result increases sharply, and, secondly, owing to the great brittleness of the material there may not be any crack-free impressions, as, for example, when studying the microhardness of InAs−InP alloys.

However, if we use the rectilinear relationship between the microhardness and brittleness number, we may obtain the true microhardness value, corresponding to zero brittleness by graphical extrapolation (Fig. 128). Clearly, the most reliable results will be obtained for smaller loads, when there is a fairly large number of impressions with a small brittleness number.

Figure 129 shows the results of a study of microhardness in the InAs−InP system using the recommended method. The results obtained indicate the existence of a maximum on the curve representing the concentration dependence of the microhardness in this system.

The maximum microhardness in the InAs−InP system for a load of 20 g is 25 kg/mm² greater than the microhardness of the harder compound, InP. On the basis of these data and also the results obtained by studying the Ge−Si system (Chapter 6) we find that there is no longer any reason to reject the generality of Kurakov's laws (Fig. 32) governing the changes in physicomechanical properties (and hardness in particular) with composition in systems with a continuous series of solid solutions.

The microhardness of solid solutions of the InAs−GaAs system was studied as a function of composition in [337] and an attempt was made at estimating the brittleness quantitatively.

The microhardness was measured over the cross sections of bars obtained by directional crystallization. The bars were cut into plates 2 mm thick. Some plates were used for x-ray diffraction and chemicospectral analysis, others for microhardness measurements.

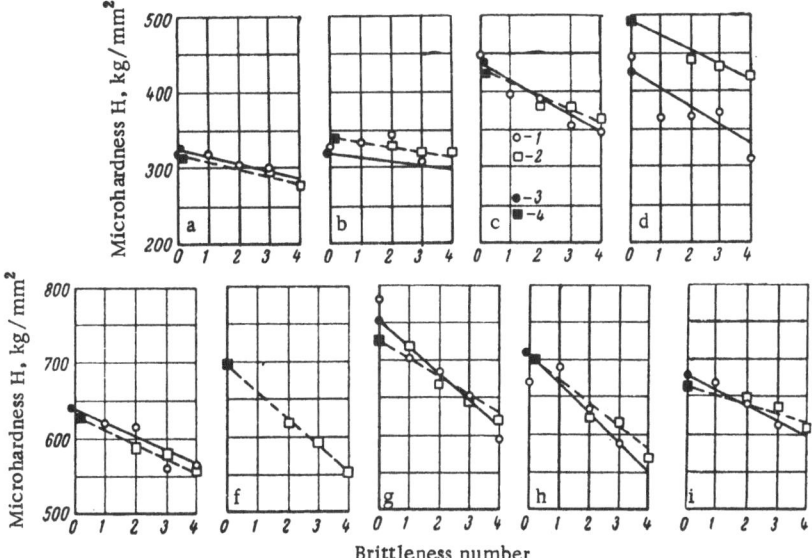

Fig. 130. Analysis of the data relating microhardness to micro-brittleness number, with the aim of securing a true extrapolated value of microhardness for solid solutions of various compositions in the InAs−GaAs system. 1) For a load of 20 g; 2) 50 g; 3, 4) extrapolated values: a) InAs; b) 9.55% GaAs; c) 14.10% GaAs; d) 22.20% GaAs; e) 45.23% GaAs; f) 64.32% GaAs; g) 74.70% GaAs; h) 87.97% GaAs; i) GaAs.

After mechanical polishing with M-1 powder the selected samples were chemically polished for several seconds in an HNO_3:HCl:H_2O = 1:3:2 mixture, after which they were carefully washed in deionized water and wiped with ethyl alcohol.

The microhardness was measured in a PMT-3. The loading time was 10 sec, the time under load 10 sec, and the time of removing the load 5 sec. Forty impressions were made on each sample, 20 at each of the loads chosen (these were 20 and 50 g). Greater loads were not employed owing to the considerable brittleness of the samples studied. The brittleness was estimated from the impressions on a five-point scale. Table 43 shows the microhardness results and the microbrittleness characteristics of InAs−GaAs solid solutions. For a load of 20 g certain of the impressions had no traces of cracking, and the microhardness H_{20} was estimated from these.

In addition to this, impressions with cracks were grouped according to their brittleness numbers $Z_i = 0, 1,\ldots,5$, and for each group of impressions the arithmetic mean microhardness was determined. Then the results were analyzed by the method of least squares: The slope $\Delta H_P/\Delta Z_i$ of the proposed linear $H_P - Z_i$ relationship was determined together with the extrapolated true microhardness H_P' (Fig. 130). Allowance was made for the weighted-mean contribution of each group of impressions. For a load of 20 g fairly good agreement was obtained between H_P and $H_{P'}$.

The form of the curves giving the true microhardness of the solid solutions in the InAs− GaAs system agrees excellently with the Kurnakov law. The microhardness in systems with a continuous series of solid solutions varies along a curve with a maximum (Fig. 131a). This maximum exceeds the microhardness of the harder component by 67.3 kg/mm² and is displaced in the direction of the latter (corresponding to 74.7 wt.% GaAs).

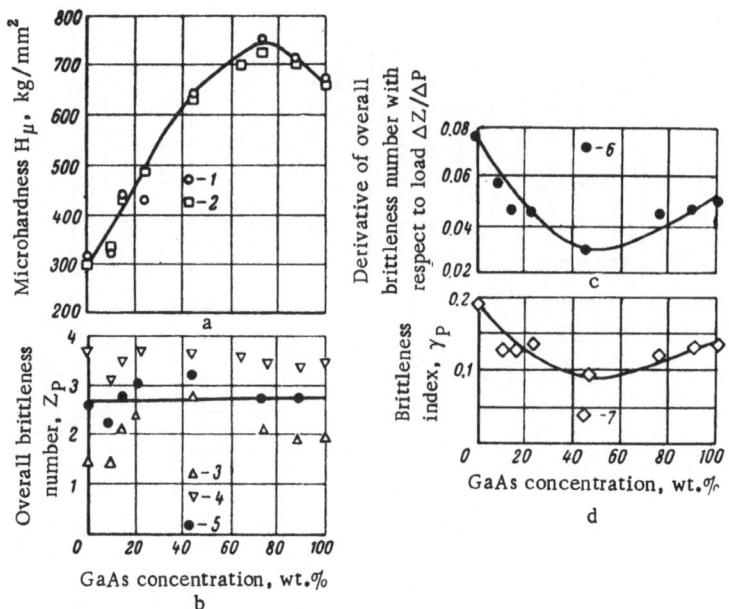

Fig. 131. Dependence of the microhardness (a), overall
brittleness number (b), derivative of the overall brittleness number with respect to the load (c), and brittleness
index (d) on the concentration of the solid solution in the
InAs−GaAs system: 1, 3) load 20 g; 2, 4) load 50 g; 5, 6, 7)
average data.

A quantitative estimate of the brittleness of InAs−GaAs solid solutions showed that the
total brittleness number $Z_P = \sum\limits_{i=o}^{i=n} N_i \cdot Z_i$ (N_i is the number of impressions with a specific
brittleness number Z_i at a load P) for loads of 20 and 50 g and also the average overall brittleness index $(Z_{20} + Z_{50})/2$ varied in a less regular manner with composition, although for a load
of 20 g there was a vague maximum (Fig. 131b). The absence of any regular dependence of Z_P
on composition may be explained as being due to the inadequacy of this criterion for a characteristic of brittleness. We may indeed note from Table 43 that there is no regularity in the
variation of the number of impressions having a specified brittleness number with composition
of the solid solutions, while with increasing load the number of impressions with a large brittleness number increases, and hence so does the overall brittleness number. Thus the proportion
of impressions with higher brittleness number affects the overall brittleness more than the
proportion of impressions with lower brittleness number for the same number of impressions.

Whereas the microhardness H_{20}, H'_{20}, and H'_{50} in the InAs−GaAs system varies with concentration along a curve with a maximum, the increment in the total brittleness number with
increasing load $(\Delta Z / \Delta P)$ and the brittleness index $\gamma = (\Delta Z / \Delta P) \cdot \Delta Z_{av}$ move along a curve
with a minimum (Fig. 131c and d). A slight rise in the overall brittleness number Z_P in the
InAs−GaAs system is reflected in a considerable fall in the brittleness index γ.

The two characteristics (Z_P and γ) characterize the brittleness of the system quantitatively in different ways. We see that the tendency toward crack formation in impressions made
with a load of 50 g is greater than in those made with a load of 20 g, and practically the same
for all concentrations; for 20 g loads applied to alloys containing 30 to 50 wt.% GaAs it is
slighly smaller than for 50 g, but larger than that obtained with the same load in the two compounds composing the system (InAs and GaAs respectively). It follows from this that a rise in

TABLE 43. Estimates of Microhardness and Microbrittleness Characteristics of InAs–GaAs Solid Solutions

Concentration, wt.%	P	H_P, kg/mm²	N_i for Z, equal to 0	1	2	3	4	5	z_P	$-\dfrac{\Delta H_P}{\Delta z_i}$	H_P, kg/mm²	$\dfrac{\Delta z}{\Delta P}$	z_{av}	γ
0.00	20	318.8	2	10	6	2	—	—	1.40	7.80	323.0	0.077	2.55	0.19
	50	—	—	—	—	6	14	—	3.70	3.80	314.3			
9.55	20	327.0	7	4	3	6	—	—	1.40	7.57	317.7	0.057	2.25	0.13
	50	—	—	—	2	16	3	—	3.10	4.89	333.0			
14.10	20	445.0	2	3	11	2	4	—	2.10	23.63	438.6	0.047	2.80	0.13
	50	—	—	—	2	7	11	—	3.50	18.50	430.0			
22.20	20	444.0	1	6	4	3	6	—	2.35	24.30	429.5	0.045	3.05	0.14
	50	—	—	—	1	7	12	—	3.70	21.40	494.6			
45.23	20	—	—	4	2	10	5	—	2.76	19.66	638.5	0.030	3.20	0.10
	50	—	—	—	1	5	14	—	3.65	18.30	629.5			
64.32	20	—	—	—	—	—	—	—	—	—	—	—	—	—
	50	—	—	—	2	5	13	—	3.55	32.83	698.8			
74.70	20	784.0	1	6	6	7	1	—	2.05	36.50	754.2	0.045	2.72	0.12
	50	—	—	1	3	3	13	—	3.40	23.10	725.0			
87.97	20	668.0	2	4	7	8	—	—	1.90	42.40	710.0	0.047	2.71	0.13
	50	—	—	—	2	7	10	—	3.42	31.80	703.9			
100.0	20	677.0	6	1	2	11	—	—	1.90	22.07	678.8	0.048	2.62	0.13
	50	—	—	—	3	9	11	—	3.35	16.30	666.1			

load from 20 to 50 g has a much greater effect in increasing the number of cracks appearing when testing the individual compounds for microhardness than when testing solid solutions based on these compounds. Solid solutions containing 30 to 50 wt.% GaAs crack so considerably on impressing a diamond pyramid with only 20 g that the picture changes very little on using 50 g. However, the compounds InAs and GaAs have a lower tendency to crack near the impressions for a load of 20 g but a greater tendency for a load of 50 g. The reason for the differences observed in the numerical estimation of brittleness in InAs−GaAs alloys evidently lies in the different values of the critical shearing stresses for the compounds InAs and GaAs and their solid solutions.

A considerable amount of experimental material has now been gathered together in relation to the dependence of the microhardness and other properties of solid solutions based on semiconducting compounds of the $A^{III}B^{V}$ type on composition. Qualitative consideration of these relationships for isovalent solid solutions of diamond-like substances leads to the following conclusion: The relationship corresponds to a smooth curve passing through a maximum, in complete accord with the Kurnakov law. The maximum is usually displaced in the direction of the harder component. Quantitative consideration of this relationship shows that for a considerable number of such systems is may be expressed [338] as

$$H_\mu^{A-B} = (H_\mu^A - H_\mu^B) N_A + H_\mu^B + K N_A (1 - N_A) ,$$

where H_μ^A and H_μ^B are the microhardnesses of the original components A and B, referred to the same number of bonds; N_A is the molar proportion of components A in the alloy; K is a constant; H_μ^{A-B} is the reduced microhardness of the alloy.

V. N. Ivanov-Omskii and B. T. Kolomiets [339] studied the InSb−GaSb system, measuring the microhardness of three alloys with loads of 20, 50, and 100 g. For loads of 50 and 100 g the samples showed great brittleness, and this led to a fall in microhardness, particularly in alloys containing over 50 mol.% GaSb ($K = 330$ kg/mm^2).

B. V. Baranov and N. A. Goryunova [340] studied the InSb−AlSb system, which was very difficult to homogenize. The curve relating microhardness to composition showed two maxima, which the authors explained as being due to the fact that the samples had not been adequately brought into a state of equilibrium.

I. I. Burdiyan and A. S. Borshchevskii* [328] studied the GaSb−AlSb system. The microhardness was measured for five samples obtained by zone melting and a relationship differing from linear was obtained.

A. S. Borshchevskii and D. N. Tret'yakov [332, pp. 14-18] studied the InAs−AlAs system and in three samples of intermediate compositions observed a rise in microhardness, exceeding the microhardness of the original compounds.

N. N. Sirota and V. V. Rosov [341] and N. A. Goryunova and V. I. Sokolova [342] presented some data relating to the microhardness/composition characteristics of the InP−GaP system. This relationship constituted a smooth curve with a maximum of 1140 kg/mm^2 close to the compound GaP.

N. A. Goryunova, A. S. Borshchevskii, and D. N. Tret'yakov showed that in the InSb−InAs system the microhardness/composition curve had a maximum (430 ± 19 kg/mm^2) exceeding the value corresponding to the original compounds (InSb 220 ± 10 kg/mm^2 and InAs 330 ± 12 kg/mm^2).

N. N. Sirota and L. A. Makovitskaya [343] established that the microhardness of the InP−GaAs system varied smoothly, having a maximum at the composition InP:GaAs = 1:1.

---

*A. S. Borshchevskii, Candidate's Dissertation, M. I. Kalinin Leningrad Polytechnic Institute (1962).

There are also some data relating to the microhardness of $A^{III}B^V - A^{II}B^{VI}$ systems. One of the first cases in which the microhardness of this type of system was studied is [327], in which microhardness was employed as a method of monitoring the homogeneity of alloys.

N. A. Goryunova and her colleagues [344, 345] studied the InSb − CdTe system by the microhardness method and established a region of limited solubility within which the microhardness increased; A. V. Voitsekhovskii [346, 347, 332, p.12] and L. I. Ksishchenskii, É. N. Khabarov, and P. V. Sharavskii [348] used the microhardness method in studying the region of limited solubility in the InAs − CdTe system and found a rise in microhardness as CdTe dissolved in InAs. N. A. Goryunova and colleagues [332, p. 7] studied the InAs − HgTe system and found a maximum on the smooth curve relating microhardness to composition at the point InAs:HgTe = 3:2.

There are also some data regarding the microhardness of systems $A^{III}B^V - A^{II}B^{IV}C_2^V$, $A^{III}B^V - A^IB^{III}C_3^{VI}$, and so on.

A. V. Voitsekhovskii [332, p. 12, 346, 347] studied the InAs − CdSnSb$_2$ (the microhardness/composition curve has a maximum; the compound CdSnSb$_2$ is regarded as nonexistent), InSb − CdSnSb$_2$ (the maximum occurs at 60% of the compound 2 · InSb), InAs − ZnGeAs$_2$ (the maximum is displaced in the direction of 2 · InAs), InAs − CdSnAs$_2$ (the maximum is displaced in the direction of CdSnAs$_2$), InSb − AgInTe$_2$, and InAs − CuInTe$_2$ systems (in the latter K = 460 kg/mm^2 and the maximum is displaced in the direction of 2 · $A^{III}B^V$).

Microhardness has also been studied in connection with solid solutions of the $A^{III}B^V - A_2^{III}B_3^{VI}$ type [349-353]. The variations in microhardness in one such case were very complicated in view of the formation of the compound In$_4$SbTe$_3$ [349-351]. In the InAs − In$_2$Te$_3$ and InAs − In$_2$Sb$_3$ systems [352, 353] the maximum is displaced in the direction of the harder compound InAs. Analogous results were obtained for the systems InP − In$_2$Sb$_3$, GaAs − Ga$_2$S$_3$, and GaP − Ga$_2$S$_3$ [354-356]. Only in the GaP − Ga$_2$Sb$_3$ system [357] did the microhardness/composition relationship lack a maximum, apparently as a result of the failure of the alloys to achieve an equilibrium state.

The microhardness of alloys of the Cu − Ge − As − Se system was studied in [358], it being considered that along the Cu$_2$GeSe$_3$ − CuGe$_2$As$_3$ section there was a phase of variable composition with tetrahedral coordination of the atoms in the lattice. A PMT-3 tester was employed and 24 impressions were made with a load of 50 g. According to microscope analysis, the homogeneous region lay between the compositions Cu$_5$Ge$_4$As$_3$Se$_6$ − Cu$_9$Ge$_6$As$_3$Se$_{12}$. Within these limits a heterovalent replacement of selenium atoms with arsenic took place in the section under consideration, the chalcopyrite lattice transforming into the lattice of zinc blende; however, the lattice constant and microhardness hardly changed (the microhardness measurements remained at a level of 430 ± 10 kg/mm^2). The excess phases appearing beyond the limits of the homogeneous range had a microhardness of 190 and 360 kg/mm^2 for a load of 20 g, i.e., they differed in properties from the tetrahedral phase.

V. F. Bankina and N. Kh. Abrikosov [359] studied the Bi$_2$Se$_3$ system. Thermal analysis, x-ray diffraction, and microscope examination established unlimited solubility in this system over the whole range of compositions. The curves relating thermo-emf thermal conductivity, carrier mobility, hardness, and microhardness to composition (obtained after annealing the samples in the solid state at 300°C) showed singular points indicating an ordering process and the formation of a compound Bi$_2$Te$_2$Se at a temperature below 500°C. Measurements of the electrical conductivity of this compound between room temperature and 500°C showed that it had semiconducting properties. The microhardness of the compounds was Bi$_2$Te$_3$ 64 kg/mm^2, Bi$_2$Te$_2$Se 60 kg/mm^2, and Bi$_2$Se$_3$ 85 kg/mm^2.

The Microhardness of Chemical Elements and
Compounds and Their Crystal-Chemical Properties

The principal laws relating hardness to crystal-chemical properties were initially formulated for minerals, artificial crystals, and metal-like compounds; very recently further information has been added as a result of using the microhardness method when studying the properties of semiconducting materials [326-363]. However, many of the laws thus derived relate almost exclusively to the Mohs hardness procedure and cannot readily be applied to the microhardness method based on the impression of a diamond pyramid.

It has been established that for crystals possessing similar simple structures the hardness diminishes with increasing interatomic distance and rises with increasing valence and coordination numbers of the atoms. In this it is assumed that the chemical bond forces in the crystals under consideration are purely ionic in nature.

V. M. Gol'dshmidt [361] suggested without any theoretical foundation that an increase in the polarization of the ions in a crystal led to a fall in hardness. This was how the necessity of allowing for a possible continuous transformation from the ionic to the covalent type of chemical bond first became evident, indicating the existence of crystals with an intermediate type of chemical bond. These principles were developed in relation to the linkage between hardness and crystal-chemical properties in [364-366].

A. S. Povarennykh [366], analyzing data from mineralogical handbooks [367, 368] and microhardness measurements [4, 369, 370], found that crystals with essentially covalent bonds were harder than those in which the ionic type of bond predominated, the relative difference increasing with increasing valence of the constituent atoms and diminishing interatomic distances. It was also found that the rise in the hardness of the crystals associated with increasing valence of the constituent cations tended to increase with increasing degree of covalence in the corresponding bonds.

In addition to this, in cases in which the covalence of the bond diminished with increasing interatomic distances (this occurs in series of analogous compounds on substituting other cations), the hardness tended to fall more rapidly than in cases in which the covalence of the bond rose (this occurs in series of analogous compounds on substituting other anions). In a number of cases these laws were exhibited both for chemical elements and for intermetallic and metal-like compounds.

Let us follow the changes in microhardness for some so-called isoelectron series of compounds of the semiconducting type [326] (values in kg/mm^2):

Ge	GaAs	Ga_2Se_3	ZnSe	CuBr
~992	721	316	137	21.2
Ge+α-Sn	GaSb	Ga_2Te_3	ZnTe	Cu I
2	469	237	82	19.2
α-Sn	InSb	In_2Te_3	CdTe	AgI
10	225	166	56	7.3

It is not difficult to see that the well-known fact of the intensification of the ionic bond in isoelectron series of the $A^{III}B^V - A^I B^{VII}$ type is accompanied by a fall in microhardness.

The chemical bond in the monocarbides of transition metals and its effect on the hardness of these compounds were studied in [371]. It was suggested that in these carbides there was a superposition of the two types of bond (ionic and covalent); the hardness of the carbides was mainly determined by the covalent bonds. It was further asserted that the deformation processes associated with hardness measurements were of the same nature as those involved in melting. For a numerical characteristic of microhardness it was therefore proposed to use

TABLE 44. Energy of Interatomic Interaction (U, kcal/mole) and Microhardness (H_μ, kg/mm^2) of Some Metal-Like Compounds [319]

Com-pound	U	H_μ	Com-pound	U	H_μ	Com-pound	U	H_μ
TiC	3980	2988	TiB$_2$	3260	3370	TiN	3900	2160
ZrC	3470	2925	ZrB$_2$	2540	2252	ZrN	3540	1988
HfC	2800	2800	HfB$_2$	—	—	HfN	2840	—
VC	3900	2094	VB$_2$	2880	2077	VN	3820	—
NbC	3216	1961	NbB$_2$	3060	2200	NbN	3560	—
Nb$_2$C	2570	2123	—	—	—	—	—	—
TaC	2770	1599	TaB$_2$	2940	2500	TaN	3320	3236
Ta$_2$C	2390	1714	—	—	—	—	—	—
Mo$_2$C	2280	1499	Mo$_2$B	2620	2790	Mo$_2$N	2670	—
WC	2760	1780	—	—	—	—	—	—
W$_2$C	2000	3000	W$_2$B	2470	2350	W$_2$N	2390	—

TABLE 45. Characteristic Frequency (ν) of Some Metal-Like Compounds and Their Microhardness (H_μ) [319]

Compound	H_μ, kg/mm^2	ν, cm$^{-1} \cdot 10^{-12}$		
		from results of [388]	calculated from melting points	calculated from elastic constants
TiC	2988	5.6	5.55	15.6
ZrC	2925	4.2	3.94	10.3
VC	2800	5.3	5.20	12.0
NbC	1961	3.8	4.26	9.9
TaC	1599	3.4	3.33	6.7

the Lindemann expression for the characteristic frequency of atomic vibrations:

$$\nu = 2.8 \cdot 10^{12} \sqrt{\frac{T_m - T}{MV^{2/3}}},$$

where T_m is the melting point; T is the test temperature; M is the molecular weight; V is the molecular volume.

The applicability of this relationshp simply to carbides of transition metals with the same principal quantum number of the incomplete electron d level was indicated by G. V. Samsonov and V. S. Neshpor [319].

The systematic consideration of physical metal-like compounds presented in [318, 319] led to the conclusion that the microhardness of metal-like compounds was mainly determined by the bond forces between the metal and the metalloid, whereas the bond between the metal atoms themselves exerted more influence on properties associated with the dynamics of the crystal lattice (melting point, elastic constants, mean-square amplitude of the thermal vibrations of the structural complexes in the lattice).

The assertion is primarily supported by the fact that the microhardness of the metal-like compounds does not vary in any regular way if the compounds are placed in order of the atomic numbers of the metal components; however, the microhardness shows a clear tendency to fall as the number of the metal component increases within each group.

As indicated by P. A. Rebinder [372, 373] and V. D. Kuznetsov [374], the hardness is determined by the surface energy of the solid and hence should be related to the energy of interatomic interaction.

TABLE 46. Microhardness (H_μ) of Certain Metal-Like Compounds and Shear Modulus (G) of the Corresponding Metals [319]

Metallic component	G kg/mm²	H_μ, kg/mm²		
		MeC	MeB₂	MeN
Ti	3 870	2988	3370	2160
Zr	2 540	2925	2252	1988
Hf	3 100	2800	—	—
V	5 500	2095	2077	—
Nb	6 000	1961	2200	—
Ta	7 000	1599	2500	—
Cr	7 300	1300 (Cr₂C₃)	1785	—
Mo	12 200	1449 (Mo₂C)	1380	—
W	15 140	1780	1350 (W₂B)	

The data presented in Table 44 (taken from [319]) show that in each group of compounds (nitrides, borides, carbides) a larger microhardness corresponds to a greater value of the energy of interatomic interaction.

In each group of the Periodic System the microhardness of the compounds falls regularly with increasing atomic number of the metal component, in accordance with the fall in the energy of interatomic interaction and the rise in the degree of screening of the electron d shell of the transition metal.

However, as indicated by G. V. Samsonov and V. S. Neshpor [100], the dependence of the surface energy and hence the microhardness of the metal-like compounds on the interaction energy is not unambiguous.

We obtain some interesting results on comparing the microhardness numbers of metal-like compounds with the characteristic atomic vibration frequency (Table 45) and with the shear modulus of the corresponding metals (Table 46).

The shear modulus is determined from the formula

$$G = \frac{E}{2(\mu + 1)},$$

where E is the elastic modulus ($E = \sigma/\delta$; σ, normal stress; δ, relative elongation); μ is the Poisson ratio [$\mu = (\Delta a/a)/(\Delta l/l)$; $\Delta a/a$ and $\Delta l/l$ are the relative changes in the transverse and longitudinal dimensions of a body of prismatic shape].

We see that the microhardness increases as the characteristic vibration frequency falls, and falls with increasing shear modulus of the corresponding metal, i.e., the microhardness increases when shear strains of the metal become more difficult or (what is the same thing) the interatomic bond between the metal atoms becomes stronger.

The establishment of a quantitative relationship between the microhardness and the crystal or crystal-chemical structure of matter would be an extremely important step.

Such a relationship was first proposed by V. M. Gol'dshmidt [323, 360, 361]. According to this relationship, the hardness determined for salt crystals on the Mohs scale is determined by the expression

$$H = S \cdot e_a \cdot e_c \cdot r^{-m},$$

where e_a and e_c are the valences of the ions or atoms of the anions and cations; r is the interatomic distance; S and m are constants.

Measuring the microhardness in the Knoop manner [330] for elements of the silicon carbide group and compounds of the $A^{III}B^V$ and $A^{II}B^{VI}$ type revealed the following relation:

$$H = \text{const} \cdot r^{-m}$$

or

$$\log H = \text{const}' - m \log r,$$

where m = 5 for elements and m = 9 for compounds of the $A^{III}B^V$ type.

This relation received a new confirmation in E. Krucheanu's dissertation* [375]. The results of Knoop microhardness measurements [330] for compounds of the $A^{II}B^{VI}$ type (ZnS, CdS, and CdSe) were considered together with Krucheanu's own measurements on the PMT-3 for the same group of compounds from the results of 50 tests [375] with loads of 30 and 100 g [ZnSe 183 ± 6 kg/mm² for the (0001) plane, ZnTe 78 ± 3 kg/mm² for the (110) plane, HgSe 28 ± 3 kg/mm² for the (100) plane, and HgTe 21 ± 1 kg/mm² for the (110) plane]. As a result of this a value of the constant m = 13 was recommended for the $A^{II}B^{VI}$ compounds.

Since the energy of the crystal lattice may be defined by the expression

$$U = \text{const}'' \cdot r^{-n},$$

while the microhardness may be regarded as a quantity directly proportional to the energy of deformation of the volume

$$H = \text{const}''' \cdot U \cdot r^{-3},$$

we have the following relation [330]:

$$m = n + 3.$$

* E. Krucheanu, Dissertation, M. I. Kalinin Institute of Nonferrous Metals (1960).

REFERENCES

1. V. M. Glazov and V. N. Vigdorovich, Microhardness of Metals, Metallurgizdat (1962).
2. M. M. Khrushchov and E. S. Berkovich, Microhardness Determined by the Indentation Method, Izd. AN SSSR, Moscow – Leningrad (1943).
3. M. M. Khrushchov and E. S. Berkovich, The PMT-2 and PMT-3 Microhardness Testers, Institute of Engineering, Izd. AN SSSR (1950).
4. Microhardness. Transactions of a Conference on Microhardness, November 21 to 23, 1950, Institute of Engineering and All-Union Scientific Engineering Society of Apparatus Construction, Izd. AN SSSR (1951).
5. Methods of Testing Microhardness, Izd. Nauka (1965).
6. N. S. Kurnakov, Selected Works, Vols. 1 and 2, ONTI-NKTP, Moscow – Leningrad (1938 and 1939).
7. N. S. Kurnakov, Introduction to Physicochemical Analysis, Izd. AN SSSR (1940).
8. A. A. Bochvar, Izv. Akad. Nauk SSSR, Otd. Tekhn. Nauk, No. 10, 1369-1384 (1947).
9. A. A. Bochvar, Technology of Metals, Selection of Scientific Works (M. I. Kalinin Moscow Inst. of Nonferrous Metals and Gold), No. 18, pp. 5-17 (1947).
10. A. A. Bochvar, Metallography, Metallurgizdat (1956).
11. V. K. Grigorovich, Tr. Inst. Met. im A. A. Baikova Akad. Nauk SSSR, No. 5, Izd. AN SSSR (1960), pp. 244-249.
12. G. A. Il'inskii, Zavod. Lab., No. 3, 366-367 (1958).
13. L. G. Kharitonov, Determination of Microhardness. Method of Testing, Measurement of Impressions, Nomograms and Tables for Determining Microhardness, Izd. Metallurgiya (1967).
14. N. A. Sologub, Zavod. Lab., No. 12, 1521 (1958).
15. V. D. Lisitsyn, Zavod. Lab., No. 6, 711-715 (1957).
16. E. S. Berkovich and A. D. Kuritsyna, Zavod. Lab., No. 7, 868-869 (1949).
17. I. L. Mirkin and É. P. Rikman, Zavod. Lab., No. 11, 1338-1341 (1957).
18. I. L. Mirkin and É. P. Rikman, Lit. Proizv., No. 12, 13-16 (1957).
19. G. V. Bokuchava, Izv. Vysshikh Uchebn. Zavedenii, Mashinostroenie, No. 5, 184-187 (1959).
20. M. G. Lozinskii, Structure and Properties of Metals and Alloys at High Temperatures, Metallurgizdat (1963).
21. M. G. Lozinskii and V. S. Mirotvorskii, Izv. Akad. Nauk SSSR, Otd. Tekhn. Nauk., Metallurgiya i Toplivo, No. 3, 135 (1959).
22. M. G. Lozinskii and V. S. Mirotvorskii, The IMASh-9 Device for Studying the Microhardness of Metals and Alloys on Heating to 1300°C and Straining in Vacuum, Izd. TsITÉIN, Subject 32, No. P-61-16/4 (1961).
23. M. M. Khrushchov and E. S. Berkovich, Study of the Hardness of Ice, Izd. AN SSSR (1960).
24. V. I. Trefilov and Yu. V. Mil'man, Zavod. Lab., No. 4, 484-485 (1964).

25. M. M. Khrushchov and E. S. Berkovich, Zavod. Lab., No. 3, 345-347 (1950).

26. G. V. Akimov, Dokl. Akad. Nauk SSSR, 51(3):205-208 (1946).

27. V. N. Novogrudskii and I. G. Fakidov, Fiz. Met. Metalloved., 7(6):903-906 (1959).

28. E. M. Strug and E. V. Panchenko, Nauchn. Dokl. Vysshei Shkoly, Met., No. 2, 252-255 (1959).

29. V. M. Glazov and A. N. Krestovnikov, Zavod. Lab., No. 4, 416-419 (1961).

30. V. Z. Bengus-Olevskii and V. I. Startsev, Transactions of the Second Conference on the Physics of Alkali Halide Crystals, June 19-24, 1961, Izd. Latv. Gos. Univ., Riga (1962).

31. Ya. Krylov and V. I. Iveronova, Kristallografiya, 6(5):784-786 (1961).

32. W. H. Vanghau and J. W. Davisson, Acta Metallurgia, 6:554-559 (1958).

33. Yu. S. Boyarskii et al., Zavod. Lab., No. 4, 477-480 (1960).

34. Yu. S. Boyarskii, Uch. Zap. Kishinevsk. Gos. Univ., 17 (Fiz.-Mat.):159 (1956).

35. Yu. S. Boyarskii and R. P. Rukatskaya, Uch. Zap. Kishinevsk. Gos. Univ. 55(Fiz.-Mat.): 67-76 (1960).

36. M. V. Klassen-Neklyudova and A. A. Urusovskaya, Tr. Inst. Kristallogr. Akad. Nauk SSSR (1955), p. 146.

37. Yu. S. Boyarskaya, Kristallografiya, 2(5):709-712 (1957).

38. Yu. S. Boyarskaya and M. I. Val'kovskaya, Kristallografiya, 7(2):261-265 (1962).

39. Semiconductor Studies, Izd. Karte Moldavenské, Kishinev (1964), pp. 64 and 131.

40. V. A. Mokievskii, Kristallografiya, 4(3):410-413 (1959).

41. V. K. Grigorovich, Zavod. Lab., No. 5, 601-605 (1959).

42. Yu. S. Boyarskaya et al., Kristallografiya, 4(4):597-602 (1959).

43. V. N. Lange and T. I. Lange, Izv. Akad. Nauk Mold. SSR, No. 7, 23-28 and 29-34 (1963).

44. V. N. Lange and T. I. Lange, Kristallografiya, 10(2):260-262 (1965).

45. V. N. Lange and T. I. Lange, Fiz. Tverd. Tela, 5(7):2029-2031 (1963).

46. V. N. Lange et al., Izv. Akad. Nauk Mold.SSR, No. 12, 61-67 (1964).

47. V. N. Lange and T. I. Lange, Fiz. Tverd. Tela, 1(4):559-561 (1959).

48. P. Ludwick, Die Kegeldruckprobe, Ein Neues Verfahren zur Nartebestimmung von Materiallen, Berlin, Fulins, Springer (1908).

49. R. L. Smith and G. E. Sandland, J. Iron and Steel Inst., 3:285-294 (1925).

50. N. W. Thibault and H. L. Niquist, Trans. ASM, 38:271-330 (1947).

51. R. F. Campbell et al., Trans. ASM, 40:954-982 (1948).

52. A. A. Bochvar and O. S. Zhadaeva, Izv. Akad. Nauk SSSR, Otd. Tekhn. Nauk, No. 3, 341-348 (1947).

53. D. B. Gogoberidze and N. A. Kopatsky, Collection from the All-Union Scientific Engineering Society of Metallurgists, Vol. 2, Metallurgizdat (1954), p. 156.

54. B. I. Kostetsky and P. K. Topakha, Zavod. Lab., No. 8, 972 (1948).

55. H. Hanemann and E. O. Bernhardt, Z. Metallkunde, 32(2):35-38 (1940).

56. F. Schultz and H. Hanemann, Z. Metallkunde, 33(3):124-135 (1940).

57. A. F. Ioffe, Physics of Crystals, GIZ, Leningrad (1930).

58. I. V. Grebenshchikov et al., Treatise on Optics, Gostekhizdat (1946).

59. A. R. G. Brown and E. Ineson, J. Iron and Steel Inst., 169, Pt. 4, December, 376-388 (1951).

60. H. Buckle, Metall, 9(13/14):549-554 (1955).

61. B. V. Mott, Hardness Testing by Microimpression, Metallurgizdat (1960).

62. B. Ya. Petrenko, Zavod. Lab., No. 7, 869 (1957).

63. V. M. Glazov and V. A. Borisov, Zavod. Lab., No. 12, 1420-1422 (1960).

64. V. N. Vigdorovich and I. F. Chernomordin, in: New Machines and Devices for Testing Metals, Metallurgizdat (1963), p. 87.

65. E. S. Berkovich, Zavod. Lab., No. 10, 1250 (1963).

66.   M. S. Ablova and A. A. Averkin, Zavod. Lab., No. 8, 1015-1017 (1965).

67.   M. S. Ablova and A. A. Averkin, Methods of Testing the Techological, Mechanical, and Physical Properties of Materials and Substances, Leading Scientific-Technical and Industrial Experience, No. 2-64-1195/27, GOSINTI (1964).

68.   G. N. Kalei, Zavod. Lab., No. 2, 224-227 (1967).

69.   H. Buckle, L'Essai de Microdureté et ses Applications, Paris (1960).

70.   V. K. Grigorovich, Zavod. Lab., No. 2, 196-202 (1951).

71.   V. K. Grigorovich, Zavod. Lab., No. 4, 457-464 (1954).

72.   V. K. Grigorovich, Zavod. Lab., No. 5, 601-605 (1959).

73.   V. V. Nalimov, Use of Mathematical Statistics in Analyzing Matter, Fizmatgiz (1960).

74.   I. N. Bronshtein and K. A. Semendyaev, Handbook of Mathematics, Gostekhteoretizdat (1953).

75.   V. M. Glazov et al., Zavod. Lab., No. 11, 1343-1348 (1950).

76.   H. Buckle, Metall, 9(23/24):1067-1074 (1955).

77.   G. Ya. Vasil'ev, Zavod. Lab., No. 2, 223-227 (1947).

78.   W. Boas, and M. E. Hargreaves, Proc. Roy. Soc., 193(1032):89-97 (1948).

79.   A. A. Bochvar and O. S. Zhadaeva, Izv. Akad. Nauk SSSR, Otd. Tekhn. Nauk, No. 4, 41-48 (1947).

80.   E. M. Shvetsova and T. V. Lebedeva, Zavod. Lab., No. 7, 850-857 (1950).

81.   D. D. Paishev, Zavod. Lab., No. 2, 231 (1954).

82.   I. L. Mirkin and I. I. Trunin, Zavod. Lab., No. 2, 229-235 (1957).

83.   V. D. Lisitsyn, Zavod. Lab., No. 4, 474-478 (1954).

84.   V. D. Lisitsyn, Zavod. Lab., No. 12, 1490-1494 (1958).

85.   V. N. Vigdorovich and A. E. Vol'pyan, Zavod. Lab., No. 6, 762-764 (1958).

86.   R. I. Garber et al., Fiz. Metal. i Metalloved., 9(2):274-278 (1960).

87.   M. S. Ablova, Fiz. Tverd. Tela, 3(6):1815-1820 (1961).

88.   M. S. Ablova, Fiz. Tverd. Tela, 3(10):3133-3136 (1961).

89.   M. S. Ablova and A. R. Regel', Fiz. Tverd. Tela, 3(7):1052-1054 (1962).

90.   M. S. Ablova, Fiz. Tverd. Tela, 6(10):3159-3161 (1964).

91.   G. V. Kukuladze and M. S. Mirgalovskaya, Izv. Akad. Nauk SSSR, Neorg. Mat., 1(7):1025-1026 (1965).

92.   A. V. Sandulova and V. M. Rybak, Fiz. Tverd. Tela, 5(9):2587-2590 (1963).

93.   M. S. Ablova and N. N. Feoktistova, Fiz. Tverd. Tela, 5(2):364-366 (1963).

94.   V. N. Lange and A. R. Regel', Fiz. Tverd. Tela, 1(4):562-565 (1959).

95.   R. Willardson and H. Goring (Editors), Semiconducting Compounds $A^{III}B^V$, Izd. Metallurgiya (1967).

96.   V. A. Kokoshkin, Tr. Inst. Met. im A. A. Baikova Akad. Nauk SSSR, Metallurgy, Metallography, and Physicochemical Test. Methods, No. 11, Izd. AN SSSR, 114-119 and 120-123 (1962).

97.   J. R. Dale and J. C. Brice, Solid-State Electronics, 3(2):105-109 (1961).

98.   M. G. Mil'vidskii and L. V. Lainer, Scientific Transactions of the State Inst. of Rare Metals (Work of 1959), Vol. 6, Metallurgizdat (1962), pp. 149-155.

99.   L. A. Shreider, in: Analysis of Methods of Measuring Hardness and Quantitative Hardness Scales, Izd. AN SSSR, Moscow–Leningrad (1949).

100. G. V. Samsonov and V. S. Neshpor, Dokl. Akad. Nauk SSSR, 104(3):405 (1955).

101. H. Koto et al., J. Japan Inst. Metal, 21(7):429 (1956).

102. S. Palmquist, Jernrontorets Annaler (Switzerland), 141(5):300 (1957).

103. I. N. Frantsevich and A. N. Pilyankevich, Inzh.-Fiz. Zh., 1(10):47-54 (1958).

104. E. V. Tsinzerling, Dokl. Akad. Nauk SSSR, 60(6):1033-1036 (1948).

105. I. N. Frantsevich and A. N. Pilyankevich, Transactions of a Seminar on Heat-Resistant Materials, No. 5, Izd. AN Ukr.SSR, Kiev (1960), p. 28.

106. A. N. Pilyankevich, Zavod. Lab., No. 1, 88-90 (1960).

107. G. V. Samsonov et al., Fiz. Metal. i Metalloved., 8(4):622-630 (1959).

108. N. S. Kurnakov and S. F. Zhemchuzhnyi, Izv. Leningrad. Politekh. Inst., No. 9, 393 (1908).

109. S. A. Pogodin et al., Soob. Fiz.-Khim. Analiz., 17:868 (1949).

110. V. M. Glazov et al., Zh. Fiz. Khim., 3(8):1891-1897 (1957).

111. V. M. Glazov et al., in: Technology of Nonferrous Metals, Coll. of Works of the M. I. Kalinin Moscow Inst. of Nonferrous Metals and Gold and the All-Union Scientific-Research Inst. of the Heat Treatment of Metals, No. 29, Metallurgizdat (1958), pp. 135-142.

112. I. I. Novikov, in: Technology of Nonferrous Metals and Alloys, Coll. of Works of the Moscow Inst. of Nonferrous Metals and Gold and the All-Union Scientific-Research Inst. of the Heat Treatment of Metals, No. 23, Metallurgizdat (1952), pp. 16-34.

113. V. M. Glazov et al., Izv. Akad. Nauk SSSR, Otd. Tekhn. Nauk, No. 12, 131-135 (1955).

114. H. Buckle, Zeissnachr., 5:93 (1944).

115. H. Buckle, Z. Metallkunde, 34:130 (1942).

116. H. Buckle, Z. Metallforschung, 1:43 (1946).

117. H. Buckle and I. Descamps, Compt. Rend. Acad. Sci., 230:752 (1950).

118. H. Buckle and I. Descamps, Rev. Metall., 48:569 (1951).

119. L. Mondolfo, Metallography of Aluminum Alloys, Inst. of Metals, London (1943).

120. M. Hansen and K. Anderko, Constitution of Binary Alloys, McGraw-Hill Co., New York—Toronto—London (1958).

121. M. V. Mal'tsev et al., in: Metallurgical Fundamentals of Casting Light Alloys, Oborongiz (1957), pp. 140-154.

122. V. M. Glazov et al., Izv. Akad. Nauk SSSR, Otd. Tekhn. Nauk, No. 4, 131-136 (1956).

123. N. N. Glagoleva et al., Izv. Akad. Nauk SSSR, Otd. Tekhn. Nauk, No. 8, 89-94 (1957).

124. V. M. Glazov et al., Zh. Neorg. Khim., 4(7):1620-1624 (1959).

125. V. M. Glazov et al., Metallovedenie i Term. Obrabot. Metallov., No. 10, 48-50 (1959).

126. V. N. Vigodorovich et al., Izv. VUZ, Tsvetnaya Met., No. 2, 143-146 (1960).

127. V. P. Elyutin and Z. F. Funke, Izv. Akad. Nauk SSSR, Otd. Tekhn. Nauk, No. 3, 68-76 (1956).

128. V. N. Vigdorovich et al., Izv. Akad. Nauk SSSR, Otd. Tekhn. Nauk, No. 2, 145-148 (1958).

129. E. M. Savitskii et al., Zh. Fiz. Khim., 3(9):2138-2142 (1958).

130. V. N. Vigodorovich and A. Ya. Nashel'skii, Zh. Neorg. Khim., 4(9):2034-2038 (1959).

131. E. E. Schumacher and G. M. Bonton, Metals and Alloys, Vol. 1 (1930), p. 405.

132. V. F. Terekhova et al., Zh. Neorg. Khim., 6(5):1252-1253 (1961).

133. V. M. Glazov and Chen Yüan-lu, Izv. Akad. Nauk SSSR, Otd. Tekhn. Nauk, Metallurgiya i Toplivo, No. 4, 150-155 (1960).

134. V. M. Glazov et al., Zh. Neorg. Khim., 7(3):576-581 (1962).

135. V. M. Glazov et al., Zh. Neorg. Khim., 7(4):831-835 (1962).

136. B. G. Zhurkin et al., Izv. Akad. Nauk SSSR, Otd. Tekhn. Nauk, Metallurgiya i Toplivo, No. 5, 86-90 (1959).

137. V. M. Glazov et al., Izv. Akad. Nauk SSSR, Otd. Tekhn. Nauk, No. 1, 162-164 (1956).

138. A. A. Bochvar and I. I. Novikov, in: Technology of Nonferrous Metals and Alloys, Coll. of Works of the M. I. Kalinin Moscow Institute of Nonferrous Metals and Gold and the All-Union Scientific-Research Inst. of the Heat Treatment of Metals, No. 23, Metallurgizdat (1952), pp. 5-15.

139. A. A. Bochvar and I. I. Novikov, Izv. Akad. Nauk SSSR, Otd. Tekhn. Nauk, No. 2, 217-224 (1952).

140. E. M. Savitskii and V. V. Baron, Tr. Inst. Met. im. A. A. Baikova Akad. Nauk SSSR, No. 3, Izd. AN SSSR (1958), pp. 191-194.

141.   H. Spengler, Metall, 9, I–G 5/6 (1955).

142.   F. A. Trumbore et al., J. Phys. Chem. Solids, 2:239 (1959).

143.   V. M. Glazov, Izv. Akad. Nauk SSSR, Otd. Tekhn. Nauk, Metallurgiya i Toplivo, No. 5, 39–42 (1961).

144.   D. A. Petrov, Zh. Fiz. Khim., 21(12):1449–1452 (1947).

145.   D. A. Petrov and A. A. Bukhanova, Zh. Fiz. Khim., 28(1):161–165 (1954).

146.   V. M. Glazov et al., Izv. Akad. Nauk SSSR, Otd. Tekhn. Nauk, No. 10, 143–146 (1955).

147.   V. M. Glazov and Chen Yüan-lu, Izv. Akad. Nauk SSSR, Otd. Tekhn. Nauk, Metallurgiya i Toplivo, No. 2, 99–107 (1961).

148.   I. I. Novikov, Hot Shortness of Nonferrous Metals and Alloys, Izd. Nauka (1966).

149.   Ya. E. Geguzin and B. Ya. Pines, Dokl. Akad. Nauk SSSR, 75(4):535–538 (1950).

150.   Silicon [Russian translation], IL (1959).

151.   V. M. Glazov et al., Coll. of Works of the A. A. Baikov Inst. of Metallurgy of the Academy of Sciences of the USSR, Vol. 14, Izd. AN SSSR (1963), pp. 108–119.

152.   V. M. Glazov and G. L. Malyutina, Zh. Neorg. Khim., 8(8):1921–1927 (1963).

153.   V. M. Glazov and G. L. Malyutina, Zh. Neorg. Khim., 8(10):2372–2375 (1963).

154.   W. Hume-Rothery, Introduction to Physical Metallography [Russian translation], Izd. Metallurgiya (1965).

155.   V. N. Lange and A. R. Regel', Fiz. Tverd. Tela, 1(4):560–561 (1959).

156.   V. I. Veraksa et al., Izv. Vysshikh Uchebn. Zavedenii, Fizika, No. 3, 124–126 (1962).

157.   D. A. Petrov, in: Theory and Study of Semiconductors and Processes of Semiconductor Metallurgy, Izd. AN SSSR (1955), pp. 53–92.

158.   B. A. Kolachev and D. A. Petrov, in: Growth of Crystals, Vol. 1, Consultants Bureau, New York (1958), pp. 126–134.

159.   D. A. Petrov, in: Light Alloys, Izd. AN SSSR (1958), pp. 5–16.

160.   R. F. Mehl, Metals and Alloys, 12:41 (1941).

161.   H. Buckle, Z. Elektrochemie, 49:238 (1913).

162.   H. Buckle and A. Keil, Metaux et Corros., 24:59 (1949).

163.   H. Buckle and A. Keil, Mikroskopie, 4:266 (1949).

164.   D. P. Jenson, Iron Age, 161:66 (1948).

165.   A. A. Bochvar and A. S. Titova, Izv. Vysshikh Uchebn. Zavedenii, Tsvetnaya Met., No. 1, 127–132 (1958).

166.   H. Buckle and J. Blin, J. Metals, 80:365 (1952).

167.   H. Buckle, Z. Naturforschung, 1:81 (1946).

168.   H. Buckle, J. Inst. Metals, 81:742 (1953).

169.   M. I. Parfenova and N. A. Izgaryshev, Zh. Prikl. Khim., 25(6):757–761 (1952).

170.   U. Zwicker, Metalloberflache, 6A;31 (1952).

171.   F. Hanser, Enclides, 47:24 (1945).

172.   E. Raub, Z. Naturforschung, 1:71 (1946).

173.   E. Raub, Z. Naturforschung, 2:281 (1947).

174.   G. G. Maksimovich, in: Mechanics of Real Solids, Vol. 1, Izd. AN Ukr.SSR, Kiev (1962), pp. 76–79.

175.   A. A. Bochvar and Z. A. Sviderskaya, Izv. Akad. Nauk SSSR, Otd. Tekhn. Nauk, No. 10, 125 (1958).

176.   R. A. Akopyan and L. V. Tuzov, Fiz. Metal. i Metalloved., 20(3):361–367 (1965).

177.   R. A. Akopyan, Izv. Akad. Nauk Azerbaidzhan SSR, Ser. Fiz.-Tekhn. i Mat. Nauk, No. 3, 81–86 (1965).

178.   R. Graf, Compt. Rend. Acad. Sci., 246:1544 (1958).

179.   S. A. Pogodin and L. M. Kefeli, Izv. Soob. Fiz. Khim. Anal., 18:86–116 (1949).

180.   V. M. Glazov et al., Metallovedenie i Obrabot. Met., No. 3, 23–27 (1957).

181.   M. V. Zakharov et al., Metallovedenie i Obrabot. Met., No. 3, 23–28 (1956).

182. M. V. Mal'tsev and Van-Bok Yan, Izv. Vysshikh Uchebn. Zavedenii, Tsvetnaya Met., No. 2, 130-142 (1958).

183. V. N. Vigdorovich et al., Izv. Akad. Nauk SSSR, Otd. Tekhn. Nauk, No. 3, 110-113 (1958).

184. V. N. Vigdorovich et al., Izv. Vysshikh Uchebn. Zavedenii, Tsvetnaya Met., No. 2, 142-152 (1958).

185. T. A. Badaeva and R. I. Kuznetsova, Tr. Inst. Met. im A. A. Baikova Akad. Nauk SSSR, Vol. 3, Izd. AN SSSR (1958), pp. 216-231.

186. V. M. Glazov and M. V. Stepanova, Izv. Vysshikh Uchebn. Zavedenii, Tsvetnaya Met., No. 4, 129-131 (1967).

187. V. M. Glazov et al., Izv. Akad. Nauk SSSR, Otd. Tekhn. Nauk, Metallurgiya i Toplivo, No. 4, 153-155 (1959).

188. V. M. Glazov and N. N. Glagoleva, Zavod. Lab., No. 12, 1481-1484 (1957).

189. A. A. Bochvar et al., Tr. Mosk. Inst. Tsvetn. Met. i. Zol. im. M. I. Kalinina, No. 1, Izd. ONTI-Metallurgizdat, Moscow—Leningrad—Sverdlovsk (1933), pp. 5-18.

190. N. N. Glagoleva and V. M. Glazov, Izv. Akad. Nauk SSSR, Otd. Tekhn. Nauk, No. 1, 130-134 (1958).

191. V. Ya. Anosov, Geometry of the Chemical Diagrams of Binary Systems, Izd. AN SSSR (1959).

192. V. Ya. Anosov and S. A. Pogodin, Origins of Physicochemical Analysis, Izd. AN SSSR, Moscow—Leningrad (1947).

193. V. N. Vigdorovich, Dokl. Akad. Nauk SSSR, 120(5):1027-1030 (1958).

194. V. N. Vigdorovich and V. M. Glazov, Izv. Vysshikh Uchebn. Zavedenii, Tsvetnaya Met., No. 3, 122-126 (1958).

195. V. M. Glazov and V. N. Vigdorovich, Zavod. Lab., No. 1, 57-62 (1959).

196. D. A. Petrov, Tr. Inst. Met. im A. A. Baikova Akad. Nauk SSSR, No. 5, Izd. AN SSSR (1960), pp. 174-177.

197. A. A. Bochvar, Izv. Akad. Nauk SSSR, Otd. Tekhn. Nauk, No. 1, 136-138 (1957).

198. T. A. Badaeva, Dokl. Akad. Nauk SSSR, 64(4):533 (1949).

199. G. G. Urazov and T. N. Shushpanova, Izv. Akad. Nauk SSSR, Ser. Khim., No. 2, 321 (1939).

200. V. G. Kuznetsov and E. S. Makarov, Dokl. Akad. Nauk SSSR, 23(3):245-248 (1939).

201. R. B. Hill et al., J. Inst. Met., 83(7):321-368 (1955).

202. N. N. Sirota, Coll. of Annotations to the Works of the M. I. Kalinin Moscow Inst. of Nonferrous Metals and Gold, Metallurgizdat (1960).

203. V. M. Glazov and M. V. Stepanova, Izv. Akad. Nauk SSSR, Otd. Tekhn. Nauk, Metallurgiya i Toplivo, No. 6, 61-64 (1960).

204. Chang Bao-chan, Izv. Vysshikh Uchebn. Zavedenii, Tsvetnaya Met., No. 1, 138 (1958).

205. V. M. Glazov et al., Izv. Akad. Nauk SSSR, Otd. Tekhn. Nauk, No. 9, 123 (1957).

206. V. M. Glazov et al., Izv. Akad. Nauk SSSR, Otd. Tekhn. Nauk, Metallurgiya i Toplivo, No. 3, 58-62 (1962).

207. V. M. Glazov and M. V. Stepanova, Dokl. Akad. Nauk SSSR, 144(3):565-567 (1962).

208. V. M. Glazov and S. N. Chizhevskaya, Dokl. Akad. Nauk SSSR, 129(4):869-872 (1959).

209. V. M. Glazov and Lu Chen-yüan, Zh. Neorg. Khim., 7(3):582-589 (1962).

210. V. M. Glazov and D. A. Petrov, Izv. Akad. Nauk SSSR, Otd. Tekhn. Nauk, No. 4, 125-129 (1958).

211. V. M. Glazov and A. A. Vertman, Dokl. Akad. Nauk SSSR, 123(3):492-494 (1958).

212. A. A. Vertman and V. M. Glazov, Izv. Akad. Nauk SSSR, Otd. Tekhn. Nauk, No. 1, 60-63 (1959).

213. V. M. Glazov, Zavod. Lab., No. 7, 824-828 (1958).

214. V. M. Glazov and V. N. Vigodorovich, Zh. Fiz. Khim., 33(10):2164-2168 (1959).

215.   V. M. Glazov, Izv. Akad. Nauk SSSR, Otd. Tekhn. Nauk, Metallurgiya i Toplivo, No. 1, 89-93 (1962).

216.   A. M. Zakharov, Izv. Vysshikh Uchebn. Zavedenii, Tsvetnaya Met., No. 1, 124-127 (1961).

217.   A. M. Zakharov, Izv. Vysshikh Uchebn. Zavedenii, Tsvetnaya Met., No. 3, 121-126 (1965).

218.   A. M. Zakharov et al., Zh. Neorg. Khim., 6(5):1165-1171 (1961).

219.   E. M. Savitskii and A. M. Zakharov, Zh. Neorg. Khim., 9(10):2424-2432 (1964).

220.   A. M. Zakharov, Izv. VUZ, Tsvetnaya, Met., No. 2, 117-122 (1966).

221.   E. M. Savitskii and A. M. Zakharov, Zh. Neorg. Khim., 7(11):2575-2580 (1962).

222.   I. I. Novikov et al., in: Structure and Properties of Nonferrous Metals and Alloys, Vol. 3, Izd. AN SSSR (1961), pp. 136-142.

223.   D. A. Petrov, Questions in the Theory of Aluminum Alloys, Metallurgizdat (1951).

224.   I. V. Gorbachev, Tr. Dal'nevost. Politekh. Inst. im. V. V. Kuibysheva, No. 26 (1941).

225.   E. Scheil, Z. Metallkunde, 34:3 (1942).

226.   A. A. Bochvar and O. S. Zhadaeva, Izv. Akad. Nauk SSSR, Otd. Tekhn. Nauk, No. 4, 419-424 (1947).

227.   B. S. Ioffe, Zh. Tekhn. Fiz., 19(8):1089-1092 (1949).

228.   D. A. Petrov and L. A. Raikovskaya, Tr. Mosk. Aviats. Tekhnol. Inst., No. 7, Oborongiz (1949), pp. 20-32.

229.   D. A. Petrov and L. A. Raikovskaya, Izv. Akad. Nauk SSSR, Ser. Khim., No. 8, 225 (1952).

230.   H. Buckle, Plansee Proc. (1952).

231.   P. Brenner and H. Kostron, Metallurgia (Manchr.), 41:219 (1950).

232.   D. Jaffe and M. Bever, J. Metals, 8(8):972-975 (sec.2) (1956).

233.   I. N. Golikov, Dendritic Liquation in Steel, Metallurgizdat (1958).

234.   A. A. Presnyakov and N. S. Sakharova, Fiz. Metal. i Metalloved., 6(5):886-892 (1958).

235.   V. S. Zolotorevskii et al., Izv. Vysshikh Uchebn. Zavedenii, Tsvetnaya Met., No. 5, 129-134 (1966).

236.   I. F. Kolobnev, Heat Treatment of Aluminum Alloys, Metallurgizdat (1960).

237.   I. I. Novkov and M. V. Zakharov, Heat Treatment of Metals and Alloys, Metallurgizdat (1962).

238.   A. A. Bochvar, Fundaments of Heat Treatment, ONTI (1931).

239.   K. P. Bunin and Ya. N. Malinochka, Introduction to Metallography, Metallurgizdat (1954).

240.   M. Hesselblatt, Z. Phys. Chemie, 83:1 (1913).

241.   A. A. Popov, in: Problems of Metallography and Heat Treatment, Mashgiz, Moscow−Sverdlovsk (1956), pp. 5-22.

242.   W. T. Olsen and R. Hultgren, J. Metals, 188(11):1323 (1950).

243.   Z. A. Sviderskaya et al., Metallovedenie i Obrabot. Metallov., No. 5, 23-28 (1957).

244.   M. E. Drits et al., Autoradiography in Metallography, Metallurgizdat (1962).

245.   G. A. Korol'kov and I. I. Novikov, Izv. Akad. Nauk SSSR, Otd. Tekhn. Nauk, Metallurgiya, i Toplivo, No. 3, 70-74 (1959).

246.   I. I. Novikov and F. S. Novik, Zavod. Lab., No. 10, 1195-1198 (1959).

247.   I. I. Novikov and V. S. Zolotorevskii, Izv. Akad. Nauk SSSR, Otd. Tekhn. Nauk, Metallurgiya i Toplivo, No. 1, 39-43 (1961).

248.   I. I. Novikov et al., Fiz. Metal. i Metalloved., 16(2):241-250 (1963).

249.   I. I. Novikov et al., Izv. Akad. Nauk SSSR, Metallurgiya i Gornoe Delo, No. 5, 121-125 (1963).

250.   I. I. Novikov and V. S. Zolotorevskii, Dendritic Liquation in Alloys, Izd. Nauka (1966).

251. V. G. Lyutsau, Zavod. Lab., No. 3, 311-315 (1959).
252. B. M. Rovinskii et al., Izv. Akad. Nauk SSSR, Ser. Fiz., 23(5):541-551 (1959).
253. L. Ya. Koshevnik and G. A. Kashchenko, Izv. Vysshikh Uchebn. Zavedenii, Tsvetnaya Met., No. 5, 165-169 (1961).
254. I. I. Novikov and V. S. Zolotorevskii, Lit. Proizv., No. 4, 13-18 (1962).
255. H. W. Phillips, Annotated Equilibrium Diagrams of Some Aluminum Alloy Systems, London (1959).
256. Yu. A. Krishtal, Izv. Vysshikh Uchebn. Zavedenii, Chernaya Met., No. 3, 110-117 (1960).
257. A. R. Regel', in: Structure and Properties of Molten Metals, Inst. Met. im. A. A. Baikova, AN SSSR (1959), pp. 3-50.
258. V. M. Glazov, Izv. Akad. Nauk SSSR, Otd. Tekhn. Nauk, Metallurgiya i Toplivo, No. 6, 111-116 (1960).
259. V. I. Danilov, Structure and Crystallization of Liquids, Izd. AN Ukr.SSR, Kiev (1956).
260. N. B. Kondrat'ev and I. N. Fridlyander, in: Metallurgical Fundaments of Casting Light Alloys, Oborongiz, Moscow (1957), p. 414.
261. A. G. Spasskii et al., Lit. Proizv., No. 10, 35-37 (1959).
262. M. V. Pikunov and A. I. Desipri, Izv. Akad. Nauk SSSR, Metallurgiya i Gornoe Delo, No. 6, 133-138 (1963).
263. V. S. Zolotorevskii and I. I. Novikov, Fiz. Metal. i Metalloved., 18(6):862-868 (1964).
264. V. G. Smith et al., Canad. Journ. Phys., 33:723 (1955).
265. Ya. I. Frenkel', Kinetic Theory of Liquids, Izd. AN SSSR (1945).
266. B. Frost, Advances in the Physics of Metals [Russian translation], Vol. 2, Metallurgizdat (1958), p. 126.
267. I. Z. Fisher, Statistical Theory of Liquids, Fizmatgiz (1961), pp. 25-100.
268. V. M. Lomer, in: Vacancies and Point Defects, Metallurgizdat (1961), pp. 99-122.
269. B. A. Movchan, Izv. Akad. Nauk SSSR, Otd. Tekhn. Nauk, No. 4, 122-123 (1958).
270. B. A. Movchan and I. Ya. Dzykovich, Dokl. Akad. Nauk SSSR, 125(2):354-355 (1959).
271. Ya. N. Malinochka, Izv. Akad. Nauk SSSR, Otd. Tekhn. Nauk, Metallurgiya i Toplivo, No. 2 (1961).
272. V. M. Glazov, Izv. Akad. Nauk SSSR, Otd. Tekhn. Nauk, Metallurgiya i Toplivo, No. 2, 65-70 (1962).
273. A. A. Bochvar and O. S. Zhadaeva, Izv. Akad. Nauk SSSR, Otd. Tekhn. Nauk, No. 10-11, 1089-1092 (1945).
274. M. Rosbaund and H. Schmid, Z. Physik, 32:197 (1925).
275. M. Straumanis, Z. Anorg. und Allg. Chemie, 180(1):1-10 (1929).
276. A. A. Bochvar and R. G. Kozolupova, Coll. of Works, Moscow Evening Metallurgical Inst., No. 1, Metallurgizdat (1955), pp. 118-135.
277. V. P. Shishokin and V. A. Ageeva, Tsvetn. Metally, No. 11, 1434-1470 (1930).
278. V. M. Vozdvizhenskii, Izv. Vysshikh Uchebn. Zavedenii, Tsvetnaya Met., No. 5, 116-120 (1960).
279. V. M. Vozdvizhenskii, Fiz. Metal. i Metalloved., 11(2):309-311 (1960).
280. V. M. Vozdvizhenskii and A. A. Dushan, Zavod. Lab., No. 2, 197-198 (1961).
281. V. M. Vozdvizhenskii et al., Zavod. Lab., No. 12, 1473-1475 (1965).
282. V. M. Glazov and G. A. Korol'kov, Metallovedenie i Obrabot. Met., No. 7, 18-23 (1957).
283. A. A. Vertman and V. M. Glazov, Izv. Akad. Nauk SSSR, Otd. Tekhn. Nauk, Metallurgiya i Gornoe Delo, No. 6, 148-151 (1964).
284. A. A. Vertman et al., Izv. Akad. Nauk SSSR, Metallurgiya i Toplivo, No. 6, 37-42 (1962).
285. V. M. Glazov and V. N. Vigdorovich, Dokl. Akad. Nauk SSSR, 118(5):924-927 (1958).
286. G. V. Kurdyumov, Zh. Tekhn. Fiz., 18(3):999-1025 (1948).
287. G. V. Kurdyumov, in: Problems of Metallography and the Physics of Metals, Metallurgizdat (1949).

288. G. V. Kurdyumov, in: Problems of Metallography and the Physics of Metals, Metallurgizdat (1952).

289. A. A. Bochvar, Metallurg., No. 8, 52-53 (1932).

290. D. D. Saratovkin, Dendritic Crystallization, Metallurgizdat (1953).

291. V. D. Kuznetsov, Crystals and Crystallization, Izd. AN SSSR (1954).

292. V. N. Vigdorovich and V. M. Glazov, Kolloidn. Zh., 21(4):405-412 (1959).

293. V. M. Glazov and V. N. Vigdorovich, Kolloidn. Zh., 21(1):18-24 (1959).

294. A. V. Rakovskii, Introduction to Physical Chemistry, GONTI-NKTP SSSR (1938).

295. R. Berrer, Diffusion in Solids [Russian translation], IL (1948).

296. I. F. Shreder, Gornyi Zh., 4(11):272-327 (1890).

297. S. A. Pogodin et al., Dokl. Akad. Nauk SSSR, 27(7):670-672 (1940).

298. S. T. Konobeevskii, in: X-ray Diffraction as Applied to the Study of Materials, ONTI-NKTP, Moscow—Leningrad (1936), pp. 193-216.

299. S. T. Konobeevskii, Izv. Akad. Nauk SSSR, Ser. Khim., No. 5, 1209-1244 (1937).

300. S. T. Konobeevskii, Zh. Éksp. Teor. Fiz., 13(11-12):418-427 (1943).

301. B. N. Finkel'shtein, Zh. Éksp. Teor. Fiz., 10:341-345 (1940).

302. M. E. Drits, Friction and Wear in Machines, Coll. VII, Izd. AN SSSR (1952).

303. A. M. Korol'kov and É. S. Kanader, Dokl. Akad. Nauk SSSR, 74(2):271 (1950).

304. E. M. Savitskii, Dokl. Akad. Nauk SSSR, 62(3):349-352 (1948).

305. E. M. Savitskii and M. A. Tylkina, Dokl. Akad. Nauk SSSR, 63(1):49-52 (1948).

306. K. N. Portnoi and A. A. Lebedev, in: Magnesium Alloys, Metallurgizdat (1952), p. 151.

307. G. V. Zakharova, in: Methods of Testing Metals and Alloys, Oborongiz (1951), pp. 92-99.

308. M. E. Drits et al., Izv. Akad. Nauk SSSR, Otd. Tekhn. Nauk, Metallurgiya i Toplivo, No. 4, 111-119 (1960).

309. A. A. Bochvar and Z. A. Sviderskaya, in: Study of the Alloys of Nonferrous Metals, Vol. 2, Izd. AN SSSR (1960), pp. 3-8.

310. A. N. Rozanov, Dokl. Akad. Nauk SSSR, 115(4):721-722 (1957).

311. A. A. Zhukov, Lit. Proizvod., No. 12, 16-20 (1957).

312. A. A. Zhukov, in: Questions in the Theory of Casting Processes, Mashgiz (1960), pp. 163-252.

313. A. N. Zelikman and S. S. Loseva, Tsvetn. Metally, No. 4, 52-55 (1956).

314. A. E. Koval'skii and L. A. Kanova, Zavod. Lab., No. 11, 1362-1365 (1950).

315. G. A. Meerson et al., Zh. Neorg. Khim., 3(4):898-903 (1958).

316. A. E. Koval'skii and T. G. Makarenko, Zh. Tekhn. Fiz., 23(2):265-279 (1953).

317. G. V. Samsonov and V. B. Rukina, Dopovidi Akad. Nauk Ukr.SSR, No. 3, 247-250 (1957).

318. G. V. Samsonov, Izv. Soob. Fiz.-Khim. Anal., 27:97-125 (1956).

319. G. V. Samsonov and V. S. Neshpor, in: Questions of Powder Metallurgy and the Strength of Materials, No. 5, Izd. AN Ukr.SSR, Kiev (1958), pp. 3-35.

320. Ya. S. Umanskii, Carbides of Hard Alloys, Metallurgizdat (1947).

321. R. Schwarzkopf and R. Kieffer, Hard Metals, New York (1953).

322. G. A. Meerson et al., Izv. Soob. Fiz.-Khim. Anal., 25:89 and 98 (1954).

323. V. M. Gol'dshmidt, Usp. Fiz. Nauk, 9(6):811-829 (1929).

324. T. S. Liu and E. A. Peretti, Trans. ASM, 45:677 (1952).

325. T. S. Liu and E. A. Peretti, Trans. ASM, 44:539 (1952).

326. N. A. Goryuniva et al., Zh. Tekhn. Fiz., 27(7):1408-1413 (1957).

327. N. A. Goryunova and N. N. Fedorova, Fiz. Tverd. Tela, 1(2):344-345 (1959).

328. I. I. Burdiyan and A. S. Borshchevskii, Zh. Tekhn. Fiz., 28(12):2684-2688 (1958).

329. V. N. Vigdorovich et al., Scientific Works of the State Inst. of Rare Metals (Work of 1959), Vol. VI, Metallurgizdat (1962), pp. 180-193.

330. G. A. Wolff et al., Semiconductors and Phosphors, New Jersey (1958).

331. V. M. Glazov et al., Izv. Akad. Nauk SSSR, Otd. Tekhn. Nauk, No. 10, 68-70 (1957).

332. In: Physics. Contributions to the Twentieth Scientific Conference of the Leningrad Constructional Engineering Institute, Leningrad (1962).

333. L. A. Baidakov et al., Zh. Prikl. Khim., 33(11):2486 (1961).

334. V. N. Vigdorovich and A. Ya. Nashel'skii, Poroshkovaya Met., No. 2 (14), 43-48 (1963).

335. Ya. A. Ugai, Introduction to the Chemistry of Semiconductors, Izd. Vysshaya Shkola (1965).

336. G. G. Wang and B. N. Alexander, Acta Metallurgica, 3:515 (1955).

337. V. N. Vigdorovich et al., Poroshkovaya Met., No. 9, 36-39 (1969).

338. A. S. Borshchevskii et al., Trans. of the Third Scientific Conference of Young Scientists of Moldavia, Izd. Kartya Moldovenyaské, Kishinev (1964), p. 10.

339. V. N. Ivanov-Omskii and B. N. Kolomiets, Fiz. Tverd. Tela, 1(6):913-918 (1959).

340. B. V. Baranov and N. A. Goryunova, Fiz. Tverd. Tela, 2(2):284-287 (1960).

341. N. N. Sirota and V. V. Rozov, Dokl. Akad. Nauk Beloruss.SSR, 7(7):446-448 (1963).

342. N. A. Goryunova and V. I. Sokolova, Izv. Moldav. Fil. Akad. Nauk SSSR, No. 3(68), 31-35, 97-98 (1960).

343. N. N. Sirota and L. A. Makovitskaya, Dokl. Akad. Nauk Beloruss. SSR, 7(4):230 (1963).

344. N. A. Goryunova, Chemistry of Diamond-Like Semiconductors, LGU, Leningrad (1963).

345. N. A. Goryunova et al., in: Physics and Chemistry, Contributions to the Nineteenth Scientific Conference of the Leningrad Constructional Engineering Institute, Leningrad (1961), p. 22.

346. A. V. Voitekhovskii, Contributions to the All-Union Congress on Semiconducting Compounds, Izd. AN SSSR, Leningrad (1961), p. 13.

347. N. A. Goryunova et al., Vestnik. Leningrad. Univ., No. 10(2), 156 (1961).

348. L. I. Ksishchenskii et al., in: Physics, Contributions to the Twenty-Second Scientific Conference of the Leningrad Constructional Engineering Institute, Leningrad (1964), p. 12.

349. N. A. Goryunova et al., Fiz. Tverd. Tela, 1(12):1858-1860 (1959).

350. S. I. Radautsan and I. P. Molodyan, Izv. Akad. Nauk Moldav. SSR, No. 3(68), 37-47 (1960).

351. S. I. Radautsan et al., Izv. Akad. Nauk Moldav. SSR, No. 10(88), 512-514 (1961).

352. N. A. Goryunova and S. I. Radautsan, Dokl. Akad. Nauk SSSR, 121(5):848-849 (1958).

353. N. A. Goryunova et al., Fiz. Tverd. Tela, 1(3):512-514 (1959).

354. S. I. Radautsan et al., Izv. Akad. Nauk Moldav. SSR, No. 3(69),107-109 (1960).

355. I. I. Kozhina et al., Vestnik. Leningrad. Gos. Univ., No. 4, 122-127 (1962).

356. V. V. Negreskul, Trans. of the Third Scientific Conference of Young Moldavian Scientists, Izd. Martya Moldavenyaské, Kishinev (1964), p. 35.

357. S. I. Radautsan et al., Izv. Akad. Nauk Moldav. SSR, No. 10(88), 98-101 (1961).

358. N. A. Goryunova et al., Vestnik Leningrad. Gos. Univ., Ser. Fiz. i Khim., No. 22(4), 97-101 (1961).

359. B. F. Bankina and N. Kh. Abrikosov, Zh. Neorg. Khim., 9(4):931-936 (1964).

360. V. M. Gol'dshmidt, in: Fundamental Ideas of Geochemistry, Vol. 1, A. E. Fersman, ed., Khimteoretizdat (1933).

361. V. M. Gol'dshmidt, Crystal Chemistry, ONTI-NKTP, Moscow—Leningrad (1937).

362. V. S. Sobolev, Introduction to the Mineralogy of Silicates, Izd. L'vov. Gos. Univ., L'vov (1949).

363. A. E. Fersman, Geochemistry, Vol. 3, Geolitizdat (1937).

364. A. S. Povarennykh, Zap. Vses. Mineralog. Obshch., Pt. 78, No. 4, 242 (1949).

365. A. S. Povarennykh, Zap. Vses. Mineralog. Obshch., Pt. 84, No. 4, 469 (1955).

366. A. S. Povarennykh, Dokl. Akad. Nauk SSSR, 112(6):1098-1100 (1957).

367. A. G. Betekhtin, Mineralogy, Gosgeolizdat (1950).

368. C. Palache et al., Dana's System of Mineralogy, New York, 1-2 (1944-1951).

369. E. O. Bernhardt, Z. Metallkunde, 33(3):135 (1941).

370. S. D. Dmitriev, Priroda, No. 8, 57 (1950).

371. H. Howotny and F. Vitovec, Pulvermetallurgee, I. Plansee Seminar: De Re Metallica, Vienna (1953).

372. P. A. Rebinder, Hardness, Technical Encyclopedia, Vol. 22 (1933), p. 703.

373. V. I. Likhtman et al., Physicochemical Mechanics of Metals, Izd. AN SSSR (1962).

374. V. D. Kuznetsov, Surface Energy of Solids, GITTL (1954).

375. E. Gruceanu and S. Ionescu-Bujor, Studii si Cercetari de Metalurgie [in Russian], VIII(4):391-395 (1963).

376. V. M. Glazov, N. N. Glagoleva, and N. M. Makhmudova, Izv. Akad. Nauk SSSR, Neorg. Mat., 5(9):1508 (1969).

377. A. N. Krestovnikov, N. M. Makhmudova, and V. M. Glazov, Izv. Akad. Nauk SSSR, Neorg. Mat., 4(4):615 (1968).

378. I. P. McHugh and W. A. Tiller, Trans. Metallurg. Soc., AIME, 218:187 (1960).

379. R. F. Brebrick, J. Phys. Chem. Solids, 24:127 (1963).

380. L. E. Shelimova and N. Kh. Abrikosov, Zh. Neorg. Khim., 9:1879 (1964).

381. L. E. Shelimova, N. Kh. Abrikosov, and V. V. Zhdanova, Zh. Neorg. Khim., 10:1200 (1965).

382. R. F. Brebrick, J. Phys. Chem. Solids, 27:1495 (1966).

383. R. F. Brebrick and A. J. Strauss, Bull. Amer. Phys. Soc., 7:203 (1962).

384. R. C. Allgaier and P. O. Scheil, Bull. Amer. Phys. Soc., 6:436 (1961).

385. N. V. Kolomoets, E. Ya. Lev, and L. M. Sysoeva, Fiz. Tverd. Tela, 6:706 (1964).

386. N. V. Kolomoets, E. Ya. Lev, and L. M. Sysoeva, Fiz. Tverd. Tela, 7:652 (1965).

387. N. V. Kolomoets, E. Ya. Lev, and L. M. Sysoeva, Fiz. Tverd. Tela, 7:2223 (1965).

388. N. V. Kolomoets, E. Ya. Lev, and L. M. Sysoeva, Fiz. Tverd. Tela, 7:2558 (1965).

389. N. V. Kolomoets, E. Ya. Lev, and L. M. Sysoeva, Fiz. Tverd. Tela, 8:1212 (1966).

390. N. V. Kolomoets, E. Ya. Lev, and L. M. Sysoeva, Fiz. Tverd. Tela, 8:2925 (1966).

391. I. A. Kafalas, R. F. Brebrick, and A. J. Strauss, Appl. Phys. Letters, 4:93 (1964).

392. V. M. Glazov, N. N. Glagoleva, and K. V. Papoyan, in: Chemical Bond in Semiconductors and Solids, Nauka i Tekhnika, Minsk (1965), p. 135.

393. A. N. Krestovnikov, V. M. Glazov, V. A. Evseev, and A. A. Aivazov, Zh. Neorg. Mat., 2:850 (1966).

394. L. V. Poretskaya, N. Kh. Abrikosov, and V. M. Glazov, Zh. Neorg. Khim., 8:1196 (1963).